AF692595

84

Advances in Biochemical Engineering/Biotechnology

Series Editor: T. Scheper

Editorial Board:

W. Babel · H. W. Blanch · I. Endo · S.-O. Enfors
A. Fiechter · M. Hoare · B. Mattiasson · H. Sahm
K. Schügerl · G. Stephanopoulos · U. von Stockar
D. T. Tsao · J. Villadsen · C. Wandrey · J.-J. Zhong

Springer
Berlin
Heidelberg
New York
Hong Kong
London
Milan
Paris
Tokyo

Biotechnology in India I

Volume Editors: T. K. Ghose · P. Ghosh

With contributions by
S. K. Basu, D. Bhardwaj, V. S. Chauhan, N. Dhar,
B. N. Johri, J. P. Khurana, A. Mukhopadhyay,
G. Padmanaban, A. Sharma, M. Sharma, R. Swarup,
A. K. Tyagi, J. S. Virdi

Springer

Advances in Biochemical Engineering/Biotechnology reviews actual trends in modern biotechnology. Its aim is to cover all aspects of this interdisciplinary technology where knowledge, methods and expertise are required for chemistry, biochemistry, microbiology, genetics, chemical engineering and computer science. Special volumes are dedicated to selected topics which focus on new biotechnological products and new processes for their synthesis and purification. They give the state-of-the-art of a topic in a comprehensive way thus being a valuable source for the next 3–5 years. It also discusses new discoveries and applications.

In general, special volumes are edited by well known guest editors. The series editor and publisher will however always be pleased to receive suggestions and supplementary information. Manuscripts are accepted in English.

In references Advances in Biochemical Engineering/Biotechnology is abbreviated as *Adv Biochem Engin/Biotechnol* as a journal.

Visit the ABE home page at
http://www.springerlink.com/series/abe/

ISSN 0724-6145
ISBN 3-540-00609-5
DOI 10.1007/b11029
Springer-Verlag Berlin Heidelberg New York

Library of Congress Catalog Card Number 72-152360

This work is subject to copyright. All rights are reserved, whether the whole or part of the material is concerned, specifically the rights of translation, reprinting, reuse of illustrations, recitation, broadcasting, reproduction on microfilm or in any other way, and storage in data banks. Duplication of this publication or parts thereof is permitted only under the provisions of the German Copyright Law of September 9, 1965, in its current version, and permission for use must always be obtained from Springer-Verlag. Violations are liable for prosecution under the German Copyright Law.

Springer-Verlag Berlin Heidelberg New York
a member of BertelsmannSpringer Science+Business Media GmbH

http://www.springer.de

© Springer-Verlag Berlin Heidelberg 2003
Printed in Germany

The use of general descriptive names, registered names, trademarks, etc. in this publication does not imply, even in the absence of a specific statement, that such names are exempt from the relevant protective laws and regulations and therefore free for general use.

Typesetting: Fotosatz-Service Köhler GmbH, Würzburg
Cover: Künkellopka GmbH, Heidelberg; design & production GmbH, Heidelberg

Printed on acid-free paper 02/3020mh – 5 4 3 2 1 0

Series Editor

Professor Dr. T. Scheper
Institute of Technical Chemistry
University of Hannover
Callinstraße 3
30167 Hannover, Germany
E-mail: scheper@iftc.uni-hannover.de

Volume Editors

Professor Tarun K. Ghose

281 SFS Apartments
Hauz Khas
New Delhi 110 016, India
E-mail: tarunghose@attbi.com

Professor Purnendu Ghosh

Birla Institute of Scientific Research
Statue Circle
Jaipur-302 001, India
E-mail: pghosh@bisrjaipur.com

Editorial Board

Prof. Dr. W. Babel
Section of Environmental Microbiology
Leipzig-Halle GmbH
Permoserstraße 15
04318 Leipzig, Germany
E-mail: babel@umb.ufz.de

Prof. Dr. H.W. Blanch
Department of Chemical Engineering
University of California
Berkely, CA 94720-9989, USA
E-mail: blanch@socrates.berkeley.edu

Prof. Dr. I. Endo
Faculty of Agriculture
Dept. of Bioproductive Science
Laboratory of Applied Microbiology
Utsunomiya University
Mine-cho 350, Utsunomiya-shi
Tochigi 321-8505, Japan
E-mail: endo@cel.riken.go.jp

Prof. Dr. S.-O. Enfors
Department of Biochemistry and Biotechnology
Royal Institute of Technology
Teknikringen 34,
100 44 Stockholm, Sweden
E-mail: enfors@biotech.kth.se

Prof. Dr. A. Fiechter
Institute of Biotechnology
Eidgenössische Technische Hochschule
ETH-Hönggerberg
8093 Zürich, Switzerland
E-mail: ae.fiechter@bluewin.ch

Prof. Dr. M. Hoare
Department of Biochemical Engineering
University College London
Torrington Place
London, WC1E 7JE, UK
E-mail: m.hoare@ucl.ac.uk

Prof. Dr. B. Mattiasson
Department of Biotechnology
Chemical Center, Lund University
P.O. Box 124, 221 00 Lund, Sweden
E-mail: bo.mattiasson@biotek.lu.se

Prof. Dr. H. Sahm
Institute of Biotechnolgy
Forschungszentrum Jülich GmbH
52425 Jülich, Germany
E-mail: h.sahm@fz-juelich.de

Prof. Dr. K. Schügerl
Institute of Technical Chemistry
University of Hannover
Callinstraße 3
30167 Hannover, Germany
E-mail: schuegerl@iftc.uni-hannover.de

Prof. Dr. G. Stephanopoulos
Department of Chemical Engineering
Massachusetts Institute of Technology
Cambridge, MA 02139-4307, USA
E-mail: gregstep@mit.edu

Prof. Dr. U. von Stockar
Laboratoire de Génie Chimique et
Biologique (LGCB)
Départment de Chimie
Swiss Federal Institute
of Technology Lausanne
1015 Lausanne, Switzerland
E-mail: urs.vonstockar@epfl.ch

Prof. Dr. G. T. Tsao
Director
Lab. of Renewable Resources Eng.
A. A. Potter Eng. Center
Purdue University
West Lafayette, IN 47907, USA
E-mail: tsaogt@ecn.purdue.edu

Prof. Dr. J. Villadsen
Center for Process of Biotechnology
Technical University of Denmark
Building 223
2800 Lyngby, Denmark
E-mail: john.villadsen@biocentrum.dtu.dk

Prof. Dr. C. Wandrey
Institute of Biotechnology
Forschungszentrum Jülich GmbH
52425 Jülich, Germany
E-mail: c.wandrey@fz-juelich.de

Prof. Dr. J.-J. Zhong
State Key Laboratory
of Bioreactor Engineering
East China University of Science
and Technology
130 Meilong Road
Shanghai 200237, China
E-mail: jjzhong@ecust.edu.cn

Advances in Biochemical Engineering/Biotechnology also Available Electronically

For all customers with a standing order for Advances in Biochemical Engineering/Biotechnology we offer the electronic form via SpringerLink free of charge. Please contact your librarian who can receive a password for free access to the full articles. By registration at:

http://www.springer.de/series/abe/reg_form.htm

If you do not have a standard order you can nevertheless browse through the table of contents of the volumes and the abstracts of each article at:

http://www.springerlink.com/series/abe/

There you will find also information about the

- Editorial Board
- Aims and Scope
- Instructions for Authors

Attention all Users of the "Springer Handbook of Enzymes"

Information on this handbook can be found on the internet at
http://www.springer.de/enzymes/

A complete list of all enzyme entries either as an alphabetical Name Index or as the EC-Number Index is available at the above mentioned URL. You can download and print them free of charge.

A complete list of all synonyms (more than 25,000 entries) used for the enyzmes is available in print form (ISBN 3-540-41830-X).

Save 15%

We recommend a standing order for the series to ensure you automatically receive all volumes and all supplements and save 15% on the list price.

Preface

The biotechnology business in India with an increase from USD 500 million in 1997 and reaching an estimated USD 1 billion next year – health related products accounting for 60%, agro and veterinary products together 15%, and contract R&D, reagents, devices and supplies adding up to the remaining 25% of which the diagnostics share was about 10% of the total – surely presented an encouraging picture even five years ago. While volumes have increased, the pattern has not. According to a report, prepared by McKinsey & Co, India's Pharmaceutical industry including domestic and export sales and contract services totals nearly USD 5 billion. Furthermore, the company optimistically projects the growth to a factor of five-fold only if both the industry and the government are able to put in place achievable solutions that must take care of the formidable obstacles preventing further growth. If this assessment is correct, then the established transformation made by IT growth should also provide the confidence required by the high expectations for biotechnology which have arisen in the country in recent years. Some contributors to this are overenthusiastic – these are bureaucrats, some retired scientists and of course the complacent politicians who have the least knowledge of what the new biotechnology is all about. However, there are clear indications of biotechnology growth demonstrated by a few but rapidly expanding biotech companies such as Biocon Ltd, Shantha Biotech (P) Ltd, Dr. Reddy's Laboratory, Bharat Biotech Ltd, Ranbuxy Ltd, Lupin, Torrent, RPG Life Sciences, United Breweries, BCG Laboratory, Godrej Biotech, Zydus-Cadila, Bengal Immunity, EID Parry, Wockhardt and several other small but fast growing units. In addition, there are dozens of joint ventures among these companies and other biotechnological start-ups with each other or with multinational corporations having a presence in India to sell products produced abroad or those from joint R&D operations in India.

Biotechnology entrepreneurship in India is severely handicapped by the lack of critical service sectors such as expression hosts, vectors, genome libraries, large scale sterile fluid transport systems, animal cell bioreactors and their components for clean and prolonged performance and an almost total absence of a biosystems – engineering database for downstream purification of drugs and therapeutic proteins. There are often complaints that government agencies are the only sources for funding research by academics with little or nothing coming from industry. While the statement is basically true, Indian industries, like others almost anywhere, have to have clear-cut goals and because they are not in a position to control the distribution of tax money nor are they charitable institutions, they can only invest funds in new ideas based on good scientific logic and those having a reasonable chance of success.

Cutting edge research with high investment possibilities such as physiomics, metabolic engineering and pathway management, search for or synthesis of new genes for useful applications, in vitro synthesis of short chain peptides, cell-surface engineering , product design models, separation of large molecules like DNA, RNA, etc, to name a few have yet to become established in India. Difficulties of obtaining very large funding, absence of sophisticated analytical infrastructures and indeed top scholars continuing to look for better opportunities elsewhere continue to handicap project leaders and universities. Despite these difficulties, the fast growing biopharma industry is making gains in international business with comparatively priced life-saving drugs that are challenging products of multinational pharmaceutical companies operating outside India.

Parliament has intensively debated the problems Indian biotechnology is faced with in order to be able to match its ability in IPR, TRIPS and GATT Agreements applicable to GM agro and transgenic animal and medical products, protection of the nation's eco balance and the unique biodiversity endowment. It defined its services and obligations to the people. However, except for the availability of high quality human resources, few of the legislative steps taken so far can match with those existing in most parts of the developed globe. In the midst of innumerable problems faced by the largest section of the Indian population, the scientific contributions made to lift life sciences from their observed empirical domain to the level of exactness is very clearly visible in the quality of students graduating in these sciences with the support of biochemical engineering. It is certainly an excellent achievement. Notwithstanding, one source predicts the slow but steady progress with a lower number of highly trained people moving to the USA and Europe and the rate of development in the rich nations slowing down, India's growth rate will reach a high social peak in about twenty years when most certainly teaching will not be conducted in the classroom but in cyber space. Given India's strong IT base, the information cross over appears to be the obvious choice. The Department of Biotechnology plans to award USD 120 million for supercomputer networking a dozen centers to allow research with qualified access to the recently discovered genome sequences and its extensions into proteomics and physiomics data bases. The usual hype associated with anything new in biotechnology is also visible in genome science. Most of those involved have inadequate understanding of what products are needed and how must they be designed and where the huge amount of money has to come from to bring the drug to the market place following approval.

These two volumes with fourteen chapters give accounts of several areas of active research pursuits and many project supports provided by the DBT. Two important sectors of significant importance namely, bioinformatics and politics of management and protection of GM products appearing for human consumption could not be reported. The science and engineering of bioinformatics covering areas like genomics, proteomics and physiomics are new and expensive for biotechnology applications. However, after the supercomputer network connecting a dozen centers in the country becomes fully functional, contributions in these areas at least in few places are expected to be seen. The other area somewhat nebulous in nature relates to the IPR in biotechnology and its legal jurisdictions exercising control over GM agro crops and transgenic animal and plant products for human consumption. Even amongst the active participants, there exist important gaps of knowledge and general understanding of this technology is poor causing the debates to generate more heat than light. A forum for professional interactions between legislators, lawyers, arbitrators, judges and social science activists is currently absent. Until this is in place, knowledge-based and useful contributions with balanced views will have to wait.

The editors were pleased to see that biochemical engineering training and research pioneered and promoted at the IIT Delhi over the last 25 years with bioethanol as one of its substantive, graduate research programs must have helped formulate the recently declared national policy on the use of bioethanol as a transport fuel. This is a big step forward in support of India's advances in biotechnology. The editors are grateful to all authors for submitting the reviewed and edited final manuscripts. The senior editor (TKG) would like to give special thanks to the Managing Editor, Prof. T. Scheper for prompt acknowledgement of the properly edited final manuscripts.

Jaipur, May 2003

Tarun K. Ghose
Purnendu Ghosh

Contents

The Way Ahead – The New Technology in an Old Society
M. Sharma, R. Swarup 1

Rhizobacterial Diversity in India and Its Influence on Soil and Plant Health
B. N. Johri, A. Sharma, J. S. Virdi 49

Plant Molecular Biology and Biotechnology Research in the Post-Recombinant DNA Era
A.K. Tyagi, J.P. Khurana 91

Drug Targets in Malaria Parasites
G. Padmanaban 123

Current Status of Malaria Vaccine Development
V. S. Chauhan, D. Bhardwaj 143

Intracellular Delivery of Drugs to Macrophages
A. Mukhopadhyay, S.K. Basu 183

Recent Advances in Tuberculosis Research in India
A.K. Tyagi, N. Dhar 211

Author Index Volumes 51 – 84 275

Subject Index 289

Contents of Volume 85

Biotechnology in India II

Volume Editors: T. K. Ghose · P. Ghosh

Bioethanol in India: Recent Past and Emerging Future
P. Ghosh, T. K. Ghose

Commercialization of a Novel Fermentation Concept
K.Mazumdar-Shaw, S. Suryanarayan

Bioprocessing of Therapeutic Proteins from the Inclusion Bodies of *Escherichia coli*
A. K. Panda

Research and Application of Microbial Enzymes – India's Contribution
S. Chand, P. Mishra

Molecular Mechanisms of Energy Transduction in Cells: Engineering Applications and Biological Implications
S. Nath

Status of Immunodiagnosis and Immunocontraceptive Vaccines in India
S. K. Gupta

Medical Biotechnology in India
B. B. Lohray

Adv Biochem Engin/Biotechnol (2003) 84: 1–48
DOI 10.1007/b11035

The Way Ahead – The New Technology in an Old Society

Manju Sharma[1] · Renu Swarup[2]

[1] Department of Biotechnology, Ministry of Science & Technology, Government of India, New Delhi, India. *E-mail: manju@dbt.nic.in*
[2] Department of Biotechnology, Ministry of Science & Technology, Government of India, New Delhi, India. *E-mail: swarup@dbt.nic.in*

Biotechnology is one of the most important scientific and technological revolutions of the last century and has greatly benefited various aspects of human life. The potentials are enormous and many breakthroughs have already been achieved in the area of healthcare, food, agricultural products and environmental production. The developments in this important area provide immediate benefits to mankind and offer environmentally friendly technologies for sustainable development. The Department of Biotechnology, Government of India, set up in 1986, has played an important catalytic role in promoting this revolutionary field. Research and development, technology validation and demonstration, technology transfer, human resource development, setting up of Centers of Excellence and promoting industry-academia interactions have been some of the major achievements during the last 15 years. A unique feature of this Department is the strong interaction with scientists and institutes across the country to promote biotechnology research and development efforts for commercialization and also to benefit the rural population for socio-economic development. A large number of research institutes/universities and organizations across the country have been supported in the areas of agriculture, healthcare, environment and industry. In addition, basic research has also been an important thrust area. In order to ensure that the benefits of biotechnology reach the masses at large, a very stringent biosafety mechanism has been adopted. India is a country rich in biodiversity with two hot spots and has a strong base of expertise available in nearly all fields – thus biotechnology could flourish leading to a Bioindustrial Revolution. We are today poised to be the leaders in this field in the 21st Century.

Key words. Biotechnology, Biodiversity, Bioprospecting, Bioresources, Transgenics, Plant Tissue Culture, Healthcare, Environment, Human Genetics, Biosafety, Bioinformatics

1 **Introduction** 4

2 **Biotechnology in India** 5

2.1 Agriculture 9
2.1.1 Transgenic Research 11
2.1.2 Plant Tissue Culture 12
2.1.2.1 Horticultural and Plantation Crops 16
2.1.2.2 Forestry 17
2.1.2.3 Tissue Culture Pilot Plants 17
2.1.2.4 Hardening Units 18
2.1.2.5 Micropropagation Technology Parks 18

© Springer-Verlag Berlin Heidelberg 2003

2.1.3 Medicinal and Aromatic Plants . . . 19
2.1.4 Biological Software . . . 20
2.1.4.1 Biofertilizer . . . 21
2.1.4.2 Biological Pesticides . . . 22
2.1.5 Animal Biotechnology . . . 22
2.1.6 Aquaculture and Marine Biotechnology . . . 23
2.2 Biodiversity Conservation and Bioprospecting . . . 24
2.3 Biofuels . . . 24
2.3.1 Production and Demonstration . . . 26
2.3.2 Alternative Sources of Energy/Fuel . . . 26
2.3.3 Bioengineering . . . 26
2.4 Environment . . . 26
2.5 Healthcare . . . 27
2.5.1 Human Genetics . . . 29
2.6 Industry . . . 30
2.7 Human Resource Development . . . 31
2.8 Bioinformatics . . . 33
2.9 Centers of Excellence . . . 34
2.9.1 Repositories . . . 36
2.9.2 Programme Support . . . 36
2.10 Biosafety . . . 38
2.10.1 Regulatory Mechanisms in India . . . 38
2.10.2 The Recombinant DNA Advisory Committee (RDAC) . . . 39
2.10.3 Bt. Cotton . . . 43
2.10.4 Public Awareness . . . 44
2.11 Socio-Economic Development . . . 44

3 Future Perspective . . . 45

4 Conclusion . . . 47

5 References . . . 48

Abbreviations

AIIMS	All India Institute of Medical Sciences
BERPDC	Biochemical Engineering Research and Process Development
BGA	Blue green algae
BHU	Banaras Hindu University
BTIC	Biotechnology Information Center
BTISnet	Biotechnology Information System Network
CBSE	Central Board of Secondary Education
CBT	Center for Biotechnology
CDRI	Central Drug Research Institute
CIMAP	Central Institute of Medicinal and Aromatic Plants
DBCC	District level Biotechnology Coordination Committee
DBT	Department of Biotechnology

DIC	Distributed Information Centers
DMD	Duchenne Muscular Dystrophy
DNA	Deoxyribonucleic acid
DST	Department of Science and Technology
ELISA	Enzyme-linked immunosorbent assay
FAO	Food and Agriculture Organization
FACS	Flourescence Activated Cell Sorter
GDP	Gross Domestic Product
GEAC	Genetic Engineering Approval Committee
GMO	Genetically Modified Organisms
HAL	Hindustan Antibiotics Limited
HIV	Human immuno-deficiency virus
IADP	Intensive Area Development Programme
IARI	Indian Agricultural Research Institute
IBA	Integrated Biotechnological Approach
IBSC	Institutional Biosafety Committee
ICAR	Indian Council of Agriculture Research
IFPRI	International Food Policy Research Institute
IHBT	Institute of Himalayan Bioresource Technology
IICT	Indian Institute of Chemical Technology
IIHR	Indian Institute of Horticulture Research
IISc	Indian Institute of Science
IIT	Indian Institute of Technology
IMTECH	Institute of Microbial Technology
INM	Integrated Nutrient Management
IPM	Integrated Pest Management
IRGSP	International Rice Genome Sequencing Programme
JEV	Japanese encephalitis virus
MALDI-TOF	Matrix-assisted laser desorption ionization-time of flight
MGIMS	Mahatma Gandhi Institute of Medical Sciences
MKU	Madurai Kamaraj University
MTP	Micropropagation Technology Park
NAP	National Agricultural Policy
NBPGR	National Bureau of Plant Genetic Resources
NCL	National Chemical Laboratory
NFMC	National Facility for Marine Cyanobacteria
NII	National Institute of Immunology
PCR	Polymerase chain reaction
R & D	Research and Development
RCGM	Review Committee of Genetic Manipulation
RDAC	Recombinant DNA Advisory Committee
RIS	Research and Information Survey
SAARC	South Asian Association for Regional Cooperation
SBCC	State level Biotechnology Coordination Committee
TBGRI	Tropical Botanic Garden Research Institute
TC	Tissue culture
TCPP	Tissue culture pilot plant

TERI	Tata Energy Research Institute
UAS	University of Agricultural Sciences
UDSC	University of Delhi, South Campus
UNESCO	United Nations Education, Science and Cultural Organization
VPN	Virtual Private Network
WSSV	White spot shrimp virus
WTO	World Trade Organization
WUE	Water Use Efficiency

1 Introduction

"It is science alone that can solve the problem of hunger and poverty, of insanitation and illiteracy, of superstition and deadening customs and traditions, of vast resources running to waste, of a rich country inhabited by starving people... Even more than the present, the future belongs to science and those who make friends with science". 'Jawaharlal Nehru'

Since India gained independence in 1947, science and technology have been an integral part of the national development. Science in India can be divided into phases – the first up to the 1960s, concentrated on infrastructure development. The second phase up to the mid 1980s concentrated on capacity development. Thereafter, there has been a strong emphasis on globalization which increases international competitiveness. Science and technology has moved in the direction which offers great strengths not only to the globalization and industrialization of the country but also for the societal up-liftment and socio-economic development. There has been concerted effort to develop a suitable linkage between long-term basic research and strategic science that offers immediate returns. In reality, modern science is still comparatively young, dating from the time of Copernicus at the beginning of the 16th century. The pace of technological advances today is certainly not slacking. Basic sciences have a very major role to play. Some everyday artifacts depend on basic sciences. The revolutionary advances witnessed during the last century world-wide have built up a very strong scientific temper and the scientific community is actively engaged in advanced thrust, directions and goals as we move in this new millennium.

A better understanding of what is life and life processes has come from advances in molecular biology starting with the major breakthrough discovery of the structure of DNA by Watson and Crick in 1953. The genetic material, i.e., the DNA, can be isolated, cut, cloned, sliced, transferred and expressed in different organisms. With these techniques, the genes of interest from microbes, plants, animals and humans can be transferred and expressed in any other organism or cell lines derived from them.

The potential of biotechnology is enormous, and has already offered new breakthroughs in healthcare, food, agricultural products and environmental protection. The new biotechnologies responsible for a bioindustrial revolution are genetic engineering, cell fusion technology, bioprocess technologies and structure based molecular designs. The pharmaceutical sector has had maximum benefit but gradually, agriculture and the environment also have had profound impact, especially the areas of bioprospecting and bioremediation are becoming a money spinner.

The developments in biotechnology, particularly with the basic understanding of genetics, immunology, biochemistry, biochemical engineering and molecular biology, have paved the way for major biotechnology products and processes and have provided tools to the manufacturing and service industry. The developments in the field of molecular biology, plant and animal cell culture, immunology and related areas are now being correlated with the progress of industrial development in the world. The areas where there has been maximum impact are: pharmaceutical industry, agriculture and related areas, environment and industrial development of different types of biologicals and recombinant products. Also the field has become extremely important because it provides solutions to the most burning problems, faced in particular, by the developing countries.

2
Biotechnology in India

In order to give a systematic impetus to the development of this very important field of science – biotechnology – during, 1981-1982 the then Scientific Advisory Committee to the Cabinet, had detailed deliberations with the scientific community and a National Biotechnology Board (NBTB) was set up in 1982 to look into scientific programmes in the area of biotechnology, which required financial support for strengthening the indigenous capabilities.

Based on the progress in the areas of creating infrastructure, human resource development and specific research initiatives as well as the need for a major effort in this field, the Government decided to set up a separate Department of Biotechnology, in February, 1986. The mandate of the Department is to support research and development, technology validation and demonstration, set up centers of excellence, build a strong human resource and promote industry-academia interaction and technology transfer. To fulfill the mandate a number of programmes have been supported in different fields of agriculture, health, environment and industry. During the last fifteen years, since the Department has come into being, a strong infrastructure for biotechnology research and services has been created in the national laboratories and academic institutions across the country. An integrated human resource development programme, support to basic and applied and product-oriented research, both from the view point of attaining excellence and for product and process development has resulted in very good research publications and 46 technologies have also been developed and transferred to the industry (Table 1).

A unique feature of this Department has been the strong interaction with scientists and institutes across the country to promote biotechnology R & D efforts for commercialization and also to benefit the rural population for socio-economic development. Another important feature has been the interaction at the State Government level to promote biotechnology to meet the needs of the various regions. The investment in the area of Bbiotechnology has increased from Rs. 750 million in 1992–1993 to Rs. 2250 million in 2002–2003.

Table 1. List of technologies transferred [1]

S.No	Technologies	Developed by	Transferred to
1.	Pregnancy slide test	National Institute of Immunology, New Delhi	Ranbaxy Laboratories, New Delhi
2.	Latex agglutination	National Institute of Immunology, New Delhi	Ranbaxy Laboratories, New Delhi
3.	Pregnancy DOT-ELISA	National Institute of Immunology, New Delhi	Ranbaxy Laboratories, New Delhi
4.	Typhoid fever detection kit	National Institute of Immunology, New Delhi	Lupin Laboratories, Bombay
5.	Typhoid fever detection kit	All India Institute of Medical Sciences, New Delhi	Ranbaxy Laboratories, Ahemdabad
6.	Amoebic liver abscess	National Institute of Immunology, New Delhi	Cadila Laboratories, Ahemdabad
7.	Polypeptide P from bitter gourd	University of Rajasthan, Rajasthan	Lupin Laboratories, Bombay
8.	Bamboo by tissue culture	University of Delhi, Delhi	Tata Energy Research Institute, New Delhi
9.	Animal birth control injection (TALSUR)	National Institute of Immunology, New Delhi	Karnatka Antibiotics and Pharmaceuticals Ltd., Bangalore
10.	Osmotolerant and high alcohol tolerant yeast strain	Institute of Microbial Technology Chandigarh and Vittal Malaya Scientific Research Foundation, Bangalore	United Breweries, Bangalore
11.	Blood grouping monoclonals	National Institute of Immunology, New Delhi	Cadila Laboratories, Ahemdabad
12.	Microbial convention on benzaldehyde into L-phenylacetylcarbinol	Central Drug Research Institute, Lucknow	Atlus Laboratories, Ambala
13.	F-MOC derivatives of 12 amino acids	Center for Biochemical Technology, New Delhi	Atual Products, Bulsar
14.	Hepatitis B detection kit	National Institute of Immunology, New Delhi	M/s. Lupin Labs Ltd. Bhopal
15.	Leprosy immunomodulator	National Institute of Immunology, New Delhi	Cadila Laboratories, Ahmedabad
16.	Leishmaniasis detection kit	Central Drug Research Institute, Lucknow	Span Diagnostics Ltd. Surat
17.	Monoclonals to M 13 phage proteins III and VIII	University of Delhi, Sourth Campus	Pharmacia Inc. USA
18.	Liposomal amphotericin-B	Seth G.S. Medical College and Hospital, Bombay	ACE Diagnostics, New Delhi
19.	Western blot test for HIV-I and II	Cancer Research Institute, Bombay	M/s. J. Mithra and Co., New Delhi

Table 1 (continued)

S.No	Technologies	Developed by	Transferred to
20.	Development of a drug formulation for prevention of septic shock in burn patients	National Institute of Immunology, New Delhi	Gufic Health Care Ltd., Mumbai
21.	Process know-how manual for infectious bovine rhinotracheitis (IBR) vaccine as developed by BAIF Foundation, Pune	BAIF Foundation, Pune	Hoechst Roussel Vet India Ltd. (HRV)
22.	Agglutination based detection of HIV-I/II antibodies in human blood	University of Delhi, South Campus	Cadila Pharmaceuticals Limited, Ahmedabad
23.	Plant-tissue culture	TERI, New Delhi	Cadila Pharmaceuticals Limited, Ahmedabad
24.	Plant-tissue culture	NCL, Pune	Cadila Pharmaceuticals Limited, Ahmedabad
25.	Mass production of mycorrhiza	TERI, New Delhi	Cadila Pharmaceuticals Limited, Ahmedabad
26.	Lipase for food industry	UDSC, New Delhi	Techno Emo, New Delhi
27.	Mass production of rhizobial fertilizer	RRL, Jammu	M/S Prathistha Industries Ltd., Secunderabad M/S Javeri Agro industries and Investment Co. Ltd., Amravati
28.	Mass production of biopesticides – *Trichoderma*	RRL, Jammu	M/S Prathistha Industries Ltd., Secunderabad M/S Javeri Agro industries and Investment Co. Ltd., Amravati M/S Bee Zed Biotech., Gurgaon
29.	Mass production of biopesticides – *Trichogramma, Heliothis* NPV	TNAU, Coimbatore	Crop Health Products Ltd. Ghaziabad
30.	Mass production of biopesticides – *Trichoderma*	TNAU, Coimbatore	Crop Health Products Ltd. Ghaziabad Hoechst AgrEvo, Bombay Maharashtra Cooperative Oil Seed Federation, Jalgaon
31.	Mass production of biopesticides – *Aspergillus niger*	IARI, New Delhi	Cadila Pharmaceuticals Ltd., Ahmedabad
32.	*Amaranthus* protein gene for nutritionally enriched animal feed	NCPGR, New Delhi	Cadila Pharmaceuticals Limited, Ahmedabad

Table 1 (continued)

S.No	Technologies	Developed by	Transferred to
33.	The IgM Mac ELISA for the detection of dengue	National Institute of Virology, Pune	Zydus Cadila Health Care, Ahmedabad
34.	The IgM Mac ELISA for the detection of Japanese encephalitis	National Institute of Virology, Pune	Zydus Cadila Health Care, Ahmedabad
35.	The IgM Mac ELISA for the detection of West Nile fever	National Institute of Virology, Pune	Zydus Cadila Health Care, Ahmedabad
36.	ELISA system to measure alpha-fetoprotein levels in pregnant women.	Indian Institute of Chemical Biology, Kolkata	Shantha Biotecnics Hyderabad
37.	An IgM-based assay for the detection of hepatitis A virus using monoclonal/ polyclonal antibodies.	National Institute of Virology, Pune	Bharat Biotech. Ltd., Hyderabad
38.	Urine-based system (ELISA) for the detection of four reproductive hormones.	Institute for Research in Reproduction, Mumbai	Zydus Cadila Health Care, Ahmedabad
39.	Western Blot for detection of HIV-1 and -2	Cancer Research Institute, Mumbai	J. Mitra & Co., New Delhi
40.	Agglutination test for HIV-1 and -2 using recombinant reagents	University of Delhi, South Campus	Cadila Pharmaceuticals, Ahmedabad
41.	A technology utilizing *Yarrowia lipolytica* expressing hepatitis B surface and pre-S genes (yielding high level of proteins/single step purification).	University of Baroda, Baroda	Biological Evans Ltd., Hyderabad
42.	A technology for expressing hCG using *Pichia patoris* system	Indian Institute of Sciences, Bangalore	Cadila Pharmaceuticals Ltd., Ahmedabad
43.	OIL ZAPPER Technology for oil spill treatments	TERI, New Delhi	Sriram Biotech Ltd., Hyderabad BPCL, Mumbai
44.	Recombinant protective antigen (rPA) against anthrax.	Centre for Biotechnology, JNU, New Delhi	Panacea Biotec Ltd. New Delhi
45.	Diagnostic test for peste des ruminants virus.	Madras Veterinary College, TANUVAS, Chennai	Indian Immunologicals, Hyderabad
46.	Production of xanthan gum	Birla Institute of Scientific Research, Jaipur	M/s Shriram Biotech Ltd., Hyderabad

Briefly, the programmes and achievements in some of the broad areas of support are outlined.

2.1 Agriculture

About 800 million people in the developing World do not have enough to eat, another 34 million people in the industrialized nations also suffer from chronic food insecurity. Almost 2/3 of the industrialized people in the world live in Asia and Pacific. India alone has 204 million undernourished which is more than all of the sub-Saharan combined (FAO 1999). At the recent World Food Summit, all countries agreed that there should be a reduction by half in the number of hungry by 2015. This has to be translated into concrete objectives at local, national and regional level.

In India today we have approached a food production level of more than 200 milion tonnes (mt). According to ICAR, in 1998–1999, India for the first time had a record production of 202.5 mt. With rice and wheat being 84.7 and 71.0 mt, respectively, India has emerged as the second largest producer in the world. Production of pulses, oilseeds, potato, milk, egg and fish have also been high, placing India amongst the front runners in their production. Generation, testing and adoption of improved technologies have played a major role in enhancing the production and productivity. Our population is likely to range from 1.4–1.5 billion by 2050; will it be possible for our farmers to produce over 300 mt of food grains? – The answer is "yes", provided the appropriate technologies are introduced.

Biotechnology can help to solve many problems limiting crops and livestock production in developing countries. National programmes need to ensure that biotechnology targets all sectors of society including resource-poor rural populations, particularly in marginal areas where productivity increases are difficult to achieve. Biotechnology provides powerful tools for sustainable development of agriculture including fisheries, forestry as well as the food industry. Appropriately integrated with other technologies for the production of food and agricultural products, biotechnology can be of significant assistance in meeting the needs of an expanding and increasingly urbanized population in the new millennium.

The success of the "Green Revolution" has depended on creating desirable new plant and animal varieties through conventional plant-breeding techniques. Biotechnology and especially genetic engineering offers a faster route. This technology has today proved useful for increasing yield, reducing use of pesticides, providing improved fertilizers and developing varieties tolerant to biotic and abiotic stress. Agriculture biotechnology has today come of age; it is referred to under different names – genetic engineering, gene splicing, bioengineering, recombinant DNA technologies. This technology represents the latest tool that has been provided to plant breeders. It allows well characterized genes to be transferred from one organism to another, thus increasing the genetic diversity available to improve important commercial crop plants. The need to speed up the industrialization of agriculture is increasing. Due to a liberalization in international economy and globalization, the farming community is now highly competitive and in order to survive in the market they must produce more, acquire the best varieties and use the most advanced techniques. The use of the new biotechnology tools gains importance in this context.

It can be hoped that 3/4 of the increase in food production will come from intensified production on existing land through the recent advances in bioscience. For crop improvement and enhanced productivity, the exploitation of heterosis vigor and development of new hybrids including apomixis, genes for abiotic and biotic resistance, and developing planting material with desirable traits would be essential. Genetic enhancement in the case of all important crops like wheat, rice, vegetables, coarse grains, and important horticultural crops would dominate the research scenario of the future. Around the world nearly 800 million people go to bed hungry; 170 million preschool children are undernourished. Transgenic research helps in producing new varieties that have higher productivity due to resistance to biotic and abiotic stress and are also nutritionally enhanced. The transgenic crops offer a more convenient and flexible management, higher productivity and a safer environment due to decreased use of conventional pesticides. Integrated nutrient management and development of new biofertilizers and biopesticides, particularly, plant-based ones, would be important. The impact of biotechnology on human life and economic progress of various nations worldwide has given a major impetus to accelerating research, development and application of this field in the relevant socio-economic sectors.

Application of biotechnology has had a tremendous impact on the field of biomolecular research, food processing etc. In the area of agriculture biotechnology also, there are a number of success stories – use of tissue culture for producing large quantities of elite plant material, development of transgenic plant resistance to biotic and abiotic stress, effective production of improved strains of biofertilizers and biocontrol agents, improved livestock varieties and improvement in the aquaculture sector.

In India biotechnology has been applied in all these areas and the concerted support provided by the Government has today yielded fruitful results. Support for research and development and demonstration-related activities is being provided to the universities, research institutes and voluntary organizations across the country. The support of the various Ministries and Departments of the Government of India – Department of Biotechnology, Indian Council of Agricultural Research, Indian Council of Medical Research, Council of Scientific and Industrial Research, Ministry of Agriculture – has yielded fruitful results and today India is poised to move ahead in the application of these technologies for improving the livelihood and the socio-economic status of our population.

Crop improvement involves in principle increased production potential of crop plants and sustaining it, making the produce cost-effective and improving crop quality. Commercial crop improvement programmes in the past addressed all these components and contributed significantly to the overall programme of crop improvement. Recombinant DNA technology has hastened the conventional breeding programmes. The priority crops are major cereals (rice and wheat), legumes (pigeonpea, chickpea), oilseed crops – Brassicas, groundnut, cotton, millets.

Genetic improvement of major Indian crop species has been achieved mainly through classical genetics and plant breeding methods. Recent advances in cellular and molecular genetics have opened new avenues to apply biotechnology tools for augmenting these conventional approaches. Considering the urgency to enhance food productivity and ensure nutritional quality, genetic engineering methods in producing transgenics with desired traits are assuming enormous significance worldwide. The speed with which this new technology is being commercialized is amazing. During the 20th century traditional plant breeding has brought about enormous increases in crop productivity. However, plant improvement by hybridization is slow, and is restricted to a very small gene pool owing to natural barriers of crossability. Beginning in the early 1980s, advances in plant-cell culture and genetic transformation have overcome these barriers by making it possible to transfer defined genes into all major food crops.

2.1.1
Transgenic Research

India has evolved from a net importer of food grains with a domestic production of about 50 mt in 1950–1951 to a position of self sufficiency of food grains. The country has achieved a record food grain production of 206 mt in 1999–2000. The intensive area development programme (IADP), high-yielding variety programmes and multiple variety programmes and multiple river valley projects played a key role in agricultural production. The Government announced the first National Agricultural Policy (NAP) to Parliament in June, 2000. The NAP envisages a growth rate of 4% in the agriculture sector over the next 2 decades. Besides achieving food and nutritional sufficiency, the policy parameter lays down broad management of inputs, providing incentives for agriculture, promoting investment in agriculture, ensuring better risk management and introducing management reforms.

In order to achieve food and nutritional security, the new tools of biotechnology, especially the development of transgenics is gaining importance. The potential of this technology to enhance the agriculture productivity has been globally accepted, however, there are a number of concerns which need to be addressed to make this technology acceptable to the public. The successful application of biotechnology for the production of desired crops to address issues related to malnutrition, poverty, increased productivity etc. has opened vast avenues for rapid industrialization of this technology. The number of countries growing transgenic crops has increased from 1 in 1992, to 6 in 1996, to 9 in 1998, 12 in 1999 and 16 in 2002. The total area under transgenics has increased from 1.7 million ha in 1996 to more than 52 million ha in 2001. Between 1995 and 1998 the value of the global market in transgenic crops grew from US $75 million to US $1.64 billion. The market is projected to increase to $6 billion in 2005, and to $20 billion in 2010 (James, 1998). Today nearly 30 agricultural biotech products are on the market and an equal number are being tested for release.

The Department of Biotechnology has made a concerted effort to support various programmes in the area of plant biotechnology. Three major types of activities are being supported: research and development projects in identified priority crops (rice, wheat, chickpea and mustard); multi-institutional projects on development of transgenics for stress resistance and quality traits and plant molecular biology programmes in centers of excellence. In addition to the public funded research there is a major interest being shown by the private groups also. Some of the major crops being studied are tobacco, potato, brinjal, rice, tomato, cabbage, mustard, cotton and cauliflower. Limited field trials have been initiated in most of the cases (Table 2). Some salient achievements in the development of transgenics are as follows:

- Complete transgenic systems for *Arabidopsis*, rice and wheat have been established and this is now being used for developing varieties with important traits.
- Transgenic mustard with Barnase and Barstar genes have been developed and contained field trials have been conducted in more than 15 locations. Transgenic brinjal, tomato, tobacco, cabbage and cauliflower with toxin gene have also been developed and contained field trials completed.
- The AmA1 gene has been isolated from *Amaranthus* and successfully expressed in potato. The transgenic high-protein potato is now being field evaluated. A significant increase (35–45%) has been observed with respect to total protein content.
- Several genes for stress resistance have been isolated and characterized from species of the cold and coastal eco-systems.
- Transgenic cotton containing Bt toxin gene after completing field trials in 40 locations is now being tested for large-scale field trials and seed production.

A major initiative has been taken by India in the area of genomics. The Department of Biotechnology has established a National Center for Plant Genome Research in New Delhi. The main objective of the center is to take up research work and structural, functional, and plant genomics of priority crops. The emphasis at present is on chickpea and *Catharanthus*.

Considering the fact that international collaboration is essential to foster progress in generating high-precision sequence information, the International Rice Genome Sequencing Programme (IRGSP) was conceived in 1997. India has recently participated in the IRGSP along with 10 other countries, namely Japan, USA, China, Korea, Thailand, France, Taiwan, UK, Canada and Brazil. The complete rice genome sequence information obtained through this collaborative project between public funded institutions of various countries will be placed in the public domain and, hence, can be freely used by the member countries. The Indian project aims to sequence 10 Mb of the total 430 Mb rice genome over the next five years.

2.1.2 *Plant-Tissue Culture*

The biotechnology industry has in a short span of time developed into a multi-billion dollar industry. Tissue culture technology is one such technology, which is of enormous potential for commercialization for providing improved quality products and also sufficient employment opportunities there by directly providing benefit to the socio-economic sector.

Table 2. Developments in transgenic research in India [2]

Institute	Crops used for transformation	Transgenes inserted	Aim	Current Status
Central Tobacco Research Institute	Tobacco	*Bt toxin* genes	Resistance to tobacco cater pillar *(Spodoptera litura)*	One season of contained field trial completed; further evaluation under progress
Central Potato Research Institute, Shimla	Potato	*Bt toxin* gene	Resistance to potato tuber moth *(Phthorimaea opurcullella)*	Ready to under take glasshouse trials
Indian Agricultural Research Institute (IARI), New Delhi	Brinjal	*Bt toxin* gene	Resistance to shoot and fruit borer *(Leucinodes arbonailis)*	Two seasons of field trials over; further evaluation in progress
IARI, New Delhi	Rice	*Bt toxin* gene	Resistance to yellow stem borer *(Scirpophage incertulas)*	Ready for green house trials
IARI, New Delhi	Tomato	*Bt toxin* gene	Resistance to fruit borer *(Helicoverpa armigera)*	One season field trial over; evaluation in progress
IARI, New Delhi	Cabbage	*Bt toxin* gene	Resistance to diamond back *(Plutella xylostella)*	Ready for green house trial
IARI, New Delhi	Tomato	ACC synthase gene	Delayed fruit ripening	Ready for green house trials
IARI, New Delhi	*Brassica Juncea*	*Annexin* gene from *Arabidopsis*	Tolerance to moisture stress	Field trial in progress
IARI, New Delhi	Potato	Osmotin	Tolerance to moisture stress	Field trial in progress
Directorate of Rice Research (DRR), Hyderabad	Rice	*Bt toxin* gene	Resistance to yellow stem borer	Ready for green house trials
DRR, Hyderabad	Rice	Chitinase gene	Resistance to sheath blight disease	Ready for green house trials
Bose Institute, Calcutta	Rice	*Bt toxin* gene	Resistance to yellow stem borer	Greenhouse trials in progress
Delhi Univ., South Campus, Delhi	Mustard/ Rapessed	Barnase and Barstar	Pollination control for hybrid development	Ready for field trials

Table 2 (continued)

Institute	Crops used for trans-formation	Transgenes inserted	Aim	Current Status
Jawaharlal Nehru University, New Delhi	Potato	*Ama-1* gene from *Amaranthus*	To improve nutritional quality	Transgenic lines under evaluation under containment conditions
Central Institute of Cotton Research, Nagpur	Cotton	*Bt toxin* gene	Resistance to lepidopeteran insect pest	Greenhouse trials in progress
MKU, Madurai	Rice	P5cs gene	Resistance against abiotic stresses	Transgenics ready for field trials
ICAR, Research Complex, Shillong	Rice	*Bt toxin* gene	Resistance against lepidopteran insects	Transgenics ready for field trials
Rallies India Ltd., Bangalore	Rice	Snow drop Lectin gene	Resistance against lepidopteran, coleopteran in pests	Transformation being carried out
M/s Proagro PGS India Ltd., New Delhi	Mustard/Rapeseed	Barnase and Barstar	Pollination control for hybrid development	Contained field trials in more than 15 locations over; Open-field research trials in progress
M/s Proagro PGS India Ltd.	Tomato	*Bt toxin* gene	Resistance to lepidopeteran insect pests	One season contained field trial over; further trials in progress
M/s Proagro PGS India Ltd.	Brinjal	*Bt toxin* gene	Resistance to lepidopteran insect pests	Glasshouse experiments in progress
M/s Proagro PGS India Ltd.	Cauliflower	Barnase and Barstar	Pollination control for hybrid development	Glasshouse experiments in progress
M/s Proagro PGS India Ltd.	Cauliflower	*Bt toxin* gene	Resistance to insect pests	Glasshouse experiments in progress
M/s Proagro PGS India Ltd.	Cabbage	*Bt toxin* gene	Resistance to insect pests	Glasshouse experiments in progress
M/s Mahyco, Mumbai	Cotton	*Bt toxin* gene	Resistance to lepidopteran insect pests	Field trials in over 40 locations completed; Now large scale field trials and seed production underway

Current techniques for in vitro propagation of plants and the ready acceptance of tissue culture-raised plants by the commercial sector, have allowed for continued growth within the micropropagation industry. Micropropagation technology is being extensively used for large-scale production of elite planting material of desired characteristics. A number of horticultural crops, i.e., ornamentals, vegetables and fruit crops are being propagated commercially through tissue culture. Globally the commercialization of tissue-culture technology is increasing rapidly.

The recent move towards liberalization of the Indian economy has a profound impact on the future development of the biotechnology industry. Micropropagation of plants through tissue-culture techniques has proved to be one of the most successful and popular economic developments. Micropropagation techniques have been commercialized globally, especially in the nations with industrialized agriculture. It is reported that over 796 commercial companies are engaged in such activities all over the world and their number is increasing. Annual plantlet production is approximately 900 million plantlets. In India, there are nearly 70 registered companies with approximately 30 companies having immediate plans for selling tissue-cultured plants and more than a dozen already on the market with their products. Annual production is estimated at approx. 50 million plantlets; however, the installed capacity is nearly 150 million. Tissue-cultured planting material of fruit crops like banana, strawberries, grapes, papaya, cardamom and ornamental flowers such as gerbera, carnation and orchids are available besides a number of foliage ornamentals.

The tissue-culture industry today is mainly catering to the requirements of the floriculture and horticulture sectors. In fact, the tissue-culture industry forms the major backbone of the floriculture and horticulture industries. The elite quality starter material for the floriculture and horticulture crops is provided through tissue-culture regeneration which helps in multiplying in large quantities the mother explant. There is a tremendous global market for floriculture crops which stands at approx. US $ 60 billion growing steadily at a rate of 11% p.a. The flower consumption is mainly confined to the developed nations and this is largely dependent on socio-economic and cultural factors. Globally, the tissue-culture companies are also increasing and the 900 million plants being produced worldwide results in an annual turn over of US $ 250 million as against a global market of US $ 15 billion. The potential demand for planting material of agriculture, horticulture, floriculture, and forestry crops is 16 trillion units of propugules per year globally with a potential market size of US $ 4 trillion per year.

Plant micropropagation is a powerful and potential tool for producing the required number of propogules of desired characteristics at a competitive price, as and when the customer or farmer requires. Large-scale propagation of elite clones from hybrids or specific parental lines through commercial micropropagation holds the promise of alleviating problems of shortage of healthy seeds or planting material and lack of disease-resistant clones which have been affecting the economy of various countries of the world, especially the developing countries. Plant-tissue culture can virtually double the yield of clonally propagated crops such as banana, cardamom, etc. With the shrinking per capita land, it is essential to increase the productivity of the land and quality of the produce.

In India, the existing natural resources and expertise such as abundance of genetic base, diverse agroclimate zones, highly qualified and skilled manpower, agri-based economy etc. provide enough scope for expansion of the industry. In order to fully exploit the plant-tissue culture technology commercially for production of desired planting material, it is essential to work out the economics of plantlet production. It is, therefore, necessary to suitably modify the technology in order to produce plants at the least cost. Cost-effective production of desired phenotypes of plant species should be the prime concern of the tissue-culture industry.

Although there is a great potential for the tissue-culture products both in the domestic and export market this has not been fully exploited. The main reason for the decline of the tissue culture industry has been a saturation of the products mainly because the industries have been targeting a limited export market. There is an urgent need today for the industry to tap the potential that exists in the domestic market and with proper product diversification it will be possible to achieve this. The tissue-culture technology requires intensive market survey and introduction of this new technology and product demands a high purchasing power.

The Department of Biotechnology has played a pivotal role in promoting the commercialization of plant-tissue culture. A vast network programme covering nearly 80 research institutes/universities has been working on protocol development of 60 different species; 20 technology packages have been perfected. Two Micropropagation Technology Parks set up by the Department at NCL, Pune and TERI, New Delhi have successfully transferred 5 technologies to the industry. Nearly 10 million plants have been produced and field-planted over 8500 ha area.

2.1.2.1
Horticultural and Plantation Crops

India being endowed with a wide variety of agroclimatic regions holds an important position in the horticulture crops. India is the second largest producer of fruits and vegetables next to Brazil and China. Total fruit production is estimated at 40.05 mt from 3.68 million ha. There is still enormous potential to enhance the productivity per unit ha. With the recent WTO agreement, Indian horticultural produce/products will have to be competitive both in the domestic and export markets. This calls for the use of latest hi-tech horticulture technologies. Such latest technologies include genetically modified crops, tissue culture for production of superior quality planting material, integrated nutrient and pest management, molecular diagnostic kits for disease detection etc. The efforts in this direction are continuing.

The Department of Biotechnology has been supporting activities for improvement of horticultural and plantation crops through tissue culture techniques. Some of the priority crops identified are citrus, mango, banana, apple, tea, coffee, rubber and spices. R & D efforts have been supported for perfecting regeneration protocols.

Complete regeneration and micropropagation protocols have been developed for black pepper, coffee, tea, cardamom, vanilla and some other spices such as clove, kala zira, saffron etc. In order to mass multiply these spices for commercial purposes, the efficiency of the protocols is being tested and field demonstrations are being conducted to test the performance of tissue-culture raised plants vis-à-vis the control. Banana, cardamom and vanilla have already been demonstrated and good results reported. An effort is being made now to develop complete technology packages which can be adopted by the farmers.

Major crops-based network programmes have been supported for improving productivity through biotechnological approaches. The thrust has been on developing disease-resistant high-yielding varieties through transgenic and tissue culture regeneration. The main crops are: coffee, tea, black pepper, ginger, cardamom etc.

2.1.2.2
Forestry

India has a land area of 3.29 million km^2 and 0.64 million km^2 of forest area (nearly 2.5% of the world's geographical area and only 1.8% of the world's forest area). The average productivity of forests in India is 0.7 m^3/ha/year as against the world's average of 2.1 m^3/ha/year. Research efforts for increasing the productivity and conservation of elite material have been supported. The application of tissue culture offers immense potential in this area, and this is now being optimally exploited. Tissue-culture regeneration of elite varieties not only helps in producing large quantities of true-to-type planting material but also ensures that this is done in the shortest period of time. The difficult-to-root recalcitrant species for which there is a serious dearth of planting material are multiplied through tissue culture and made available for the afforestation programme.

Taking note of the urgent requirement of producing high-quality planting material in the area of forestry, the Department of Biotechnology initiated major activities during the 7th and 8th plan period. Research and development projects were supported at various universities and research institutes across the country for standardizing complete protocols for plantlet regeneration through tissue culture of nearly 20 important forest tree species. To date protocols for more than 10 different tree species have been perfected and demonstrated effectively in the field.

2.1.2.3
Tissue Culture Pilot Plants

A concerted effort has been made by the Department to demonstrate the feasibility of large-scale production of forest trees through tissue culture. It was felt necessary that, before the technology is transferred for large-scale production to the entrepreneurs and other end-users, field data should be collected. Two Tissue Culture Pilot Plant Units were set up at NCL, Pune and TERI, New Delhi as major infrastructure facilities to successfully demonstrate the feasibility of large-scale production and plantation of forest trees. Each TCPP has a production potential of approx. 1 million plants. The protocols developed at the TCPP and by other R & D groups were refined and scaled up at the TCPP. Cost-effective technology packages have been perfected. Nearly 60 lakh plants have been produced and field planted over an area of 6000 ha. The main forest tree species being multiplied are (i) *Eucalyptus tereticornis*, (ii) *Eucalyptus camaldulensis*, (iii) *Dendrocalamus strictus*, (iv) *Populus deltoides*, (v) *Tectona grandis*, and (vi) *Anogeissus pendula* [1].

The field demonstration trials have been conducted in 17 states in association with the State Forest Department, Forest Corporations and other user agencies. In addition, the agricultural universities are concentrating on field trials of agroforestry species. The main emphasis of the study is to demonstrate the field performance of the tissue culture-raised plants of the different forest tree species in the field. The preliminary data collected indicate a survival percentage of 95–98% and a better response of the growth parameters vis-à-vis the control. The preliminary cost-benefit analysis has been conducted in the farmers' field and the demonstration plot of *Eucalyptus* demonstrates higher growth rates, early rotation leading to a higher cost-benefit ratio (1:3). Wood volume of tissue culture-raised plants was 45 m^3/ha higher and fetched 38.91% more value of wood which resulted in higher net profit of 42% and approx. Rs. 15,000–20,000 additional earning per hectare in the case of TC-raised plants of *Eucalyptus*.

An important aspect of the field evaluation of different plant species, especially forest trees, is the testing of genetic fidelity of the tissue culture-raised plants. The Department has made a concerted effort to evaluate the genetic fidelity of the plants in the field in order to collect data on different parameters. For those plants which are being produced in large quantities, the complete genetic variability has been studied using different molecular markers. Research projects have been supported in different research institutes/universities.

2.1.2.4
Hardening Units

One of the main bottlenecks faced in the large-scale multiplication of forest trees and other plants through tissue culture, especially in view of their specific requirements related to the different agroclimatic regions, is the hardening procedure. In order to tackle these issues effectively the Department took up as a priority the objective of supporting smaller hardening units catering for the requirements of the specific agroclimatic regions. Six regional hardening units have been established. Through the efforts of these hardening units the protocols for various regions have been scaled-up and the tissue-culture technology is being popularized and made highly cost effective so as to reach the grass root level. The other important feature of these hardening units is that they promote effective linkage for technology dissemination at a state level.

2.1.2.5
Micropropagation Technology Parks

The two TCPPs at NCL, Pune and TERI, New Delhi, were then converted to Micropropagation Technology Parks which act as an interface between the research institutes and industry. In addition to demonstrating the large-scale production of forest tree species the MTPs are also optimizing protocols for horticulture and plantation crops and other crops of commercial interest. The MTPs have perfected technologies for nearly 20 different plant species and have already transferred to the industry 10 different technologies. There is a strong R & D group at the two MTPs and a continuous effort is being made to refine the technologies developed either at the MTP or by different research groups, so that cost-effective technology packages could be generated. The MTPs also provide training for generating skilled manpower and interact closely with the industry, state departments and other end-users so that different technologies can be made available for adoption. They also offer contract research and contract production to the industry and smaller entrepreneurs. Projects have been taken up on a turn key basis for different state departments/universities/agencies. Forest trees are being planted in association with State Forest Departments. In addition, some economically important crops such as sugarcane and potato are also being multiplied in large number. The thrust of this activity is to promote the technology at grass-root level in association with farmers. The cost-benefit analysis for the different plant species produced through tissue culture has also been worked out. In the case of sugarcane, a 1.5-fold increase in yield and a 1% increase in sugar content have been reported. A 5-fold increase in biomass yield has been reported in the case of banana.

The Department is making a concerted effort to promote the commercialization of the plant-tissue culture technology. One of the main concerns of the plant-tissue culture industry was that the tissue culture raised plants need to be certified both in terms of being virus free and also the quality. The Department has now set up a National Facility for Virus Diagnosis and Quality Control. The main facility is located at IARI, New Delhi and there are 5 satellite centers and the facility is functioning in a network manner. Virus diagnosis is being done for known and unknown pathogens. Diagnostic kits are being developed. The quality control specially for the forest trees is being certified using molecular markers and DNA fingerprinting techniques. The planting material will be indexed and certified for the industry, and other research institutes/organizations involved in the large-scale production of tissue-culture plants.

2.1.3
Medicinal and Aromatic Plants

Medicinal and aromatic plants hold enormous potential in view of the rich bioresources existing in our country. Major programmes have been launched not only for mass cultivation but also for their sustainable utilization. Applications of biotechnology for conservation, micropropagation, production of secondary metabolites, biotransformation of intermediates into pharmaceutically important products and genetic improvement of medicinal and aromatic plants have been taken up. The Department has supported programmes for mass multiplication of elite medicinal and aromatic plants, and end-to-end programmes for herbal product development.

Under this programme germplasm conservation is an important component. Four gene banks have been set up. The four gene banks set up by the DBT have collected nearly a thousand accessions from different biogeographic regions of the country and are conserving collected accessions of rare, threatened, endangered and economically important species in field gene banks, seeds banks, in vitro repository, DNA bank and under cryopreservation. The gene banks at: Tropical Botanical Garden and Research Institute, Thiruvananthapuram, National Bureau of Plant Genetic Resources, New Delhi, Central Institute of Medicinal & Aromatic Plants, Lucknow, and Regional Research Laboratory, Jammu Tawi are concentrating on medicinal and aromatic plants, critically endangered high altitude medicinal plants of the North-West Himalayas and accessions from the wild collection of different regional stations. In addition to conservation of seeds and other materials in a seed bank and through cryopreservation, the emphasis is also on field gene banks.

2.1.4
Biological Software

Indian agriculture has shifted from traditional to intensive farming with indiscriminate use of chemical fertilizers and pesticides which, in turn, has made our soils largely non-productive and has contaminated the ground water. Chemical fertilizers use, always falling short by 8–10 million tons [3], is staggering, considering the country's future demand for food. Incorporation of biofertilizers or microbial inoculants in combination with organic manure can reduce the detrimental effects of the current agricultural practices in an eco-friendly manner. Use of biofertilizers such as *Rhizobium, Azospirillum, Azotobacter*, blue green algae, *Azolla*, mycorrhiza and phosphate solubilizers can facilitate management of crop nutrients leading to long-term sustainability in crop production. Their addition also increases the fertilizer use efficiency of the crop plants. With an ever increasing population, it is necessary to increase crop productivity per unit area without causing further deterioration in soil health and maintaining micronutrient balance.

The area of biological software has today advanced very rapidly and complete practices are available for biofertilizer and biocontrol applications. Suitable strains have been identified for different crops and regions and these are now becoming highly acceptable by the farming community since they provide valuable inputs for increased productivity and sustainable agriculture.

Most of the production of biofertilizers in the country is being done in the public sector by research institutions, universities and the National Biofertilizer Development Center. A few state and co-operative fertilizer units also have ventured into this field. Surprisingly, the involvement of the private sector is extremely limited in spite of it being a low investment and high benefit technology. Almost all the production units in the country are producing only the bacterial biofertilizers and the present day annual production is estimated to be around 9,000 metric tonnes which is far below the demand of more than 700,000 mt/annum.

On the other hand, our domestic production of nitrogen fertilizers is only around 8.6 mt with an infrastructure investment of Rs. 120,000 million. This, together with an import of 2.34 mt and a total subsidy of Rs. 48,000 million is able to meet only about 2/3 of the total fertilizer requirement of the country. The consumption of fertilizers is also not uniform and seems to be directly proportional to the size of the land holdings. Only 30 percent of the total fertilizers consumed in the country are used by about 68 million farmers owning 65 percent of the total arable land. It is this group which can be easily motivated to adopt biofertilizer technology. The area of biopesticides also has enormous potential.

The Department of Biotechnology has supported a large number of programmes in the area of biopesticides and biofertilizer technology development [1]. Several packages have been developed to facilitate the use of the new technology by the farming communities. The biofertilizer technology has been demonstrated benefiting nearly 20,000 farmers. The technology for production of mycorrhizal biofertilizer has been transferred to the industry. In the area of biocontrol 75,000 ha have already been covered and 35,000 farmers have been benefited. An antifungal formulation for control of root rot disease was developed and transferred to the industry.

2.1.4.1 Biofertilizer

Blue-green algae (BGA) for rice, *Rhizobium* for legumes and oil seeds, *Mycorrhiza* for tree crops and *Acetobacter* for sugarcane have significant potential to sustain productivity and soil fertility. The concept of the biofertlizer programme is to domesticate some of these microorganisms in our agriculture production system, so that the vast natural reserve of nitrogen in atmosphere can be tapped as an additional source to meet our requirements, especially by small and marginal farmers who cannot afford costly chemical fertilizers.

A number of technologies have been developed under a mission-mode programme by the department and to incorporate these technologies a network programme on integrated nutrient management has been started. The major objective is the integrated use of biofertilizers in the nutrient management. This is being implemented at 17 centers and some R & D projects are also being supported in the gap areas.

For popularizing use of biofertilizers, 7,000 experimental demonstrations were conducted and about 6,000 farmers were trained in the use of blue-green algae and rhizobium. About 250 training programmes have been conducted. The increase in paddy yield due to application of blue-green algae ranged from 13–16.67% whereas in pulses and oilseeds application of rhizobium resulted in a yield increase in the range of 5–13.5%. Based on this, a network programme on integrated nutrient management has been started. The technologies have been developed for inoculum production for ecto- and endo-mycorrhizae [1].

Technologies have been developed under mission-mode programmes and packaged for transfer to entrepreneurs. A major programme on genetic modification of various biofertilizer strains to enhance nitrogen fixing ability by increasing the copy number of nif genes and also by incorporating the hup gene which is responsible for hydrogenase enzyme production is to be undertaken. The major emphasis will be on development of GMOs for higher nitrogen fixation and adaptability.

2.1.4.2
Biological Pesticides

The main objective is to develop new formulations and cost-effective, commercially viable technology for the production of biocontrol agents to be used under integrated pest management of key pests and diseases of economically important crops. Overall objective of the programme is also to integrate the use of biopesticides/natural enemies along with other components of IPM and INM for developing a capable module/package of practices which is cost-effective, sustainable and ecofriendly in different crop ecosystems. The programme aims at studies on microbial pesticides and other natural enemies, transgenic Bt. and molecular aspects of other biocontrol agents, insect cell lines, pheromones and botanical pesticides. Significant progress has been made in terms of development of cost-effective and commercially viable mass production technologies of various biocontrol agents/biopesticides, and field efficacy demonstration of various biocontrol agents/biopesticides in an area of 75,000 ha covering various economically important crops under different ecosystems has been carried out. An effort has been made to study the economic benefits in IPM trials by working out the cost-benefit ratio in terms of a) monetary gain, b) yield increase, c) reduction in pesticide consumption and promoting general awareness about the biocontrol-based IPM technology among the farming community and the end-users. Through the mission-mode programme the gap in the availability of a sufficient quantity of biocontrol agents was bridged to a considerable extent by setting up 20 production units. The mass production technologies for various biocontrol agents have been fine tuned to suit the various agroclimatic zones. A number of biopesticide technologies has been developed through this programme. Four fermentor-based technologies have been transferred to the industry.

2.1.5
Animal Biotechnology

Animal husbandry is an important source of self employment and subsidiary occupation in rural and semiurban areas, especially for people living in drought prone, hilly, tribal and other poorly developed areas where crop production alone may not fully sustain them. The share of animal husbandry to the total GDP of the country in 1950–1951 was 15.5% and this rose to 27% in 1995–1996 with an annual growth rate of 4.8% in the diary sector and 15% in the poultry sector. The per capita availability of milk has increased from 140 grams in 1947 to 195 grams today [4]. This is highly significant taking note of the increase in population since then. Although there has been a increase in the livestock population there is a great need today to have genetically superior stock which is more successful and productive. The application of newer technologies greatly contributes to this.

The DBT has initiated major programmes in animal biotechnology. The latest biotechnology tools are applied for improving the livestock production and productivity as also the wealth. Embryo transfer technologies have been standardized in buffalo and camel. Good leads have been obtained on indigenous breed characterization and embryo preservation. A new protocol for camel superovulation was developed for the first time. Expression of the bovine growth hormone gene in *E. coli* bacteria, a health package for infectious bovine rhinotrachititis diagnosis and control, and preparation of reconstituted collagen sheet of bovine origin for wound healing applications are other highlights. Seven different types of transgenic mice carrying antibiotic markers, hepatitis-B antigens, interleukin genes and other markers have been developed. A new rabies vaccine for animals has been produced and is being tested for technology transfer. Similarly, a pregnancy diagnosis test for cattle and buffalo as well as a vaccine for blocking fertility in female dogs will be ready soon.

2.1.6 *Aquaculture and Marine Biotechnology*

Fisheries play an important role in the Indian economy and support nearly 6 million full-time and occasional fishermen. This continues to be a thrust area of the country's development programme due to its vital contribution to employment generation, food security and foreign exchange earnings. The country has a vast coast line of more than 8000 km and 2 island territories. There is a high potential of marine fisheries from the Indian Ocean. Fish production in India reached a level of approx. 5.40 mt in 1998 and the country is now the 6th largest producer in the world. Out of this nearly 2.90 mt are from the marine sector. The earnings from export of fish and fishery products amounted to US $ 1.30 billion in 1998. The targeted production of 10 mt actually largely depends on aquaculture. There is a megadiversity of aquatic species existing in the country. Aquaculture represents an optimum use of scarce land resources and an efficient system for conversion of biomass into proteins. Today aquaculture products are among the fast-moving commodities. An integrated holistic approach towards the aquaculture is helping not only in diversification of the bread basket but also in employment generation and foreign exchange earning.

The thrust of the DBT-supported programmes is on improving the production and productivity of the major marine bioresources, developing a better improved feed, and health-related vaccines. Research projects have been supported to strengthen the gap in the areas of health and diagnostics, transgenics, cell and tissue culture, intensive prawn culture, carp culture, feed and seed production. Emphasis was laid on the development of bioactive compounds and culture technology in non-conventional aquaculture species, etc. Projects in these areas were taken up to sustain research and development needs and to help diversification in aquaculture towards production and productivity increase. The results were significant leading to product and process development thereby enabling transfer of various technologies to industry and entrepreneurs. Useful leads generated in the programme include the development of a heat-killed whole cell vibrio vaccine using a virulent strain of *Vibrio harveyi*. This has shown immune response in shrimp (*Penaeus monodon*) and the technology has been transferred to industry. New oligopeptide primers were designed for detection of white spot shrimp virus (WSSV) by PCR, useful in developing a diagnostic test. Monoclonal antibodies have been developed for diagnosis, serotyping and characterization of WSSV. Presumptive transgenic zebra fish were developed using rat growth hormone gene. Experiments have been undertaken to transfer newly designed vectors to produce transgenic fish. Efforts are being made to produce disease-resistant transgenic carp and zebra fish. A prawn complex comprising hatchery, feed mill and farming system has been set up for *Macrobrachium rosenbergii*. Prawn hatchery has an annual capacity of 10 million post larvae. A feed unit with a capacity of 250 kg pellets per hour has been established and feed production started.

2.2 Biodiversity Conservation and Bioprospecting

A multicentric project on bioprospecting of biological wealth using biotechnological tools has been supported jointly by DBT and Department of Space. The overall objective is of characterization, inventorization and conservation of biodiversity of different ecogeographical regions and prospecting of genes and biomolecules. The programme focuses on the two hot spots – the North Eastern Himalaya and Western Ghats and also on regions of the Western Himalaya. A major component of the programme is mapping of the biodiversity at landscape level through remote sensing. This activity is being undertaken by Department of Space. In addition to preparation of Biome maps, species level mapping is also being done. A biomonitoring system for Western Ghats is being prepared. The genetic profiling and cataloguing of species is being done using molecular markers. Economically and medicinally important species are being prospected for genes and bioactive molecules of therapeutic and agricultural importance, conservation strategies are being worked out depending on the species richness. Tissue culture regeneration systems are also being studied.

Through this programme the landscape level vegetation mapping has been completed for the North East, Western Ghats and parts of Western Himalayas. Through the use of molecular markers, complete genetic profiling is being done. Economically important elite varieties have been identified and the fingerprinting is being done to compare accession/collection from different geographic locations. This is helping in developing conservation strategies, especially for the endangered species. Studies have also been initiated in the Andaman and Nicobar Island regions.

Gene prospecting studies have been successful and today nearly 25 stress-tolerant genes from plant species of fragile ecosystems have been identified, characterized and cloned. Studies are also ongoing for transferring these genes to other economically important crop plants. Important leads have been obtained in prospecting of bioactive molecules for agricultural and therapeutic products.

2.3 Biofuels

Biomass is defined as any organic matter that is available on a renewable and a recurring basis. Apart from the dedicated energy crops and trees, biomass resources constitute forage grasses, oil seeds/waxes, short rotation woody crops, agricultural and forest residues, and algae/aquatic plants.

The ideal biomass should possess low inputs and maintenance, early maturity, short rotation, high fuel wood yield and easy harvesting. Domestic biomass resources include agricultural and forest wastes, municipal solid wastes, industrial wastes, and terrestrial and aquatic crops grown solely for energy purposes, known as energy crops. Biomass is used in the production of energy in the form of solid fuel (wood, charcoal for domestic cooking), liquid fuel (biodiesel and bioethanol), gaseous fuels, electricity/thermal power.

In India, about 390 million tonnes of major agricultural residues are generated annually which have a potential of producing approximately 170 billion liters of ethanol with the right technology [5]. At present, no technology at commercial scale is available to utilize agriculture residues, large fractions of MSW, microalgae etc. to produce bioethanol and biodiesel. The total annual production of ethanol in the country is about 1.3 billion liters from sugarcane molasses-based technology.

The Indian Institute of Petroleum, Dehradun started working on the development of energy crops for liquid fuels in 1979 in collaboration with the National Botanical Research Institute, Lucknow under the sponsorship of the Department of Science and Technology and later Department of Non-conventional Energy Sources. Out of the 74 resinous plants studied, 7 were selected as the most energy rich crops. They are *Euphorbia antisiphilitica, E. caducifolia, E. royleana, Calotropis procera, C. gigantea,* and *Cryptostegia grandiflora*. Hydrocracking of biocrudes of *Euphorbia neriifolia, E. royleana, C. gigantea* and *C. grandiflora* using palladium-nickel-tungsten catalysts yielded liquid fuels. *E. antisyphilitica* was the best amongst all, yielding a maximum amount of liquid fuel (8.8% of biocrude) and a minimum amount of gases and coke. Detailed studies have been carried out at the University of Rajasthan on the methods of cultivation of *E. antisyphilitica* as well as for improving its biocrude yield. As a result a crop of 160 tons/hectare/year of plant, 15% of which is biocrude has been harvested. This will yield 3.37 tons of biocrude/hectare every year. With 94.5% of biocrude converted to liquid fuel it comes to about 3 tons of liquid fuels/hectare.

The work on cultivation and breeding of hydrocarbon-yielding plants was initiated at NBRI, Lucknow, in 1979 and at the University of Rajasthan in 1982. The first plants to be taken up were *Euphorbia lathyris, E. tirucalli, E. royleana, E. caducifolia, Pedillanthus tithimaloides, Calotropis procera, C. gigantica,* and *Crytostegia grandiflora*. Although *E. lathyris* grows and yields well it failed to produce seeds and survive the semi-arid dry summer conditions due to heavy infection of soil fungi, *E. tirucalli* has great potential but its biocrude content is relatively lower.

It is evident that further research is needed to develop better sources and methods for the production of biofuels. One useful plant is *Jatropha curcas* Linn belonging to the Euphorbiaceae family. Neem and mahua oils have been tested for use as blends. Preliminary engine tests carried out with the blends of neem and mahua oils with diesel have shown that they can be substituted by neem oil and 20% by mahua oil. Other oils studied for the purpose are karanje and rice bran oil. All these studies show that blends in the ratio 1:1 with diesel perform well, but the oil has to be heated first. If not, the efficiency of the engine drops. The Ministry of Non-conventional Energy Sources has launched major research and development programmes on biomass production.

Significant work has been done in the area of bioethanol and biodiesel. A number of groups in IIT, Kharagpur; NCL, Pune; IMTECH, Chandigarh; UDCT, Mumbai and most importantly at IIT, Delhi have been working on new routes of ethanol from various biomass material. Efficient microbial strains have been identified and are now being genetically manipulated for increased ethanol production. Prototype packages are available which are now being refined for commercial utilization. Efforts are being made to convert lignocellulosics and other various forest residues for maximum utilization.

In the area of biodiesel a lot of work has also been done by NBRI, Lucknow; CSMCRI, Bhavnagar; IISc, Bangalore etc. Here again, the emphasis is on conversion of oil and hydrocarbon from some important tree species to diesel.

The DBT has initiated a major mission network programme on "Bioengineering of Crops for Biofuels & Bioenergy". The programme envisages three collaborative steps.

2.3.1 *Production and Demonstration*

This involves the production, demonstration and utilization of bioenergy crops, to serve as a feedstock for production of bioenergy and biofuels. Demonstration is the most important aspect as the success of bioengineering production depends largely on the supply of the feedstock and its proper utilization.

2.3.2 *Alternative Sources of Energy/Fuel*

a) Identification of alternative sources such as algae and other aquatic organisms; Hydrogen as alternative energy source.
b) Biofuel production (biodiesel, bioethanol etc.) from biomass and other waste material. This will have to be clearly linked to a feedstock programme for supply of raw material.

2.3.3 *Bioengineering*

The source material identified would be processed for the production of alternative sources of energy/fuel. Bioengineering tools would be required for the downstream processing of the source material. A viable process will be developed for scaling up and pilot production of alternative sources of energy involving fermentation technology with the application of identified microorganisms.

2.4 Environment

The area of environmental biotechnology, specially the bioremediation technology has enormous commercial potential. Technology for precombustion biobeneficiation of coal containing high pyritic sulfur and ash and precombustion desulfurization of gaseous fuels and emissions containing hydrogen sulfide with concomitant recovery of elemental sulfur has been developed. Another important product which is today being scaled-up for commercial production is the crude oil and oil sludge degrading bacterial consortia named "OIL ZAPPER". A two-step approach for remediation of crude oil spills has been developed. The process includes application of cost efficient adsorbent, alkali-treated sawdust, as the first step to recover most of the spilled oil. The second step involves biodegradation of remaining oil by using a bacterial consortium for removal of xenobiotics. The degradation process is cost-effective and environmentally benign.

An integrated biotechnological approach (IBA) has been developed for bioremediation of mine spoil dumps. The process leads to the development of supportive and nutritive rhizosphere in manganese and coal mine spoil dumps through appropriate blending of spoil and organic waste(s) for the establishment of the plants microflora, and inoculation of plants with specialized culture(s) of nitrogen-fixing microorganism. Strains of endomycorrhizal fungi are used for profuse root development and stress tolerance in plants. The ecological restoration technology for revegetation of degraded land has been successfully transferred to the industry. Other important technologies which are available for commercialization are development of biosensor for the detection and estimation of organophosphorus pesticide residues in water, process for palm oil mill effluent treatment, process for dye industries effluent treatment and its validation as well as tannery-based effluent treatment.

2.5 Healthcare

To provide a low-cost, affordable, easy to use health care system through the inputs of biosciences is one of the most challenging goals for the bioscientist in the 21st Century. A large proportion of the global investment in biotechnology goes to the healthcare sector – both human and animal. Biotechnology-based healthcare products are likely to dominate the market at about 40% of the consumption level. Monoclonal and polyclonal antibodies for disease immunodiagnosis, tissue typing, clinical assays and research constitute a large proportion of the market. The potential of the diagnostic market is very high and is also anticipated to be very diverse considering the growing population, the declining status of health, the emergence of infection disease etc.

In India the prospects for investments in the biotechnology-based healthcare system are enormous. Biotechnology facilitates the development of a new phase in drug production – drug detecting and targeting. The generation of new improved vaccines through use of recombinant DNA techniques is also of high commercial importance. We are well poised to meet this challenge. Research efforts across the country have succeeded in developing a number of developments of improved vaccines including edible vaccine.

Presently in India, eight units have taken steps to produce r-DNA products in the healthcare area. Recently an Indian company had launched an indigenous r-hepatitis B vaccine, first ever-r-DNA product. At present globally there are about 35 biotechnology-derived therapeutics and vaccines in use and over 500 additional drugs and vaccines are in various phases of clinical trials.

In the diagnostic area, the market is rapidly developing with the introduction of easy-to-use, reliable and cost-effective methods of diagnostic devices for detecting both communicable and non-communicable diseases. Technologies have been local as well as imported. Biotechnology has also spurred growth in diagnostics and over 600 biotechnology-based diagnostics are now available in clinical practice with a value of about US $ 20 billion. Many more are about to enter the market, the most prominent among these will be PCR-based diagnostics. In India the diagnostic sales are expected to be between Rs.1–2 billion. At present we rely on imports for many of the immunodiagnostic kits.

One of the great areas of promise for the future in healthcare is immunodiagnostics. There are a large number of diseases where the disease itself or predisposition to it is wholly or significantly genetic in origin; and there are chronic degenerative conditions. But there is also a host of afflictions, particularly in the developing countries, whose origin lies in infection or the environment. Tuberculosis is an example of such a disease. These are not often easy to diagnose; and invariably demand highly specialized skills and tests that are not easily available or are costly. It should be possible to detect through immunodiagnostics, the body's reaction to these externalities that result in disease. Such immunodiagnostics have to be easy to use, reliable and low in cost. It would enable early detection and, therefore, more specific and ultimately less expensive treatment. Morbidity and mortality that otherwise result from delayed or wrong diagnosis can be greatly reduced. Major efforts in this direction would be particularly important for developing countries such as India.

Through the efforts of the DBT, projects have been implemented in the area of major infectious and non-infectious diseases, viz., tuberculosis, HIV, malaria, leishmania, cholera, dengue, typhoid, cancer and heart diseases. Major emphasis was also given to the generation of projects on oral cancer and clinical applications of stem cells. Three diagnostic kits for detection of HIV-I and -II, a therapeutic vaccine for leprosy, the Leprovac as an immunomodulator and a drug delivery system for systemic fungal infections have been transferred to industry. Several leads are being pursued in case of hepatitis, malaria and leishmaniasis. There is an Indo-US Vaccine Action Programme with very good results. Two hepatitis C diagnostic tests and a prototype vaccine for rotaviral diarrhoea have been developed and validated. Special projects for development of edible vaccines for rabies and cholera are in an advanced stage. An ELISA-based diagnosis system has been validated for Japanese encephalitis towards technology transfer. Another diagnosis test system for *H. pylori* is being validated. Eight test systems for diagnosis of dengue, hepatitis and reproductive hormones have been transferred to the industries. A number of projects on neurosciences including developmental neurobiology, aging, neuronal death have been supported. Studies on distribution of HIV-I subtypes in the eastern parts of India revealed interesting results. Technology for 11 diagnostic kits has been developed and validated and transferred to industry. In the area of DNA vaccines also very good leads have been obtained especially in the area of cholera, rabies and JEV. Clinical trials are underway. These products will be ready for commercialization very soon.

The pharmaceutical industry in India is very strong and vibrant with expertise for chemical drugs. India is competing well in the bulk drug production market. This is largely due to the more efficient nature of the processes and manufacturing costs. The same expertise and costs can hold true for the biotech products even though the pharmaceutical industry has little experience in diagnostics and in biotech therapeutics. The pharmaceutical industry needs to invest more in R & D and concentrate on local problems, which have remained untouched by overseas biotechnology industries. It is imperative that optimal exploitation is ensured with the availability of indigenous technologies that required skilled human resources and the appropriate infrastructure.

2.5.1
Human Genetics

The Human Genome Project ushered in a new era of genomics during 20th Century in the international scenario. The technological developments that have taken place since the initiation of this international megaproject have provided opportunities for a big leap in the area of human genetics and genome analysis. Keeping in view the priorities and also to keep pace with international efforts, the department initiated major programmes in the area of human genetics and genome analysis during 1990–1991. This was mainly to provide genetic diagnosis and counseling to the families having genetic disorders prevalent in the country with the aim to develop new methods for the diagnosis of such disorders and to find out the functions of the genomic DNA sequences. Since the inception of the human genetics and genome analysis programme, several projects have been implemented including genetic diagnosis cum counseling units in different parts of the country and also major programmes in the area of functional genomics and human genome diversity.

Sixteen genetic clinics and three major projects were established for providing molecular diagnosis and counseling for the common genetic disorders such as beta-thalassemia, Duchenne muscular dystrophy (DMD) and others prevalent in the country to the affected families. About 16,500 affected families so far have benefited from these units for genetic diagnosis including prenatal diagnosis and counseling for major genetic disorders. A genomic diversity analysis of the Indian population by utilizing genotyping of 8 human specific insertion/deletion for polymorphisms in population was carried out.

Gene therapy offers the potential to correct diseases at the genetic level. There are presently about 130 approved gene therapy protocols and some of the protocols are being tested clinically.

In the programme on functional genomics, powerful computational capability for handling large-scale human genome sequence data, robotic methodologies for genotyping and PCR-based diagnostics for common genetic disorders have been developed. As a post-genome initiative, programmes are being taken up on functional genomics, pharmacogenomics, custom-made drug designing, development of molecular diagnosis methods for various infections, genetic disorders and microbial genomics with focus on identifying virulence gene and target for drug development. A National Bioethics Committee has been set up to take up the issues related to the universal declaration on human genome and human rights, in liaison with the International Bioethics Committees of UNESCO, to prepare bioethics policy and consider/monitor programmes on "Gene Therapy" research in the country.

2.6 Industry

Biotechnology is one of the most research-intensive industries in the world. The first U.S. company, Genentech, established in 1976 had a major impact on the industrial revolution and after 20 years, there are more than 1,300 companies in USA alone. Market capitalization, the amount of money invested in the U.S. biotechnology industry, increased 4 percent in 1998, from $93 billion to $97 billion. The first US approved biotechnology product, a monoclonal antibody-based diagnostic kit had reached the market in 1981 and the following year the recombinant pharmaceutical Humulin was available to human beings. Just to indicate the magnitude of the proliferation of the biotechnology companies, in 1994, in USA, the biotechnology companies had a market value of US $ 41 billion, a total R & D expenditure of US $ 7 billion and 103,000 employees. The U.S. biotech industry spent $ 9.9 billion on research and development in 1998 and US $ 11 billion in 1999. The top five biotech companies spent an average of $121,400 per employee on R & D. This compares with an average of $ 30,600 per employee for the top pharmaceutical companies. Biotech firms focusing on agriculture employed 22,000 workers and earned US $ 2.3 billion in revenues. Such a situation did not exist 20 years ago [3, 5, 6].

In India the consumption of biotechnology products increased from Rs. 71.54 billion in 1997 to Rs. 94 billion in 2000 and the projected demand for 2005 is around Rs. 145.6 billion. Major recombinant products are therapeutics, monoclonal antibodies and genetic vaccines. Today in India 13 products are approved. The market was Rs. 1.95 billion in 1997, Rs. 4.85 billion in 2000 and is expected to grow to Rs. 8.97 billion in 2005 [3, 5–8].

The sector represented by industrial products will remain primarily based on conventional biotechnology for production although recombinant microbial strains are expected to contribute substantially to the production of biocatalysts (useful for complex chemical reactions), industrial enzymes, food-grade enzymes, production of simple microbial metabolites such as organic acids and amino acids. There would be a rise in the production of speciality enzymes and oligonucleotides in molecular biology research, speciality materials including speciality plastics for specific uses, analytical materials and reagents for diverse uses, and application of biological materials in electronic devices. Among the industrial products, new investment opportunities are foreseen in industrial enzymes (over Rs. 500 million) and in amino acids production where investment of over Rs. 700 million can come up to meet not only local needs but to cater to the export needs also. Opportunities for investment in these areas are clearly linked with India's having sizable quantities of sugarcane molasses, and also other cheap agricultural substrates like various grades of starches from tapioca, maize, potato etc.; corn steep liquor (whose quality can be improved if adequate demand is created), sugar, pea/peanut/soybean meals, and various vegetable oils. In the area of baker's and brewer's yeast, opportunities for production of fresh compressed yeast do not presently exist but production of value-added NAD/NADH and speciality enzymes could be explored by using the locally available compressed yeast. In addition to the above areas of investment, there exists reasonable scope for setting up facilities for the recovery of value-added products from wastes such as proteins from milk whey, biogas and composted fertilizers from municipal or agriculture wastes, better methods for recycling of organic wastes, production of speciality biochemicals and speciality plastics that are biodegradable. These opportunities, thoughtfully explored, can provide avenues for investment of over Rs. 1 billion in these areas too.

2.7
Human Resource Development

Recognizing the need for developing adequate human resource in the multidisciplinary field of biotechnology and for training personnel for research and industrial activities, the Department of Biotechnology has been implementing a programme on human resource development. The main focus of this programme has been to generate large numbers of highly trained scientists/students. Under this programme nearly 60 universities have been covered in almost all regions of the country and a great deal of interest has been generated in many state universities also. Post-graduate programmes include specialized biotechnology courses in the area of agriculture and industrial biotechnology and neurosciences. There has been a major thrust in the area of biochemical engineering and biotechnology and this initiative was launched in the 1970s by IIT Delhi. With gradually built up facilities and excellent students, IIT Delhi drew attention from peers around the world. As a result of this the Swiss Federal Institute of Technology, Zurich established academic collaboration with IIT Delhi on biochemical engineering. The IIT, Delhi initiated a five-year integrated M.Tech. programme on biochemical engineering and biotechnology which was later adopted as a model curriculum.

The success of any bioprocess involving changes in flow patterns and consistencies of media in bioreactors, absorption and release of heat, variation in cell morphology, loss of plasmids and reversion of rDNA cells, protein instability and inactivation, release of toxins with target products, cell disruption, a variety of purification steps through which the final product must pass to meet regulatory requirements, process economics etc., is totally dependent on the nature and quality of academic training received by the engineers managing the process.

To acquire competence in these and related areas the training must impart quantitative knowledge in both new biology and chemical engineering science. The products of such a training will be engineers capable of solving difficult problems in process biotechnology. Selection of students must also be rigorous. Because of the lack of growth of the science-based biotechnology industry in India such a model was not in demand until early 1960s. Through mutual discussions between microbiologists (without engineering) and chemical engineers (without biology), management of fermentation processes (very few in number) could be done, because products were few and well defined and competition was even less. In addition PG diploma and post MD/M.Sc training courses are supported on a regular basis (Table 3).

Table 3. Institutions offering DBT supported post-graduate teaching programmes in biotechnology

Name of the University/Institution (Year of Start)	Annual intake
M.Sc. Biotechnology (2 Year Courses)	
1. Jawaharlal Nehru University, New Delhi (1985–1986)	20
2. Madurai Kamaraj University, Madurai (1985–1986)	20
3. MS University, Baroda (1985–1986)	20
4. University of Poona, Pune (1985–1986)	20
5. Banaras Hindu University, Varanasi (1986–1987)	12
6. Indian Institute of Technology, Mumbai (1987–1988)	10
7. Roorkee University, Roorkee (1991–1992)	10
8. Aligarh Muslim University, Aligarh (1991–1992)	10
9. Guru Nanak Dev University, Amritsar (1991–1992)	10
10. Devi Ahilya Viswavidyalaya, Indore (1991–1992)	10
11. University of Hyderabad, Hyderabad (1991–1992)	10
12. Himachal Pradesh University, Shimla (1994–1995)	10
13. University of Calicut, Kerala (1994–1995)	12
14. Banasthali Vidyapeeth, Banasthali, Rajasthan (1994–1995) (for girls only)	10
15. Tezpur University, Tezpur (Assam) (1998–1999)	12
16. Gulbarga University, Gulbarga (Karnataka) (1998–1999)	10
17. Jammu University, Jammu (1999–2000)	10
18. Gujarat University, Ahmedabad (1999–2000)	10
19. Mysore University, Mysore (1999–2000)	10
20. University of Allahabad, Allahabad (1999–2000)	12
21. Guru Jambheshwar University, Hisar (2000–2001)	10
22. University of Kashmir, Srinagar (2000–2001)	10
23. Kumaun University, Nainital (2000–2001)	10
24. University of North Bengal, Siliguri (2001–2002)	10
25. Lucknow University, Lucknow (2002–2003)	10
M.Sc. Agricultural Biotechnology (2 Years)	
1. Assam Agricultural University, Jorhat (1988–1989)	10
2. Tamil Nadu Agricultural University, Coimbatore (1988–1989)	15
3. GB Pant University of Agriculture and Technology, Pantnagar (1988–89)	10
4. Birsa Agricultural Univ. Ranchi (1999–2000)	12
5. Himachal Pradesh Krishi Vishvavidhalaya, Palampur (H.P.) (1999–2000)	05
6. Indira Gandhi Agricultural Univ. Raipur (2000–2001)	10
7. Marathwada Agricultural University, Parbhani (2000–2001)	10
Master in Medical Biotechnology (2 Years)	
1. All India Institute of Medical Sciences, New Delhi (1986–1987)	10
M.Sc. Marine Biotechnology (2 Years)	
1. Goa University, Goa (1988–1989)	10
M.Sc. Neurosciences (3 Years)	
1. Tata Institute of Fundamental Research, Mumbai (2000–2001)	5
M.Tech/M.Sc.(Tech) Biochemical Engineering and Biotechnology	
1. Indian Institute of Technology, New Delhi (1969–1970) (Converted into a 5 year integrated course since 1993–1994)	10
2. Indian Institute of Technology, Kharagpur (1986–1987)	10
3. Anna University, Chennai (1991–1992)	15
4. University Department of Chemical Technology, Mumbai (1993–1994)	15
Post MD/MS Certificate Course in Medical Biotechnology (1 Year)	
1. All India Institute of Medical Science, New Delhi	4
2. Post Graduate Institute of Medical Education & Research, Chandigarh	4
3. Sanjay Gandhi Post Graduate Institute of Medical Sciences, Lucknow (2000–2001)	5

During the last five years more than 2000 students have been trained and have been placed for research and training in institutes/industries in the country and abroad. There is also a post doctoral research programme being supported. A special feature of this human resource development programme is the industrial training which is provided to the post M.Sc/M.Tech students. The in-house training provided by the industry for a period of six months enhances their career opportunities. The DBT also encourages mid-term career scientists from academic institutes and scientists and technologists from the industrial R & D laboratories for training in advanced and emerging research techniques in the field of biotechnology. These programmes are supported as short-term training courses, national and overseas associateships.

There is also an emphasis on popularization of biotechnology for which popular lectures are supported in different colleges, a number of biotechnology publications are brought out for the benefit of the students and, in addition, the Department participates widely in exhibitions with the strong emphasis on popularization of this programme in the rural sector. As a special incentive for the 10+2 students, 25 biology scholarships are given to the children performing best in CBSE and pursuing biology studies. 5 National Bioscience Career Development awards are given to scientists below 45 years for their outstanding contribution in the field of biotechnology. Three Women Bioscientist awards are also given annually for the outstanding contribution of female scientists. Biotech process and product development and commercialization are also promoted and 5 awards given annually.

2.8 Bioinformatics

Informatics is an important factor in the biotechnology revolution. Ranging from reference to type culture collections or comparing gene sequences, access to comprehensive up-to-date biological information is crucial in almost every aspect of biotechnology. The impact of biological databases has been as great as (say) that of cloning and other breakthroughs in laboratory techniques. India was one of the first countries in the world to establish a nation-wide Bioinformatics Network in 1986–1987 supported by the Department of Biotechnology. Through this initiative the R & D activities in biotechnology and molecular biology have achieved a significant growth. The navigation through the Biotechnology Information System Network (BTISnet) has lead to the emergence of very sophisticated scientific infrastructure in the country.

The broad objectives of the Biotechnology Information System programme are:

- To provide a national bioinformation network designed to bridge the interdisciplinary gaps on biotechnology information and establish links among scientists in organizations involved in R & D and manufacturing activities in the country.
- To build up information resources, prepare databases on biotechnology and to develop relevant information handling tools and techniques.

- To continuously assess information requirements, organize creation of necessary infrastructures and to provide information and computer support services to the national community of users working in biotechnology and allied areas.
- To coordinate efforts to access biotechnology information worldwide including establishing linkages with some of the international resources of biotechnology information (e.g., data banks on genetic materials, published literature, patents, and other information of scientific and commercial value).
- Perform research into advanced methods of computer-based information processing for analyzing the structure and function of biologically important molecules.
- To evolve and implement programmes on education of users and training of information scientists responsible for handling of biotechnology information and its applications to biotechnology research and development.
- To promote international collaboration towards exchange of scientific information in biotechnology through the development of appropriate network arrangements.

The network has established a link among scientists in organizations involved in research and development activities in biotechnology. The BTISnet today offers a single major information resource in the country covering several interdisciplinary areas of biotechnology. The network consist of ten Distributed Information Centers (DICs), Forty-six Sub-Distributed Information Centres (Sub-DICs) and a Biotechnology Information Centre (BTIC) which is an apex center for coordination in bioinformatics.

Human resource development has been recognized as an important area in bioinformatics. Towards meeting the needs for trained bioinformatics professionals, long-term courses are being run as part of this network in four reputed universities. More than 20 short-term courses are conducted every year to train the researchers and scholars in bioinformatics. Six national facilities on interactive graphics are dedicated to the promotion of molecular modeling and other related activities. More than 100 databases on biotechnology have been developed by the BTISnet Centres. Several major international databases for application to genomics and proteomics have been established in the form of mirror sites as part of the bioinformatics programme. These databases are being linked through high-speed and large bandwidth as VPN to promote a faster sharing of information. These sites are designed to act as knowledge pathways for discoveries in biotechnology. A directory on institutions and industries in SAARC regions has been developed and published.

2.9 Centers of Excellence

The Department of Biotechnology has been supporting a large number of centers of excellence, facilities and repositories to promote research and development, demonstration, product development and technology transfer activities (Table 4). The Government of India, as indicated, has a well structured regulatory mechanism in place and this institutional framework has been established to

Table 4. Biotech facilities established/supported by the Department of Biotechnology at various research institutions and universities [1]

A. *National Facilities/Repositories*

1. National Facility on Microbial Type Culture Collection, IMTECH, Chandigarh
2. National Facility on Blue Green Algal collection, IARI, New Delhi
3. National Facility on Plant Tissue Culture Repository, NBPGR, New Delhi
4. National Animal House Facility, CDRI, Lucknow
5. National Animal House Facility, NIN, Hyderabad
6. National Facility on Biochemical Engineering Research and Process Development, IMTECH, Chandigarh
7. National Facility on Oligonucleotide Synthesis, CBT, Delhi
8. National Facility for Enzymes and Biochemicals, CBT, Delhi
9. Three Genetic Engineering Units at JNU, N. Delhi, BHU, Varanasi and IISc., Bangalore
10. Computational Facility for Oligonucleotides, IISc., Bangalore
11. Centre for Reproductive Biology & Molecular Endocrinology, IISc., Bangalore
12. Carbohydrate Cell Surface and Cellular Transport Facility, IISc., Bangalore
13. Protein-Peptide Sequencing Facility, IISc., Bangalore
14. National NMR Facility, AIIMS, New Delhi
15. National Facility for Marine Cyanobacteria, Bharathidasan University, Tiruchirappalli
16. Antibiotic Development Consortium, HAL, Pune
17. Repository on Cryopreservation of Blood Cells, IIH, Mumbai
18. Repository of Filarial Parasites and Reagents, MGIMS, Sevagram
19. Repository of Medicinal and Aromatic Plant Materials, CIMAP, Lucknow
20. National Gene Bank for Medicinal and Aromatic Plants, NBPGR, New Delhi
21. National Gene Bank for Medicinal and Aromatic Plants, CIMAP, Lucknow
22. National Gene Bank for Medicinal and Aromatic Plants, TBGRI, Thiruvanathapuram
23. Micropropagation Technology Park, TERI, New Delhi
24. Micropropagation Technology Park, NCL, Pune
25. Centre for Genetic Engineering and Strain Manipulation, MKU, Madurai
26. Automated DNA Sequencing Facility, IISc., Bangalore
27. Automated DNA Sequencing Facility, UDSC, New Delhi
28. MALDI-TOF Mass Spectrometer Facility, IICT, Hyderabad
29. National Facility for X-Ray Crystallography, IISc., Bangalore
30. National Facility for Transgenic Containment & Quarantine, NBPGR, New Delhi
31. National Facility for Virus Diagnosis and Quality Control in Tissue Culture raised Planting materials, IARI, N. Delhi with five satellite centers at NCL, Pune, TERI, N. Delhi, IHBT, Palampur, IIHR, Bangalore and SPIC, Chennai
32. Gene Bank on Medicinal Plants, IHBT, Palampur
33. Functional Genomics Facility, CBT, New Delhi
34. National Facility for Stable Isotope Discrimination and Molecular Marker for WUE at UAS, Bangalore
35. Drosophila Repository & Research Facility, IIT, Kanpur
36. International Depository Authority, IMTECH, Chandigarh
37. FACS Facility, CCMB, Hyderabad

B. *Hardening Units*

1. Jai Narayan Vyas University, Jodhpur
2. Haryana State Council, Hisar
3. G.B.Pant Institute of Himalayan Environment & Development, Almora
4. West Bengal Council of Science & Technology, Kolkata
5. Regional Research Laboratory, Jammu
6. Tata Energy Research Institute, Guwahati

C. *Micropropagation Technology Parks*

i) TERI, New Delhi
ii) NCL, Pune

meet the specific requirements especially with regard to the plant, animal and microbial systems.

2.9.1
Repositories

The main objective of setting up of the repositories is to promote conservation/preservation of living organisms including microbes both useful and harmful in agriculture, animal husbandry and bioindustries. Some of the major repositories established for various collections are:

- National Facility for Marine Cyanobacteria (NFMC) at Bharathidasan University, Tiruchinapalli surveyed the Little Andaman and added sixty new strains to its germplasm collection.
- The Tissue Culture and Cryopreservation Repository at NBPGR, New Delhi has total of 1107 accessions of various priority crops with more than 100 accessions of orthodox seed species and difficult to store species in its cryobank.
- Repository of Medicinal and Aromatic Plants, CIMAP, Lucknow maintains 303 pure compounds and 461 plant extracts from more than 80 medicinal and aromatic plant species to serve as reference material.
- Drosophila Repository and Research Facility, Devi Ahilya University, Indore has developed an experimental kit for demonstrating principles of genetics in schools and colleges.
- National Centre for Conservation and Utilization of Blue Green Algae (BGA), IARI, New Delhi holds more than 800 strains of blue-green algae.

2.9.2
Programme Support

The department has also provided programme support to a number of centers of excellence, universities and research institutes for strengthening some of their major programmes in the area of biotechnology, especially in basic research, technology and product development (Fig. 1). This programme essentially caters to the needs of the industry in terms of sophisticated and expensive equipment support, expertise and supply of research material. Training of human resource for handling and experimentation is also an integral part of this facility.

Main Institutes/Universities receiving support are:

- Indian Institute of Science, Bangalore
- University of Bangalore, Bangalore
- Rajiv Gandhi Centre for Biotechnology, Thiruvananthapuram

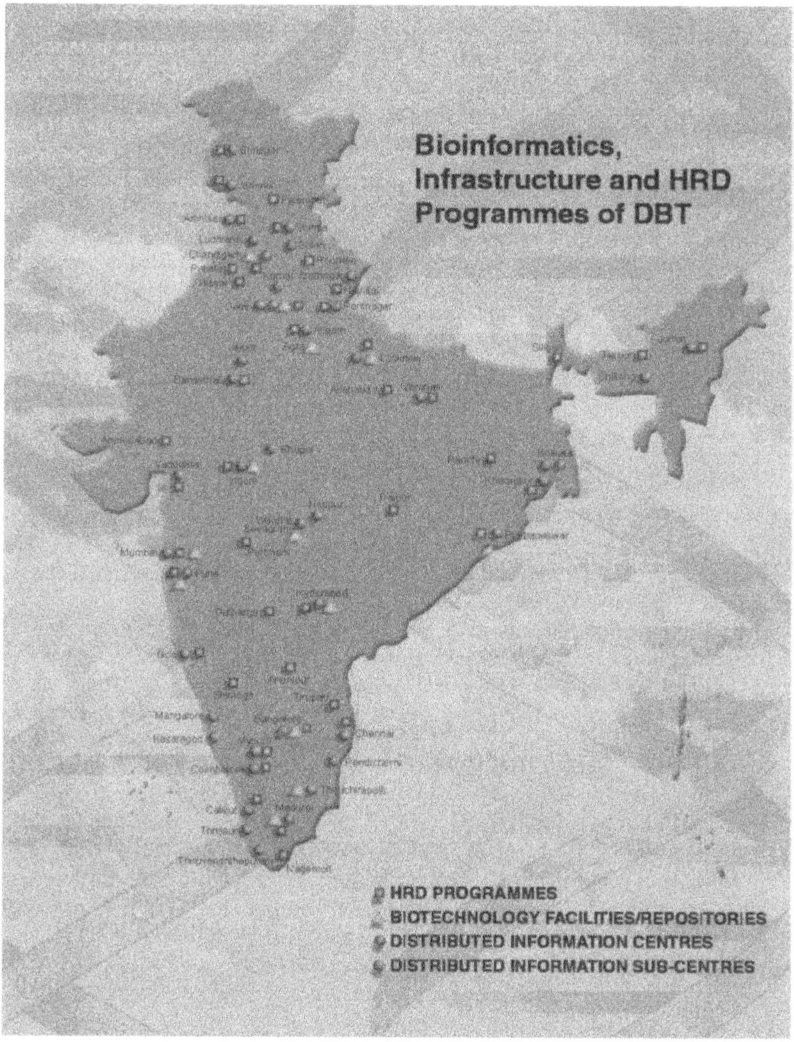
Bioinformatics,
Infrastructure and HRD
Programmes of DBT
HRD PROGRAMMES
BIOTECHNOLOGY FACILITIES/REPOSITORIES
DISTRIBUTED INFORMATION CENTRES
DISTRIBUTED INFORMATION SUB-CENTRES

Fig. 1

2.10 Biosafety

2.10.1 *Regulatory Mechanisms in India*

The new capabilities to manipulate the genetic material present tremendous potential and find use in many novel experiments and applications. These developments have generated a sense of concern among scientists working in biological areas and others to find ways how the research in the field should be carried out safely and means to regulate work involving pathogenic microorganisms and genes of virulence. Several countries have formulated safety guidelines and regulations for research in the field of recombinant DNA, large-scale use of them in production process and their applications in the environment. Considering the possible incremental risks associated with the use of new techniques in laboratory research with pathogenic microorganisms, the National Biotechnology Board issued a set of safety guidelines for India in 1983 to ensure the safety of workers in the laboratory environment. While framing the guidelines, the Committee took into account the local factors such as resistance to infection (immunity), host parasite burden in the community, laboratory environment and chances of survival and growth of altered organisms under the tropical conditions.

Remarkable developments have ensured in the last few years in the field of genetic manipulation and the scenario has shifted from the laboratories to the market place and elsewhere [3, 6]. In India there is a growing awareness of the commercial potential of biotechnology and efforts are being made to promote large-scale use of indigenously relevant biotechnologies. A large number of research institutions in Government, Universities and private R & D labs have active biotech programmes where research is being done, both in basic and applied fronts utilizing microorganisms, plant and animals, tissue culture and cell lines and on development of vaccines towards communicable diseases of both men and animals. A good deal of effort is being made in the areas of diagnostics, biofertilizers, biocides, fertility control, tissue culture of high-value crops to develop technologies and useful products. The successes in indigenous research efforts should soon be translated into commercially viable technologies through clearing houses with major R & D centers, university academic institutions and by the industry itself.

India has made significant progress towards the development of biosafety regulations and an institutionalized framework for their implementation to ensure an effective evaluation of transgenic plants before they are given clearance for release in the field. Safety guidelines for recombinant DNA research and development in India have been formulated under the environmental protection act, 1986 by the recombinant DNA Advisory Committee set up by the Department of Biotechnology in consultation with experts working in this area and representatives from different ministries. These guidelines have been extensively circulated and are now adopted across the country. These are in conformity with the international protocols.

The Government of India notified various regulations through a government order, the notification No. 1037(E) dated 5.12.1989, which was issued by the Ministry of Environment & Forests under the Environment (Protection) Act, 1986. The notification has set the rules for manufacture, use, import, export and storage of hazardous microorganisms/genetically engineered organism or cells. The notification has also set up various levels of committees considering the level of risk involved. These committees are Recombinant DNA Advisory Committee (RDAC), Institutional Biosafety Committee (IBSC), Review Committee of Genetic Manipulation (RCGM), Genetic Engineering Approval Committee (GEAC) and State and District level Biotechnology Co-ordination Committees (SBCC and DBCC)

The biosafety protocol holds great significance because it considers international law that regulates genetically modified organisms and also establishes an internationally binding framework of a minimum standard.

The area of transgenic research (plant, animal and human) is associated with many ethical issues. In the case of plants it is mainly related to the food crops and concerns issues such as food labeling etc. The regulatory mechanisms for this are covered under the Food Safety Issues. Other social and cultural concerns are related to the introduction of alien genes into the plant system. The issue of consumer knowledge and acceptance is very important here. The Government Policy is to have a very transparent system in place to keep the consumer totally informed.

2.10.2
The Recombinant DNA Advisory Committee (RDAC)

This Committee constituted by the DBT is to monitor the developments in biotechnology at national and international levels. The RDAC submits recommendations from time to time that are suitable for implementation for upholding the safety regulations in research and applications of GMOs and products thereof. This Committee prepared the Indian Recombinant DNA Biosafety Guidelines in 1990, which was adopted by the Government for conducting research and handling of GMOs in India.

The regulatory mechanism for the development and evaluation of transgenics as per the recombinant DNA safety guidelines is based on a three-tier system as given below:

a) *Institutional Biosafety Committee (IBSC).* This Committee is constituted by organizations involved in research with GMOs. The Committee requires the approval of the DBT. IBSC is an institutional committee chaired by Head of the Institution or his nominee and has members from different disciplines, the Biosafety Officer and one member nominated by DBT who oversees the activities to ensure that safety aspects in accordance with the safety guidelines are fully adhered to by the organization. Every R & D project using GMOs has to have an identified investigator who is required to inform the IBSC about the status and results of the experiments being conducted. Experiments belonging to Categories I and II risks as well as all experiments conducted with GMOs in the contained lab, contained green house conditions for plants as well as contained lab, caged or enclosed conditions for animals can be approved by the IBSC; however, the synopsis of all such experiments is required to be reported to the RCGM in the DBT in the form of reports from time to time in a prescribed format. Such information along with the progress of research work is also required to be reported to the RCGM as a mandate at least once in six months.

The committee's functions are:

- To provide half yearly reports on the ongoing projects to RCGM regarding the observance of the safety guidelines on accidents, risks and on deviations, if any.
- To review and clear project proposals falling under a restricted category that meets the requirements under the guidelines.
- To tailor a biosafety programme to the level of risk assessment.

The IBSC is a nodal point for interaction within an institute/university/commercial organization involved in r-DNA research for the implementation of the recombinant DNA guidelines. As such, in the first instance, it is necessary that the organizations intending to carry out research activities involving genetic manipulation of microorganisms, plants or animals should constitute their IBSC in accordance with the procedures in vogue and as informed to the public through the above notification. All recombinant research carried out by the organization should have a designated Principal Investigator (P.I). It would be the duty of the P.I. to appraise the IBSC about the nature of the experiments being carried out. Depending upon the category of the experiments the P.I. can inform the IBSC about the recombinant experiments, seek permission of IBSC before starting the experiments or seek the permission of the RCGM through its IBSC in cases where the risks involved in the experiments are considered to be of higher magnitude having the potential of polluting/endangering the environment, the biosphere, the eco system, animals and human beings.

b) *Review Committee on Genetic Manipulation (RCGM).* The RCGM is constituted by the DBT to monitor the safety aspects of ongoing research projects and activities involving genetically engineered organisms. The RCGM has representatives of various ministries and scientists from different disciplines. The committee is headed by an eminent biotechnologist and has the following functions:

- To establish procedural guidance manual-procedure for regulatory process with respect to activity involving genetically engineered organisms in research, production and applications related to environmental safety.
- To review the reports in all approved ongoing research projects involving high risk category and controlled field experiments, to ensure that safeguards are maintained as per guidelines.
- To recommend the type of containment facility and the special containment conditions to be followed for experimental trials and for certain experiments.
- For research in recombinant DNA work involving risks categorized as category III and above the permission of the RCGM must be obtained by the P.I. before conducting the research work.

- RCGM can authorize applicants (P.I.s) to conduct limited field trails in multi-locations in the country. The design of the trial experiments is either provided by the RCGM or it may approve the protocol designed by the P.I.
- RCGM can, if required, direct the applicants to generate toxicity, allergenicity and any other relevant data on transgenic materials in appropriate systems. RCGM may design or approve a protocol for conducting experiments to seek answers to the above.

The RCGM can put such conditions as would be required to generate long-term environmental safety data from the applicants seeking release of transgenic plants into the open environment, and who have complied with initial safety evaluation.

The Committee is mandated to bring out manuals of guidelines specifying procedures for regulatory process with respect to activities involving genetically engineered organisms in research, use and applications, including industry, with a view to ensure environmental safety. All ongoing projects involving a high risk category and controlled field experiments shall be reviewed by the RCGM to ensure that adequate precaution and containment conditions are followed. The RCGM can lay down procedures restricting or prohibiting production, sale, importation and use of GMOs.

RCGM can approve applications for generating research information on transgenic microorganisms. The growth of microorganisms under sterile conditions in a fermentation vessel under submerged conditions can be carried out in large bioreactors. However, RCGM can approve experiments in bioreactors having a geometric volume of up to 20 liters. All experiments for the use of larger bioreactors require the approval of the Genetic Engineering Approval Committee (GEAC). RCGM puts conditions of contained use of the bioreactors in order to prevent the escape of genetically modified microorganisms into the open environment. The procedures for safe handling of microorganisms have been stated in the Recombinant DNA Safety Guidelines-1990.

RCGM can also approve applications for generating research information on transgenic plants. Such information may be authorized to be generated in contained green house as well as in small plots. The small experimental field trials are limited to a total area of 20 acres in multilocations in one crop season. In one location where the experiment is conducted with transgenic plants, the land used should not be more than one acre. The design of the trial experiments requires the approval of the RCGM. The design of the experimental plot in open environment is made to seek answers to relevant and necessary questions on environmental hazards including risks to animal and human health. Data are required to be generated on economic advantages of the transgenics over the existing non-toxicity, allergenicity and any other relevant data on transgenic materials in appropriate systems.

For generating research information on transgenic animals, RCGM can authorize the investigator to conduct experiments in the lab as well as in contained and enclosed conditions so as to prevent the escape of the transgenic animals into the open environment. The experiments are designed to generate information about the growth characteristics and the health conditions of the transgenic animals, using the non-transgenic animals as controls.

The RCGM can issue clearances for import/export of etiologic agents and vectors required for producing and cloning genetically modified microorganisms, plants and animals. Clearances for import/export are also provided for transgenic microorganisms; transgenic germplasms including transformed calli, seed and plant parts, as well as transgenic animals of various kinds for research use only. All experiments using GMOs which belong to category III risks and above as elaborated in the guidelines require a permit to be issued by the DBT for authorizing such experiments. All such permits are issued on the basis of the recommendations of the RCGM.

According to the Indian classification of risks, the category I risks involve routine recombinant DNA experiments in lab and work involving defined genes/DNA of microbial, plants or animals origin which are generally considered as safe. Category II risks involve lab and contained green house experiments involving genes or DNA of microbial, plant or animal origin which are non-pathogenic to human, but can have implications on plants and insects. Fermentation experiments with GMOs conducted in fermentation vessels up to 20-liter can be category II risk experiments, if the transgenic microorganism or cell lines used are harmless and non-pathogenic. Such experiments can also belong to category III risks, if the GMOs are coding for toxins or are infective to human and animals. Other category III risk experiments involve genes/DNA of microbial, plant or animal origin, which can cause alterations in the biosphere and not fall in categories I and II. All open field experiments of GMOs, howsoever organized, are considered to belong to category III risks, although they may be carried out under reasonably contained conditions by taking all the precautions to prevent the escape of GMOs or parts thereof that have propagating traits into the uncontrolled open environment.

The RCGM has a subcommittee namely Monitoring-cum-Evaluation Committee (MEC) to design field experiments and formats for collection and evaluation of scientific data on plants grown under contained conditions as well as in limited field trials. The DBT has set up this inter-ministerial committee to field evaluate the transgenic crop trials undertaken by the investigators under the approval of RCGM. The committee has also been made responsible for monitoring the large-scale field trials on transgenic crops permitted under GEAC.

c) *Genetic Engineering Approval Committee (GEAC).* This Committee functions as a body in the Ministry of Environment and Forests and is responsible for approval of activities involving large-scale use of GMOs in research, industrial production and applications. The clearance of GEAC is only from an environmental angle under the EPA. All other relevant laws would apply even though EPA clearance is available for using GMOs and products thereof. For example, drugs made through GMOs would require separate approval for manufacture and use under the Indian Drugs Act; production of GMOs is also authorized under the Indian Industries (Development & Regulation) Act, and, therefore, these clearances are also mandatory. Large-scale experiments beyond the limits specified within the authority of RCGM are authorized by the GEAC only.

It is an interministerial committee under the Ministry of Environment and Forests and is the competent authority for release of GMOs into the environment.

It has members from different ministries and scientists drawn from different disciplines. The Ministry of Environment and Forests issues authorization to applicants on the basis of approvals accorded by the GEAC. The important functions of the committee are:

- Import, export, transport, manufacture, process, selling of any microorganism or genetically engineered substances or cells including foodstuffs and additives.
- Discharge of genetically engineered/classified organisms/cells from laboratory, hospitals and related areas into environment.
- Large-scale use of genetically engineered organisms/classified microorganisms in industrial production and applications (production shall not be commenced without approval).
- Deliberate release of genetically engineered organisms. The approval will be for a period of 4 years.

2.10.3
Bt. Cotton

Cotton is grown on approximately 9 million hectare of land in India, the highest in the world, and accounts for about 20% of the world acreage. India stands 3rd in the lint production, producing only 319 kg/ha lint yield compared to the world average of 603 kg/ha. Cotton is highly susceptible to insects and consumes over 50% of the total insecticides used in the country. India spends nearly Rs. 12 billion worth of pesticides for controlling various pests of cotton. Approximately, 60–70% of the pesticides are used only for control of the dreaded pest namely bollworm against the world average of 30%. It is estimated that in India about 20–30% of the productivity of cotton is lost due to the damage of insect pests.

Transgenic Bt. cotton is a technology that has been proposed as an alternative to overcome the problem of bollworm, reduction in pesticide use and improving the environment.

A gene from the soil borne bacterium *Bacillus thuringiensis* has been introduced into the cotton plant. This gene encodes a protein, Cry1A(c) that has specific insecticidal properties to control bollworms. This gene was transformed into cotton plants by M/s. Monsanto, USA and M/s. Maharashtra Hybrid Seeds Co. Ltd. (Mahyco), Mumbai has introduced it into Indian cotton germplasm by traditional breeding methods. This transgenic cotton on which cotton pests, namely bollworm, feeds, interferes with the digestive tract of insect and kills it. The new plant has no effect on other insects or pests.

Studies like gene flow, aggressiveness and cross-ability with wild near relatives of cotton, conducted with transgenic cotton revealed that the possibility of gene flow in nature becomes negligible and therefore there is a least chance of effecting the biodiversity and less chances of possibilities of cross-pollination in nature. The toxicity and allergenicity studies conducted on birds, fish, mammals including ruminants (goats, cows and buffaloes) on Bt. cotton established that it is as safe as non-Bt. cotton and has no impact on non-target insects and on bioflora.

The Bt. cotton hybrids were further characterized for the absence of the Terminator gene and it was confirmed that these hybrids do not have any terminator gene sequences. Composition analysis of Bt. cotton showed that it is comparable with non-Bt. cotton. Contained limited field studies and large-scale field studies in farmers' fields and also in the ICAR system indicated that Bt. cotton showed distinct agronomic advantages over non-Bt. cotton.

The economic benefit of Bt. cotton is estimated to be substantial when compared to current practices of cotton cultivation. Significantly higher benefits are expected to be derived from the yield savings of the Bt. cotton in addition to savings in the cost of pesticides. The farmers are expected to gain 50% to 70% of the economic benefits generated from the technology. Small farmers are more likely to adopt this technology as they do not have to spray every 3 to 4 days and 20 to 25 times.

In March 2002, the Genetic Engineering Approval Committee (GEAC) of Ministry of Environment and Forests examined the data generated on Bt. cotton and cleared three Bt. cotton hybrids for commercial cultivation with certain conditions. This is the first clearance for the commercial cultivation of a transgenic crop in India.

2.10.4
Public Awareness

A proper public awareness is very important with respect to the benefits and risks associated with the GM technology. There are several factors that can play an important role in public acceptance of genetically modified crops. Scientific demonstration of biosafety of transgenic crops and review of Government Agencies are extremely important in gaining public acceptance. Public acceptance is also greatly determined by the kind of information provided by the media to the general public. Misinformed public debates on key issues related to crop biotechnology can result in erosion of public confidence and can arose mistrust among the technology and its developers. Therefore, clear and understandable consumer information is a very important part of the public acceptance process. Besides, media, research organizations and scientific institutions concerned with crop improvement must also take up the responsibility of bringing awareness to the public about the application of genetic engineering in agriculture, the potential benefits as well as constraints.

The DBT has been organizing a number of roving seminars across the country. In addition some of the important NGOs and research institutes have also organized policy discussion forums, stakeholders dialogues and media debates to spread awareness to the public in general. A concerted effort is being made to conduct and promote research activities for the benefit of the public in consultation with them.

2.11
Socio-Economic Development

The developments in applied biotechnology are directed towards economic production of new and conventional biological products for widespread human use.

Importantly, biotechnology is aiming at higher yields from agriculture with reduced inputs and to provide affordable diagnostics, pharmaceuticals, vaccines and therapies for preventing and curing human diseases. Both these approaches will have significant, concomitant impacts on industrial growth.

Applications of biotechnology will be used to ameliorate the economic conditions of a sizable fraction of 55% of the Indian families engaged in agriculture, presently suffering on account of the recent decline in the competitiveness of Indian agriculture. Careful diversification of cropping patterns will enlarge India's presence in the global markets for value-added agriculture products. Utilization of genetically engineered crop varieties will improve farm level economics of agriculture.

There will be gains to all strata from the transfer of healthcare-related biotechnologies to industry engaged in the manufacture of diagnostics, vaccines and pharmaceuticals. The costs of many kinds of tests and therapies will be reduced, accessibility will be enlarged, exports and reduction in imports of the concerned materials will have a positive impact.

Several biotechnological programmes for the benefit of SC/ST populations, women and rural areas have been implemented by the DBT in the areas of biopesticides, biofertilizers, vermicompost and vermiculture, sericulture, floriculture, mushroom cultivation and medicinal plant production, poultry and fish farming and improvement of human health. Nearly 60,000 families have already benefited. Universities, national laboratories, state government institutions, Krishi Vigyan Kendras, NGOs and voluntary organizations implement these projects. A biovillage has been established at Mocha Village, Porbandar in Gujarat for the extension of technologies which are pro-poor, pro-women and pro-nature at the grassroots level.

A special programme on biotechnology for women is also being implemented. The main objective is to provide proven technology packages on plant-tissue culture, spirulina and vermicompost production, rabbit rearing, sericulture, aquaculture etc. for the benefit of women. This helps in providing employment and income generation. Training programmes are an important component. More than 20000 women have benefited so far.

A Women's Biotechnology Park has been established in Chennai. It aims to provide opportunities to professionally qualified women for setting up self-employment ventures using various biotechnologies. The park has twenty industrial modules and an equal number of land modules for agro-biotechnology activities. The park provides facilities for technology sourcing, training and marketing; 18 modules are already operational.

3
Future Perspective

Through the efforts of DBT there has been a tremendous impact on society of the advances in the front-line areas of research and development in biotechnology. The focus has been on basic research, environmental biotechnology, genomics and bioinformatics, plant and animal biotechnology, medical and environmental biotechnology, human resource development and creation of infrastructure and centers of excellence. The vision of DBT is "to attain new heights in biotechnology research, to shape biotechnology into a premier precision tool of the future for creation of wealth and to ensure social justice specially for the welfare of the poor" [9].

In order to realize the full potential of biotechnology as a front-line area of research and development with an overwhelming impact on society, the Indian biotechnological enterprises will be systematically nurtured at three distinct levels. The focus would be on basic research in modern biotechnology, including genomics and bioinformatics, agriculture, plant and animal biotechnology, medical biotechnology, environment and biodiversity, biofuels, product and process development and bioinstrumentation, human resource development, creation and strengthening of infrastructure in existing and new institutions, biotechnology for societal development, biosafety, ethical issues and biotechnology-related policy issues, conduct of cutting edge research, large scale demonstrations, partnership with private and public sector industries for commercialization and marketing of bioproducts.

Sustainable development ensuring food, nutritional, health, environmental and livelihood security of the people by harnessing the powers of biotechnology should be a dream of the scientific community. Translating these dreams into reality would give a major impetus to our socio-economic progress. Time-bound, mission-mode, result-oriented projects to be taken up include utilization of the full potential of the genomics revolution for humankind, plants, animals and microbes, development of new vaccines, diagnostics, drugs and drug delivery system, production of a large number of low-cost, affordable small proteins and therapeutics using the plants and animals as bioreactors,; to engineer crops with enhanced nutritional status, biotic and abiotic resistance and introduce precision farming with new quality traits for productivity enhancement, and to develop environmentally friendly technology packages for pollution control, biodiversity conservation and restoration of damaged ecosystems. Long-term support would continue for basic research on all aspects of molecular biology, genetics and genomics, proteomics and neurosciences. Well-defined missions to develop products, processes and technologies addressing the problems of the nation would be launched.

Research and development, demonstration and commercialization in the above areas, and implementation of some selected missions would be the Government's endeavor. To integrate the information and biotechnology revolution as a single technological and economic force and establish a large number of biological data banks would facilitate rapid progress of biotechnology. Transgenics as an important strategy in agriculture would produce bioengineered crops to become a source of enzymes, vaccines, biochemicals, etc., crops with resistance to biotic and abiotic stresses such as salinity, drought and water logging, nitrogen fixation in cereals using more efficient microbes and value addition along with nutritional enhancement of important edible crops would be a mission. In vitro mass propagation of the desired planting material, genetic control through identification and manipulation of genes and its implication in the forestry sector would help in the rapid regeneration of forests and also result in the enhanced production of industrial timber. Development of a large number of diagnostics for major diseases, genetic disorders, cancer, tuberculosis, HIV, malaria, better understanding of the human brain from molecular to systems levels to enable control and treatment of large number of currently incurable neurological disorders and development of new generation vaccines including DNA vaccine would move fast. Our inherent strength of Ayurveda and traditional systems of medicine would be optimally utilized through biotechnological interventions. Realizing that the present century would greatly depend on medicines from plant-based systems, the development of new molecules, drugs, prospecting for new genes and the whole field of pharmacogenomics would be a mission.

It is becoming increasingly obvious that systems that combine biological and chemical molecules on the one hand, and physical devices and electrodes on the other, have a huge potential for many applications.

In order to generate a critical mass of expertise, a strong infrastructure and directed support, the critical areas for investment over a ten year period have been outlined in the Report of the Working Group on Biotechnology for Tenth Five Year Plan [10].

The vision for biotechnology research, development and commercialization, in the next 10 years should thus focus on:

- Basic Research in New Biology and Biotechnology
  (a) Genomics,
  (b) Bioinformatics,
  (c) Basic biological phenomena with potential application.
- Agriculture, Plant and Animal Biotechnology
- Environment and Biodiversity
- Medical Biotechnology
- Biofuels
- Bioprocesses, Product Development, and Bioinstrumentation
- Human Resource Development
- Creation and Strengthening of Infrastructure in Existing Institutions and Setting up New Institutions
- Biotechnology for Societal Development
- Biosafety, Ethical and Proprietary Issues.

4 Conclusion

Biotechnology has clearly emerged as a field which targets all segments of society. The unique feature is that the blend of the new and conventional technologies has paved the way for an overall socio-economic development of the nation. The field of biotechnology offers enormous opportunities for research and development to generate excellence and products and processes of relevance for human kind. It is going to be one of the most important areas for scientific development in harmony with the environment as we move along in this 21st century. In India, a country rich in biodiversity and where the base of expertise is available in nearly all areas of science and technology, biotechnology could flourish leading to a bioindustrial revolution. However, its continued success depends not only on innovative research and development but also on a favorable regulatory climate and public acceptance.

The biotech industry is just coming out of its infancy. Its potential is being tested, realized and used. The public awareness and acceptance will accelerate the process. This sector is expected to expand at least 3-fold by the end of the century and will match or surpass the computer industry in size, importance and growth. Indian efforts in utilization of modern biotechnologies are modest. Currently, these are being supplemented by private individual entrepreneurs for developing appropriate goods and services for local needs as well as for the export market. Technologies are also flowing into the country due to the changed economic scenario. The country has a pool of skilled human resources, abundant agricultural materials, infrastructure and capital. Therefore with the local development of more globally competitive biotechnologies, India could become a global player in most of the areas in this emerging field.

It is clear today that there is a need to move in a focused direction to develop what we call a sustainable society – a society which has faith in science and technology as an instrument of environmentally friendly social and economic change. This sustainable society has to aim at working in partnership with the new biotechnologies, so that there is a healthy blend of environmental, social, developmental and economic importance.

5
References

1. Department of Biotechnology, Ministry of Science and Technology, Government of India, Annual Report (2000–2001)
2. Ghosh PK (2000) Research and Information Survey Development Report, vol 3 (2)
3. Confederation of Indian Industry (2000) Business Opportunities in Biotechnology report
4. IFPRI (1999) 2020 Focus – Biotechnology for developing countries. www.ifpri.cgiar.org/2020
5. Biotechnology Consortium India Ltd. (1999) SIDBI report on Technology for SSI's
6. Biotechnology Industry Organization (1999) Economic Development. www. bio.org
7. Sharma M (1999) New Biosciences opportunities and challenges as we move into the next millennium, Presidential address 86th Session Indian Science Congress Association, Chennai
8. Sharma M, Swarup R (2000): In: Tzotzos GT, Skryabin G (eds), Biotechnology in the developing world and countries in economic transition, CABI Publishing, p 312
9. Department of Biotechnology, Ministry of Science and Technology, Government of India (2001) Biotechnology – a ten year perspective vision document.
10. Department of Biotechnology, Ministry of Science and Technology, Government of India (2000–2001) Report of the working group for the 10th Plan.

Received: May 2002

Adv Biochem Engin/Biotechnol (2003) 84: 49–89
DOI 10.1007/b11036

Rhizobacterial Diversity in India and Its Influence on Soil and Plant Health

Bhavdish N. Johri[1] · A. Sharma[1] · J. S. Virdi[2]

[1] Department of Microbiology, G. B. Pant University of Agriculture and Technology, Pantnagar-263 145, India. *E-mail: bhavdishnjohri@rdiffmail.com*

[2] Department of Microbiology, University of Delhi South Campus, Benito Juarez Road, New Delhi-110 021, India

The rhizosphere or the zone of influence around roots harbors a multitude of microorganisms that are affected by both abiotic and biotic stresses. Among these are the dominant rhizobacteria that prefer living in close vicinity to the root or on its surface and play a crucial role in soil health and plant growth. Both free-living and symbiotic bacteria are involved in such specific ecological niches and help in plant matter degradation, nutrient mobilization and biocontrol of plant disease. While the rhizosphere as a domain of fierce microbial activity has been studied for over a century, the availability of modern tools in microbial ecology has now permitted the study of microbial communities associated with plant growth and development, in situ localization of important forms, as well as the monitoring of introduced bacteria as they spread in the soil and root environment. This interest is linked to environmental concerns for reduced use of chemicals for disease control as well as an appreciation for utilization of biologicals and organics in agriculture.

Indian researchers have studied the diversity of rhizobacteria in a variety of plants, cereals, legumes and others along with assessment of their functionality based on the release of enzymes (soil dehydrogenase, phosphatase, nitrogenase, etc.), metabolites (siderophores, antifungals, HCN, etc.), growth promoters (IAA, ethylene) and as inducers of systemic disease resistance (ISR). Based on such primary screening protocols, effective rhizobacteria have been field tested with success stories from various agroecological zones of the country, as reflected in the control of root- and soil-borne diseases, improved soil health and increased crop yields. Several commercial formulations, mostly based on dry powder (charcoal, lignite, farmyard manure, etc.) have been prepared and field tested, however, problems of appropriate shelf-life and cell viability are still to be solved. Also, inherent in such low cost technologies are the problems of variability in field performance and successful establishment of introduced inoculants in the root zone. In addition, most products available in the market are not properly monitored for quality before they reach the farmer. As a consequence, the acceptance of rhizobacterial formulations in the country is limited. However, several laboratories have now developed protocols for the rapid characterization of effective isolates based on molecular fingerprinting and other similar tools. Also, the use of molecular markers (*gus*, *lux*, *gfp*, etc.) makes it easy to monitor introduced inoculants in situ in soil and rhizosphere environments. The government initiative in integrated nutrient management and pest management systems has provided additional incentives to relate rhizobacterial science to other ongoing activities so that the benefit of this research leads to technologies that are environmentally and socially acceptable.

Keywords. Rhizosphere, Rhizobacteria, PGPR, FLP, Growth promotion, Biocontrol, P solubilizers, *N*-fixers, Community dynamics

© Springer-Verlag Berlin Heidelberg 2003

1	**Introduction**	52
2	**Rhizosphere**	53
2.1	Diversity of Free-Living and Symbiotic Bacteria	53
2.2	Endorhizosphere	54
2.3	Rhizoplane	56
3	**Beneficial and Deleterious Forms**	56
3.1	Community Dynamics	56
3.2	Influence of Plant and Soil Type on Diversity	57
4	**Substrates in the Rhizosphere**	57
4.1	Influence on Dominant Populations	58
5	**Root Colonization by Rhizobacteria**	58
5.1	Competition and Mechanisms	58
6	**Influence of Rhizobacteria on Soil Processes**	59
6.1	Interconversions	59
6.2	Enzymes	60
6.3	Biomolecules – Community Interactions	60
7	**Rhizobacteria as Growth Promotory and Biocontrol Agents**	62
7.1	Siderophores	62
7.2	Antibiosis	64
7.3	ACC	66
7.4	IAA	67
7.5	Phosphorus Solubilization	67
7.6	Field Evaluation	68
7.7	Technological Developments	71
7.7.1	Newer Developments: Role of rDNA Techniques	73
7.7.2	Manipulation of Genes	74
7.7.3	Dissection of Biocontrol System	74
7.7.4	Bt in Control of Plant Disease	75
7.7.5	Risk Potential of Biological Agents	75
7.7.6	Genetic Manipulation of PGPR	76
7.7.7	Monitoring with Molecular Markers	77
7.7.8	Future Scenario	78
8	**Interactions of Rhizobacteria with other Microorganisms**	79
9	**Future of Rhizobacteria as New Inoculant**	80
9.1	Problems of Shelf-Life and Field Performance	80
9.2	Acceptability by the End User	80
9.3	Influence on Non-Target Organisms	82

9.4 Community Analysis, Population Dynamics and Integrated Management 82

10 Conclusion 85

11 References 86

Abbreviations

ACC	1-Aminocyclopropane-1-carboxylate
AM	Arbuscular Mycorrhizae
ARDRA	Amplified rDNA Restriction Analysis
Az	Azotobacter
BGA	Blue Green Algae
BGB	Blue Green Bacteria
BHU	Banaras Hindu University
CLSM	Confocal Laser Scanning Microscope
CMC	Carboxymethyl Cellulose
DBT	Department of Biotechnology
DEAE	Diethylaminoethyl
DGGE	Denaturing Gradient Gel Electrophoresis
DHBA	2,3-Dihydroxybenzoic Acid
3,5-DHBA	3,5-Dihydroxybenzoic Acid
DRB	Deleterious Rhizobacteria
DRMO	Deleterious Rhizosphere Microorganisms
ELISA	Enzyme Linked Immunosorbent Assay
ESI-MS	Electrospray ionization-Mass Spectroscopy
EPR	Emergence Promoting Rhizobacteria
EPS	Exopolysaccharide
FLPs	Fluorescent Pseudomonad
FYM	Farmyard Manure
gfp	Green Fluorescent Protein
Ggt	*Gaeumannomyces graminis* var. *tritici*
GDH	Glucose Dehydrogenase
GMOs	Genetically Modified Organisms
GUS	β-Glucuronidase
ha	Hectare
HCN	Hydrogen Cyanide
IAA	Indoleacetic Acid
IARI	Indian Agriculture Research Institute
IGS	Intergeneric spacer
INM	Integrated Nutrient Management
IPM	Integrated Pest Management
IROMPs	Iron Regulated Outer Membrane Proteins
ISR	Induced Systemic Resistance
ITS	Internal Transcribed Spacer

IVET	In vitro Expression Technology
kDa	Kilodalton
LPS	Lipopolysaccharide
MBI	Michigan Biotechnology Institute
MHB	Mycorrhiza Helper Bacteria
mM	Millimole
MRF	Mussoorie Rock Phosphate
N	Nitrogen
NBRI	National Botanical Research Institute
NMR	Nuclear Magnetic Resonance
NPR	Nodulation Promoting Rhizobacteria
OM	Organic Matter
OMPs	Outer Membrane Proteins
P	Phosphorus
PPC	Phosphoenolpyruvate Carboxylase
PCR	Polymerase Chain Reaction
PGPR	Plant Growth Promoting Rhizobacteria
Phl	Phloroglucinol
PHPR	Plant Health Promoting Rhizobacteria
PQQ	Pyrroloquinoline quinone
PSB	Phosphorus Solubilizing Bacteria
PSMs	Phosphorus Solubilizing Microorganisms
QS	Quorum Sensing
RAPD	Random Amplified Polymorphic DNA
RCR	Relative Chemotactic Response
RFLP	Restriction Fragment Length Polymorphism
rhi	Rhizosphere Induced Genes
SEM	Scanning Electron Microscopy
SPIC	Southern Petrochemical Industrial Centre
SSP	Single Superphosphate
UPGMA	Unpaired Group Arithmetic Mean Analysis
USDA	United States Department of Agriculture
VAM	Vesicular Arbuscular Mycorrhizae
X-Glc A	5-Bromo-4-chloro-3-indolyl-β-D-glucoronide
YIB	Yield Increasing Bacteria

1 Introduction

The world population will cross the 10 billion mark by 2050. This population increase will create insurmountable pressure on the existing land area for food, fiber, fuel and raw materials. Utilization of improved plant varieties and technological interventions have been instrumental in meeting the demands of the growing populace in the country. While the consumption of N, P and K increased, the relative use of N has been much higher than other nutrients. However, this has had its impact on the major cropping system, viz., rice-wheat which is the dominant component (over 70%) of India's food supply; average yield increases of approximately 2 percent obtained during 1960 to 1990, are no longer being maintained [1]. In this context, soil health and plant productivity both are relevant since it is imperative that any such intervention must also be sustainable in the long run.

When one views the soil health issue, a central role of microorganisms becomes evident for the large population present in this habitat and the inherent diversity points towards the many roles of the existing communities. It has been realized that the intensive cropping of the last decades coupled to high nutrient inputs has not been able to cope with the demands of the improved varieties. Consequently there has been a greater removal of nutrients than their replenishment.

While biological inputs have been seen as a viable alternative, there are inherent problems of variability, sustenance and appropriate technologies. However, it is known that microbially produced nitrogen through the free living blue-green bacteria (BGB), symbiotic with *Azolla* and heterotrophic bacteria in the rhizosphere and soil, coupled to fixation by legumes, can contribute up to 15–20 lakh tones of nitrogen for crop production in the country [2]. Application of such biological inputs also improves the status of soil organic matter, enzymes and microbial population which act as indicators of soil health.

2 Rhizosphere

The term "rhizosphere" was originally coined to denote the existence of a relatively large microbial population around the roots of especially the legumes, however, it is now freely applied to the zone of influence around plant roots in general [3]. The zone of influence of the root harbors an approximately 10- to 100-fold greater microbial population, suggesting fierce competition for nutrients as well as the existence of species which show a variety of functional diversity and metabolic versatility [4, 5]. Available data show that about 10 to 12% of the photosynthate is exuded out of the roots during the course of growth and development of a plant species. The other sources of nutrients include various kinds of secretions, lysates, sloughed-off cells, mucigel, polysaccharides and dead biotic components [6]. Within the rich nutrient pool, are molecules that work as signals for attraction of especially bacteria which live in closer association with the roots than others.

The rhizosphere itself can be demarcated into (a) endorhizosphere, (b) rhizoplane, and (c) ectorhizosphere. Endorhizosphere refers to the internal root area extending generally to the cortical region but which now appears to harbor rather large populations of bacteria with varied functions [3]. The differentiation between the rhizoplane and rhizosphere/ectorhizosphere is difficult to make because the surface of the root and adjoining soil component often appear as a continuum.

2.1 Diversity of Free-Living and Symbiotic Bacteria

The major rhizobacterial genera include species belonging to *Acetobacter, Arthrobacer, Azospirillum, Azotobacter, Bacillus, Flavobacter, Pseudomonas, Proteus, Rhizobium, Serratia, Xanthomnonas* and others [7].

Table 1. Relative composition of rhizobacterial population of green gram (*Vigna radiata* and *V. glabrescens*) [8]

Genus	Bacterial population (%)		
	Total	Rhizosphere + Rhizoplane	Endorhizosphere
Bacillus	33	31	38
Enterobacter	29	24	41
Klebsiella	2	1	5
Proteus	3	1	3
Pseudomonas	28	36	11
Unidentified	5	7	2

The terms used to denote specific functions of rhizobacteria based on their functional attributes include PGPR (***p**lant **g**rowth **p**romotory **r**hizobacteria*), NPR (***n**odulation **p**romoting **r**hizobacteria*), EPR (***e**mergence **p**romoting **r**hizobacteria*), PHPR (***p**lant **h**ealth **p**romoting **r**hizobacteria*), DRB (***d**eleterious **r**hizobacteria*), DRMO (***d**eleterious **r**hizosphere **m**icroorganisms*) and YIB (***y**ield **i**ncreasing **b**acteria*).

Among the dominant bacterial forms in Indian soils and rhizosphere, *Bacillus, Enterobacter* and *Pseudomonas* have usually been recovered [8, 9]. The species include *Bacillus brevis, B. cereus, B. circulans, B. firmus, B. licheniformis, B. megaterium, B. mesentericus, B. mycoides, B. polymyxa, B. pumilus, B. pulvafaciens, B. subtilis*; within *Pseudomonas* these are *P. aeruginosa, P. cissicola, P. fluoresens, P. pinophilum, P. putida, P. putrefaciens, P. stutzeri* and *P. syringae*. These forms are additionally associated with strong phosphorus solubilization potential. The relative proportions of various genera in rhizosphere and endorhizosphere can vary (Table 1).

2.2 Endorhizosphere

The existence of growth promontory bacteria inside the root and shoot tissue has attracted considerable attention lately wherein rice is a major plant being studied. Rice is grown in India under varied ecological and hydrological situations and management practices vary. However, associative nitrogen fixation in the rice rhizosphere ecosystem is one of the major components of plant productivity. The contribution of associative and free-living microorganisms based on $^{15}N_2$ and ^{15}N dilution techniques is in the order 10 to 80 kg N ha^{-1} [10]. In addition, the large rhizobacterial diversity is responsible for overall nutrient transformations and the ensuing growth promoting effects.

The rice root-soil interface exhibits not only poor aeration but also a low redox potential coupled to the existence of both reduced and oxidized zones. This is manifested in the existence of diverse physiological groups of nitrogen-fixing bacteria under the flooded soil ecosystem [11]. In rice, an additional feature of interest is the presence of endophytic bacterial diversity and its capability to fix nitrogen. Some of the confirmed endophytes include *Azoarcus* spp., *Herbaspirillum seropedicae, Pseudomonas stutzeri* A 15, *Rhizobium leguminosarum* bv. *trifolii* and *Serratia marcescens* [12]. Endophytic forms have now been reported in seeds of deep-water rice varieties. Tripathi et al. [13] have reported seven types of Box-PCR fingerprints for bacteria recovered from the seeds of Jaisurya or Desaria, a free-living variety of rice. These were identified as *Pantoea agglomerans, Ochrobactrum anthropi, Pseudomonas fulva* and *Psuedomonas boreopolis.* A *gus*-tagged *P. agglomerans* strain could be traced on the root surface, root hairs, root cap, points of lateral root emergence, root cortex and the stellar region. Among the diverse diazotrophic enterobacteria of rice, the better characterized forms are *Enterobacter clocae, Erwinia herbicola/Enterobacter agglomerans* (*Pantoea agglomerans*), *Klebsiella planticola, K. oxytoca* and *Serratia marcescens* [14]. The relative preponderance of these r-strategists is a reflection of the rather large pool of organic carbon around rice roots.

Kanungo et al. [15] reported that a cultivar BKS-13 with high N-absorption efficiency harbored a greater population of *Azospirillum, Azotobacter* and anaerobic bacteria. However, when 60 kg N ha^{-1} were applied to this cultivar and BK-7–275 (low N-absorption efficiency), the rhizospheres of both cultivars depicted a nearly identical population density of nitrogen-fixing bacteria. However, methane oxidizers and autotropic ammonium oxidizers remained more crowded in the surface and rhizosphere region than in the subsurface soil [16].

Currently, groups located at the School of Biotechnology, BHU are carrying out active research on rice endophytes. Kumar and others (personal communication) have recovered 49 bacterial isolates from rice, 41 endophytes and 8 rhizospheric and analyzed them employing functional and molecular tools. Based on ARDRA and RAPD, 19 N_2 fixing strains could be placed in 5 groups. Functionally, several of these isolates were also efficient P-solubilizers and IAA producers. Gus-tagging of strains M36C1 and M36R3 showed active colonization of the root hair surface in rice variety Malviya-36.

Tripathi and coworkers (personal communication) have studied the diversity of salt-tolerant bacteria in the rhizosphere of *Oryza sativa*. Analysis of 14 salt-tolerant (3% NaCl) strains by ARDRA based on *Sau*3A, *Alu*I and *Rsa*I and RAPD, grouped them into 4 clusters which were assigned to *Alcaligenes xylosoxidans, Ochrobactrum anthropi, Pseudomonas aeruginosa* and *Serratia marcescens*. What is intriguing in this diversity is the fact that these four species are potential human pathogens that can infect immunocompromised patients. This study, therefore, highlights the necessity to consider the possible distribution of pathogenic/potential pathogens within the new found rhizobacterial diversity besides their plant growth promontory potential.

Considering the demand on improved endophytic nitrogen fixation in rice, a new approach was attempted by Gopalaswamy et al. [17] wherein the xylem of rice (*Oryza sativa*) was colonized by forced entry of *Azorhizobium caulinodans* in the presence of the flavonoid naringenin without any symptoms of disease. In this context, endophytic *Acetobacter diazotrophicus* is known to penetrate sugarcane roots intercellularly with the resultant colonization of xylem vessels. This opens up the possibility to artificially colonize rice roots with potential nitrogen fixers in the future.

2.3 Rhizoplane

Rhizoplane refers to the root surface zone wherein microorganisms can often attach themselves employing surface structures such as flagella, fimbriae or cell surface polysaccharides. The boundary between rhizoplane and rhizosphere is very thin and therefore this habitat is largely considered as a continuum. Researches carried out by Indian workers refer to all three components as separate habitats and together as an assemblage of one or the other component [5, 18].

3 Beneficial and Deleterious Forms

A relatively large population of rhizobacteria is dominated by Gram negative forms that exhibit various functions. This is attributed to the ability of PGPR to release IAA, synthesize ACC deaminase, lower ethylene levels, secrete siderophores and antibiosis, release volatiles, form hydrolytic enzymes, solubilize phosphorus, fix nitrogen and invoke induced systemic resistance [7]. Isolates that are capable of performing these functions are termed beneficial. Most research in this respect has been carried out with fluorescent pseudomonads (FLPs) wherein a single isolate can perform more than one function [9, 19, 20]. Among the rhizobacterial population is also a fraction of FLPs which is deleterious to plant growth; however, beneficial forms usually keep such populations under check in the soil environment. The deleterious action of rhizobacteria is a result of HCN production, toxins and lytic enzymes [4, 21, 22].

3.1 Community Dynamics

The distribution of various bacterial populations in the rhizosphere zone is not haphazard since a change in abiotic and/or biotic components results in a consequent change in the population structure and community profile. However, before the advent of in situ molecular tools, such changes were studied based solely on culturable populations in selective media but with our present understanding of bacterial diversity, it is difficult to interpret the data obtained through routine methods.

In a national network programme initiated by the Department of Biotechnology, Government of India on Integrated Nutrient Management, a center at Madurai Kamraj University, Madurai has begun studying the dynamics of FLPs using molecular tools. This group [23] has developed protocols for in situ DNA extraction from soil (rhizosphere/non-rhizosphere) followed by confirmation of fluorescent *Pseudomonas* strains in soil based on PCR amplification of the 16S-23S rRNA operon ITS region using the primers ITS-F (5′-AAGTCTAAGGATG-3′) and ITS-R (5′-GACCATATAACCCCAAG-3′). The loss of viability of the pseudomonad was inferred after one week from the intensity of the band.

3.2 Influence of Plant and Soil Type on Diversity

Each plant species releases root exudates that differ quantitatively and qualitatively and the rhizosphere population and community structure vary accordingly. Studies based largely on population diversity and structure in relation to plant and soil influences are reviewed by Tilak et al. and Saxena et al. [20, 22].

One of the major problem in use of PGPR for reintroduction in soil microcosms has been strain differentiation based on suitable markers. In general, intrinsic antibiotic resistance has been used but considering the problems associated with vertical and horizontal movement of such genes, data generated through this approach are not entirely satisfactory. During the last few years, SPIC, Chennai, has developed molecular methods for strain differentiation of PGPRs [24]. Hopper has used PCR-RFLP analysis for differentiation of rhizobial strains utilization amplification of the intergeneric region (IGS). Based on the digestion of PCR amplified product with *Hind*III and *Sau*3A, cluster analysis of the RFLP band pattern of chickpea, soybean, pigeonpea and groundnut nodulating rhizobial strains was carried out. According to Hopper, clusters based on UPGMA (unpaired group method of arithmetic mean) permitted observation of closeness of strains within rhizobia and helped in differentiating strains at the species level.

To assess the role of seed and root extracts of soybean (*Glycine max*) in root colonization of the rhizoplane (GRP_6), endorhizosphere (PEn-4) and rhizosphere (PRS_9) isolates, the crude material was fractionated through CM-Sephadex C-50 and DEAE-Sephadex A-50 by Rao and Johri [25]. In addition, synthetic amino acids and sugars were used as chemical attractants in a capillary assay to assess their relative chemotactic response (RCR). There was greater agglutination of the three FLPs in CM-Sepharose (C-50) treated seed and root fractions than Sephadex A-50 (A-50) generated fractions. When soybean roots pre-treated with these fractions were tested for adherence to bacteria, GRP_6 and PRS_9 were significantly better colonizers of the first 4 cm root portion than PEn-4. These isolates showed a differential agglutination reaction towards crude and ion-exchanged extracts. Whereas, a reproducible variation in adherence pattern could be seen, a significant correlation existed only between adherence and agglutination.

4 Substrates in the Rhizosphere

The common sources of nutrients in the rhizosphere are exudates, secretions, plant mucilage, mucigel and lysates. Root exudates are a crucial component of the rhizosphere microecosystem since they are rich sources of a large variety of molecules such as sugars, amino acids, organic acids, fatty acids and sterols, growth factors, nucleotides, flavonones, enzymes and a range of other miscellaneous compounds [6, 26]. A further help to rhizobacteria are the secretions from the root cap cells, epidermal cells, and root hairs besides those derived from the microbial degradation and modification of the dead epidermal cells. These are not only a source of nutrients but create microsites that permit niche exclusivity to rhizobacteria during root colonization and sustenance.

4.1 Influence on Dominant Populations

Among various rhizobacteria studied, FLPs are the most extensively and intensively examined group globally. Several research groups in India have used them in plant growth promotion and biocontrol studies [4, 20, 22]. The reasons are clear since FLPs can be recovered from a wide variety of plant rhizospheres, are easy to cultivate and a great deal is known about their metabolic and genotypic versatility [5, 27, 28]. In addition they can bring about plant growth promotion by releasing auxins, lowering the levels of ethylene through the activity of ACC deaminase (direct mechanism), chelate iron (siderophore mediated disease suppression) and release antifungals (antibiosis mediated suppression); the latter are examples of indirect mechanisms for improvement of plant health [7, 29]. In addition, the last decade has seen the involvement of ISR in FLP-associated plant disease suppression [30]. However, until now no clear cut explanation has come forth to understand the complexities involved in the interactions of PGPR with other microbial communities or about their rhizosphere competence, an attribute determining the root colonization process and the sustainability of an inoculant [28, 29].

5 Root Colonization by Rhizobacteria

The exploitation of FLPs and other PGPR as inoculants to control plant disease, use as biofertilizers, phytostimulants and as co-inoculants for bacteria-plant based bioremediation has met with only limited success at field level on account of the inconsistency and variability wherein root colonization is the first step of establishment in the rhizosphere [4]. Researches in this direction have usually been targeted towards long-term survival rather than short-term in situ colonization. Srivastava et al. [31] have studied root colonization of wheat by FLPs, GRP_3 and PRS_9 in non-sterile soil employing culturable method and SEM and reported that both the strains followed an identical pattern of root and rhizosphere colonization; the population level increased during the early root expansion phase, reaching a constant level, followed by a decline. For example, on a single root, the bacterial density declined from stem base to the root tip. SEM studies confirmed abundant bacterial colonization of the proximal parts of wheat root surface. Nautiyal [29] has reviewed other developments including use of molecular markers in the study of root colonization.

5.1 Competition and Mechanisms

In a recent study, Rainey [27] has described the genetic complexity of the system in *Pseudomonas fluorescens* employing IVET based on promoter-tapping. He has identified 20 genes that were induced during the rhizosphere colonization and of which fourteen were found to be involved in nutrient uptake, stress response or secretion processes; the other six did not show homology to the existing sequences. Seven of the rhizosphere-induced (*rhi*) genes showed homology to known *Pseudomonas* genes; of these, one (*hrc*C) is a component of a type III secretion pathway which is not known to exist in saprophytic bacteria.

Based on the molecular dissection of the rhizosphere, it is apparent that the association between FLPs and other rhizobacteria with the plant roots is much more complex and intimate than previously perceived. Furthermore, a new development is that whereas antifungal metabolites produced by FLPs influence plant health through an indirect mechanism, i.e., elimination of phytopathogens in the rhizosphere, the pathogen itself is no longer a moot spectator. Smith et al. [32] have reported that in *Pseudomonas fluorescens* F 113, that protects the roots of sugarbeet from *Pythium ultimum* by producing antifungal metabolites, the pathogen possesses the ability to downregulate the expression of genes necessary in rhizosphere competence of the PGPR strain. Schnider-Keel et al. [33] have now reported and confirmed earlier observations that the toxin, fusaric acid produced by *Fusarium oxysporum* can inhibit Phl production by *Pseudomonas fluorescens*. In addition, rhizosphere competence is a key character since it determines root colonization potential, competitive ability and sustenance in the crowded rhizosphere environment [4, 29].

6 Influence of Rhizobacteria on Soil Processes

Rhizobacteria influence soil processes through nutrient acquisition and release, interconversions through enzymatic processes, mobilization and immobilization, influencing the physical structure through aggregation and by producing dead cell mass. Thus, their influence on soil processes is both direct and indirect with consequential effects on plant growth.

6.1 Interconversions

In an interesting study of microbial populations and diversity in Alfisols and Vertisols subjected to erosion and other soil and crop related factors in Andhra Pradesh, Venkateswarlu [34] has found that the microbial population and diversity decline in the surface layer (0–10 cm) of Alfisols could be correlated with the physical and chemical properties whereas those caused by seasonal erosion in cropped fields were largely reversible. Fungal population and diversity were more severely affected than actinomycetes and bacteria. Among bacteria, the autotrophs, *Nitrosomonas* and *Nitrobacter*, showed greater decline compared to total heterotrophs. Whereas the decline in microbial population was related to reduction in organic carbon, the change in diversity appeared more closely related to the humic and non-humic fractions of the organic matter. In low fertility marginal fields of the Ujhani area in District Badaun, we have found [35] that, at the crown root initiation stage in wheat, the FLPs and siderophore-producing populations in the rhizosphere and rhizoplane were comparable. However, phosphate solubilization predominated in low input, high yield fields.

In a recent study of bacterial diversity of organically grown crops (Jhangora and Mandua) of the Central Himalayan region of Choukhutia, Chandra [36] has reported a greater population in rhizosphere soil compared to bulk and uncultivated soils. The population of copiotrophs increased with plant growth but those of oligophiles declined. About 16% of a pool of 382 bacterial isolates was capable of solubilizing phosphorus and produced siderophores and rhamnolipids. Several of these isolates have now been found to inhibit the damping-off pathogen under in vitro assay conditions (Sharma, personal communication). Therefore, it appears that the long history of organic agriculture in these fields (approx. 100 years) has permitted the indigenous microbial communities to develop strong interconversion capability for plant matter and competitive abilities based on the release of secondary metabolites.

6.2 Enzymes

Enzymes in the rhizosphere and bulk soil have been used as indicators of biological activity by several groups associated with the National Network Programme on INM of the Department of Biotechnology, Government of India. In assessing the role of BGA, VAM, PSB and *Azospirillum* in a rice-wheat-mung bean cropping system, Kaushik [37] has reported a beneficial residual influence of biofertilizers on build up of nitrogen and phosphorus; this resulted also in enhanced dehydrogenase activity. In sugarcane-ratoon-wheat rotation, soils became biologically active 45 days after plantation followed by a decline [38]. On the other hand, in sugarcane-ratoon-mentha rotation, there was continuous fluctuation in dehydrogenase activity; in contrast, the dehydrogenase level declined in sugarcane-ratoon-lentil rotation. Adholeya [39] has examined the influence of chemical fertilizers, organic manures (FYM/compost) and biofertilizes in a poplar-*Eucalyptus* agroforestry system intercropped with wheat-pulse rotation and has reported that soil dehydrogenase activity and total culturable microbial counts were influenced by fertility level and the previous history of microbial inoculation. Talukdar [40] has assessed change in dehydrogenase activity to monitor soil health in rice-legume-rice and rice-rice cropping systems operative in the Lower and Central Brahmaputra Valley Zone along with microbial population and microbial biomass and has found it to be a good indicator of soil health. Rao et al. [41] studied acid and alkaline phosphatase and dehydrogenase after three years of crop rotation in an arid soil and found considerable differences in enzyme levels (Table 2).

6.3 Biomolecules – Community Interactions

The major molecules that appear to alter the pace of microbial community interactions which are not associated directly with competition and root colonization include cell surface polysaccharides. Glycolipids are produced by a va-

Table 2. Changes in enzyme status at 0–15 cm soil depth after 3 years of crop rotation in an arid soil [41]

Rotation[a]	Acid phosphatase (n Kat 100 g^{-1} dry soil)	Alkaline phosphatase (n Kat 100 g^{-1} dry soil)	Dehydrogenase (n Kat 100 g^{-1} dry soil)
F-PM-F	2.42	6.75	9.31
PM-PM-PM	3.20	8.99	10.16
MB-PM-MB	3.18	9.02	10.63
CB-PM-CB	3.63	9.71	10.26
F-MB-MB	3.80	9.41	11.72
F-CB-CB	3.97	9.54	11.10
MB-MB-MB	3.85	9.63	13.17
CB-CB-CB	4.25	11.64	12.60
LSD (P=0.05)	0.16	0.36	0.45

[a] F- Fallow; PM- Pearl millet – *Pennisetum americanum* (L.) Leeke; PM-PM-PM; MB (Mung bean – *Vigna radiata* (L.) R. Wilczek); PM-MB; CB (Cluster bean – *Cyamopsis tetragonoloba* (L.) Taubert) – PM-CB; F-MB-MB; F-CB-CB-CB, MB-MB-MB and CB-CB-CB. The cropping system preceding the experiment was pearl millet.

riety of rhizobacteria including FLPs [42]. Rhamnolipids from FLPs can alter membrane permeability of zoosporic forms of *Pythium, Phytophthora* and *Plasmopara* and thus break completion of an effective life cycle, indirectly influencing their spread in the root environment [43]. In our ongoing studies of rhizobacteria, we have recovered a large population of rhamnolipid producing isolates by employing a selective plate assay. Isolate GRP_3 from rhizoplane of soybean produced rhamnolipids of varying chain length, the major being C_{10}–C_{12} (Table 3); the total glycolipid content based on ^{13}C-NMR is approx. 18% of crude exocelllular products (Sharma et al. unpublished).

Table 3. Glycolipids found in the crude ethyl acetate extract of FLP isolate GRP_3 by positive ion mode EMI-MS

Structure	Mol. mass ($M+H^+$)	Relative amount
Rha-Rha-C_{10}-C_{10}	673	100
Rha-C_{10}-C_{10}	527	21
Rha-Rha-C_{10}-C_{12}	701	21
Rha-Rha-C_{12}-C_{10}	701	7
Rha-Rha-C_{10}-C_{12}	699	17
Rha-Rha C_{12}-C_{10} (with double bond in C_{12} unit)	699	1

7 Rhizobacteria as Growth Promotory and Biocontrol Agents

The ability of rhizobacteria associated with various plant species to proliferate in the crowded rhizosphere milieu requires special metabolic and functional attributes in order to evolve into "ecologically fit" species. Among the more important attributes are, the ability to colonize roots successfully (colonization potential), sustain competition through release of bioactive molecules, build up large populations during the growth stage of a plant and maintain the threshold level to meet the demands of a proliferating plant species in soil. Most successful rhizobacteria, including the dominant FLPs, achieve these objectives by releasing one or more of the following groups of chemicals, i.e., siderophores, antibiotics, HCN, IAA, polysaccharolytic enzymes, phosphatases and lipopolysaccharides [7]. Coupled to these features is the ability of several PGPR to release cytotoxic principles that make them possible pathogenic agents in human disease, viz., *Pseudomonas aeruginosa* and *Serratia marcescens.*

7.1 Siderophores

Rhizosphere inhabiting bacteria usually live in microcolonies where the transient concentration of available iron can vary greatly from that in bulk soil solution. Further influences on the availability of iron are brought about by organic acids released by plant roots and the presence of microbial siderophores and phytosiderophores [44]. Bacteria tend to produce siderophores under iron limiting conditions. The dominant rhizobacterial population in most plant rhizospheres in Indian soils consists of FLPs that produce pyoverdines [45]. Pyoverdines chelate ferric iron in the environment via the hydroxamate and hydroxy acid groups present within the peptide moiety of the molecule [46]. Bacterial strains possess outer membrane receptor proteins that can transport the ferric iron complex to the respective cognate pyoverdine into the bacterial cell; iron thus becomes available for metabolic processes [45]. The iron-regulated outer membrane proteins of various siderophore producing bacteria have been characterized (Table 4).

The genes involved in the biosynthesis of pyoverdines and their functions are known (Table 5) [45].

Several workers in India have studied the distribution and activity of siderophore producing FLPs from the viewpoint of plant disease control since the production of pyoverdines scavenges available iron in the rhizosphere and creates an environment inimical to the growth of especially phytopathogenic fungi [19, 22]. The multitude of siderophores and their specificity with respect to the producer strain has permitted "siderotyping" for characterization of pseudomonads and their monitoring in soil [51]. Early reports of siderophores from growth promoting FLPs relate to production under in vitro conditions [52]. In subsequent work from this group, rhizobacteria of chilli, cotton, groundnut and soybean were found to secrete trihydroxymate type siderophores under iron deficient conditions [52, 53]. *Azospirillum lipoferum* M, an isolate from surface

Table 4. Iron regulated outer membrane proteins (IROMPs) involved in iron siderophore uptake

IROMP	Gene	Receptor protein	Reference
FpvA	*fpv*A	Ferripyoverdine receptor	[46]
FptA	*fpt*A	Ferripyochelin receptor	[47]
PupA	*pup*A	Pseudobactin 358 receptor	[48]
PupB	*pup*B	Pseudobactin BN7 and BN8 receptor	[49]
FhuE	*fhu*E	Ferrioxamine E Receptor	[50]

Table 5. Genes of *P. aeruginosa* ATCC 15692 (PAO 1 strain) involved in the biosynthesis of pyoverdine [45]

Pyoverdine biosynthesis genes	Expected function	Number of amino acid residues	Size (Da)
pvcA	Diaminobutyric acid condensation	327	37,019
pvcB	Oxygenase	289	33,165
pvcC	Hydroxylase	500	55,812
pvcD	Cyt C-related e-transfer	215	23,076
pvdA	L-Orn-N^5-oxygenase	426	47,700
pvdD	Peptide synthetase	2448	273,000

sterilized roots of maize (*Zea mays*) was found to produce catechol type siderophores under iron starved conditions [54]. The active components were characterized as salicylic acid, DHBA and 3,5-DHBA conjugated to threonine and lysine. Similar siderophore moieties were detected also in culture filtrates of a *Rhizobium* GN 1 isolate, nodulating pea [55]. Two iron repressive OMPs of approx. 80 and 70 kDa were detected in iron starved cells; the sid^- mutant was unable to grow in a medium containing synthetic iron chelators unless exogenous iron was supplied [56]. Based on colony blot and dot blot hybridization, Pandya and Desai [57] have shown that *nod* (common *nod* genes), *nif* (*nif*KDH) and EPS (epoxypolysaccharide, P^{ss}) determinants reside on the large plasmid. Root exudate components provided a chemotactic response [58]. Saikia and Bezbaruah [59] studied siderophore production by *Azotobacter chroococcum* RRLJ 203 and found that azotobactin produced in cultures containing Fe, Cu, Ni, Co, Mn and V was able to increase seed germination of *Cicer arietinum, Phaseolus mungo, Vigna catjung* and *Zea mays*. The production of siderophore in four fluorescent pseudomonads (FPC 32, FPM 14, FPM 22, FPO 4) isolated from the rhizosphere of *Capsicum* spp., *Zea mays* L., and *Oryza sativa* L., was influenced by amino acids, organic acids and sugars under in vitro growth conditions [60, 61]. Sharma [62] had studied the influence of carbon sources and metal ions on siderophore biosynthesis in a PGPR isolate GRP3A and found standard succinate medium to be superior to others; siderophore production was stimulated by Zn^{2+}, Cu^{2+} and Mn^{2+} in this bacterium. In a field trial conducted with GRP_3 at Malegaon Farm, Sharadnagar, Dist. Pune in 1999 with groundnut, a considerable reduction in iron chlorosis was achieved through seed dressing. The yield in the treated plot was 16.5 q ha^{-1} compared to 11.0 q ha^{-1} in an untreated control (Johri et al., unpublished).

7.2 Antibiosis

Many rhizobacteria including FLPs secrete a variety of antifungal molecules under in vitro and in situ conditions [63]. The diversity in the type of antibiotics is rather broad and compounds such as pyrrolnitrin, pyoluteorin, tropolone, pyocyanin, phenazines and 2,4-diacetylphloroglucinol are fully characterized [64]; this has helped in the development of suitable probes for detection of antibiotic producing strains within the diverse PGPR populations. There are several reports of in vitro antagonism of pathogenic fungi and field performance by bacteria recovered from the rhizosphere of plants in India [19, 22].

The rhizosphere and root zone of tea (*Camellia sinensis*) is a good habitat for PGPR strains represented by *Bacillus, Proteus* and *Pseudomonas*, inhibitory to phytopathogenic fungi in vitro, viz., *Fusarium oxysporum* f. sp. *lycospersici*, *Fusarium oxysporum* f. sp. *ciceri*, *F. udum*, *F. solani*, *F. moniliformae*, *F. semitectum*, *Fomes lamiensis* and *Ustulina zonata* [65–67]. One of the molecules was tentatively identified as *N,N*-dimethyl-2-phenazineamine ($C_{14}H_{13}N$) [67]. Most isolates were able to enhance plant growth of tea, pigeon pea, chickpea and maize when used as seed bacterization preparations. Bacterization with *Bacillus* strain SR 2 (peanut rhizosphere) reduced chickpea wilt in wilt-sick soil [68]. In a subsequent study, Dileep Kumar [69] utilized a siderophroe producing fluorescent *Pseudomonas* strain RBT 13 (tomato rhizoplane) and antibiotic-producing *Bacillus subtilis* strain AF1 (pigeonpea rhizosphere) for bacterization of chickpea seeds and assessment of plant growth and spread of wilt caused by *Fusarium oxysporum* f. sp. *ciceris*. Both the strains were active under in situ conditions, however, influence of RBT 13 (Sid^+) but not AF 1 (ant^+) could be eliminated by addition of iron. Dileep and Dileep Kumar [70] have reported recovery of a *Pseudomonas* strain from the rhizoplane of paddy root that was effective in suppression of collar rot disease of peanut caused by *Aspergillus niger*. Rao et al. [27] examined growth and nodulation of lentil (*Lens esculentus*) in *Fusarium* infested soil employing five (GRP_3, GRP_6, PRS_9, PEn-4, RBP_2) FLPs recovered from rhizoplane and rhizosphere of pea and soybean. In a pot experiment, disease reduction was correlated with the population dynamics of the introduced bacterium. Isolates PRS_9 and GRP_3 showed strong survival in the rhizosphere and rhizoplane up to 45 d under heavy inoculum pressure of the wilt pathogen *Fusarium oxysporum* f. sp. *lini*. In subsequent work from our group, Tripathi [71] has reported control of anthracnose of soybean caused by *Colletotrichum dematium* and sheath blight of maize caused by *Rhizoctonia solani* under in vitro and field conditions by FLPs, GRP_3 (rhizoplane of soybean), PEn-4 (endorhizosphere of pea), PRS_1 (rhizosphere of pea) and WRS_{24} (rhizosphere of wheat); besides the role of antifungal antibiotics and siderophores, involvement of induced systemic response also appears as a possible explanation for disease suppression. All FLPs were compatible with the commonly used fungicides, bavistin and vitavax (Table 6).

Table 6. Compatibility of fluorescent pseudomonad GRP_3 with commonly used fungicides and inhibition of phytopathogenic fungi [71]

Treatment	% inhibition		
	Colletotrichum dematium	*Rhizoctonia solani*	*Sclerotium roflsii*
GRP_3	72.7	0.0	59.4
Bavistin (80 ppm)	25.4	22.6	45.6
Bavistin + GRP_3	73.8	44.8	61.6
Vitavax (80 ppm)	39.4	51.7	43.3
Vitavax+GRP_3	80.7	71.9	83.5

Pandey et al. [72] have described two strains of *Pseudomonas corrugata* (1 and 7) from subtropical and temperate soils in the Sikkim Himalaya that were effective in situ against three major pathogens of maize, viz., *Pythium ultimum, Pythium arrhenomanes* and *Fusarium graminearum*. Fluorescent pseudomonad isolate EM 85 and two *Bacillus* isolates [MR-11(2), MRF] recovered from the maize rhizosphere were found by Pal et al. [73] to be strongly antagonistic towards maize pathogens, *Fusarium moniliforme* (foot rot and wilt), *Fusarium graminearum* (collar stalk/root rot/wilt) and *Macrophomina phaseolina* (charcoal rot). These isolates were endowed with various growth promotory properties (IAA, P-solubilization, nitrogen fixation) besides secretion of antifungals (volatile and non-volatile) and siderophores. Under in situ conditions, the two bacilli significantly (P=0.05) reduced the charcoal rot of maize; disease caused by the two fusaria was effectively checked by isolate MRF and a *Tn*5:*lac*Z mutant (M 23) of fluorescent *Pseudomonas* EM85. *Pseudomnonas* spp. EM85 was strongly inhibitory towards *Rhizoctonia solani*, the causal agent of damping-off of cotton (Table 7, [74]). Considering its potency towards a variety of soil borne plant pathogenic fungi, the molecular basis of this character was analyzed by Anith et al. [75], employing a functional complementation analysis; the mutant was, however, unable to produce the antifungal antibiotic and therefore failed to check fungal growth [76]. Pal et al. [74] monitored this isolate employing the *Tn*5:*lac*Z molecular marker. Introduction of this trait into the chromosome of EM85 provided antibiotic deficient and over-producing mutants with different antifungal traits; in general, deficient mutants (Afu^-, Sid^-, Flu^-, HCN^-) built up higher populations than the wild type, isogenic and over-producing mutants. The siderophore mediated antagonism of rhizoplane isolates of *Pseudomonas* from chilli, cotton, groundnut and soybean against species of *Aspergillus, Fusarium, Mucor, Penicillium, Pythium aphanidermatum, Rhizopus oryzae, Rhizoctonia solani, Sclerotium rolfsii* and *Sclerospora graminicola* was reported by Yeole and Dubey [77].

Several new exhaustive and intensive searches for biocontrol PGPRs for specific crop pathogens have been undertaken in the country. Kishore et al. [78] recovered 150 bacterial strains from the rhizosphere of groundnut from different soils in Andhra Pradesh of which ten inhibited conidial germination of *Phaeosariopsis personatum* (Berk and Curt) V. Arx under in vitro conditions and as a spray for leaf; two chitinase-producing, Gram-positive strains in particular

Table 7. In situ disease suppression of cotton caused by *Rhizoctonia solanii** [74]

Treatments	Mean disease rating/plant at 14 days after sowing**	Per cent disease reduction
Soil control	0.0^{h}	–
Pathogen control (P)	2.97^{a}	–
P+EM85	1.80^{c}	39.15^{abc}
P+M18	2.56^{bc}	20.71^{cd}
P+M58	1.10^{g}	52.48^{a}
P+M3	1.68^{ef}	39.45^{abc}
P+M54	2.69^{ab}	13.21^{d}
P+M42	1.35^{fg}	48.26^{ab}
P+M30	2.24^{cd}	29.08^{bcd}
P+M55	2.25^{bc}	11.31^{d}
P+M43	1.28^{g}	49.00^{ab}
P+M23	1.95^{de}	35.70^{abc}

* Means within the same column and followed by the same letter(s) are not significantly different at P=0.05 according to Dunnett and Duncan multiple range tests.

** Results based on 100 plants in each treatment with five replications, having 20 plants in each replication. The experiment was repeated twice.

were very effective. Vasudevan and Gnanamanickam [79] found *Pseudomonas fluorescens* PF 7–14, a rice rhizosphere isolate, to suppress the leaf and neck blast disease of rice by 78 and 82%, respectively.

Besides suppression of disease through the siderophore-mediated system and the release of antifungals, rhizobacteria are now known to act through ISR as well wherein a pathogen can be checked away from the place of application. In our laboratory (Pathak and Johri, unpublished), isolate GRP_3 has been found to induce systemic resistance in rice against sheath blight caused by *Rhizoctonia solani*; the level of peroxidase and phenolics was higher in the diseased tissue and resulted in considerable disease reduction.

7.3 ACC

Among the mechanisms operative in PGPR, Glick [7] has reported stimulation of plant growth through the activity of the enzyme ACC deaminase. His group has subsequently proposed a model according to which the IAA produced by a PGPR bacterium attached to either seed or root surface, is taken up by the plant [7]. Depending upon the endogenous level of IAA in plant tissue, two things can happen, either cell proliferation and (or) elongation or induction of the enzyme ACC synthase. Should the latter event take shape, the level of ACC in plant increases; some component of ACC, exuded out by plant roots or seeds, is taken up by the bacterium while the remainder is cleaved by ACC deaminase to ammonia and α-ketobutyrate. ACC is the immediate precursor of ethylene in plants and therefore lowering of the level decreases ethylene inhibition of plant seedling, resulting in root elongation.

Shah et al. [80] have analyzed the genes of ACC deaminase producing PGPR, *Enterobacter cloacae* CAL 2 and UW 4 and found them homologous to three *Pseudomonas* strains (6G5, F17, ACP). Based on Southern hybridization, these workers have reported the presence of a single copy of the ACC deaminase gene in strains AL 2 and UW but other bacteria may possess different ACC deaminase genes. Rhizobacerial diversity reported from India in general has not been exclusively screened for this property although direct growth promotion has been reported [20, 74]. In our ongoing work with wheat rhizobacteria (Gaur et al., unpublished) over 100 isolates were screened for ACC deaminase of which 15 were positive. Molecular dissection of ACC genes in these isolates is currently underway.

7.4 IAA

Bacterial biosynthesis of phytohormone, indole-3-acetic acid (IAA) is known in many rhizobacteria. It is believed that approximately 80% of rhizosphere bacteria can secrete IAA [81] and the phytohormone thus produced can help promote growth or pathogenesis in plants. Among the first group are included bacteria within the genera *Azotobacter, Azospirillum, Pseudomonas, Rhizobium, Xanthomonas* and others whereas the latter group consists of *Agrobacterium rhizogenes, A. tumefaciens* and pathovars of *Pseudomonas syringae*. Several workers in India have reported IAA production by rhizobaceria but detailed analysis of this characteristic has not been worked out [18].

7.5 Phosphorus Solubilization

Indian soils are normally deficient in available phosphorus even though the bound component may be sufficiently abundant. Therefore, the use of PSM is very common. However, immobilization of added P in soils is a serious problem [9]. In systems such as banana, up to 90% of P can often be immobilized. For example, in the major banana growing area of Jalgaon in Maharashtra (Kothari, personal communication) where the soils are calcareous and are irrigated with saline groundwater, there is large scale immobilization of phosphate as calcium phosphate; in neutral to alkaline pH soils, the solubility goes down to as low as 0.1%. Application of PSM therefore is a necessity to solubilize immobile P into the primary orthophosphate ion ($H_2PO_4^-$) available for the plants.

Rhizobacteria of various kinds can solubilize not only calcium phosphate but also rock phosphate, iron phosphate and aluminum phosphate by secretion of organic acids. Limited solubilization of organic phosphates by secretion of extracellular phosphatases is also known [9]. Furthermore, in addition to rhizobacteria, several fungi such as species of *Aspergillus* can efficiently solubilize P [82]. This is based on the premise that nearly all PSM when multiplied on a simple "C" source, produce formic, acetic, lactic, oxalic, succinic, malic, maleic, gluconic, glycolic and 2-ketogluconic acids in the medium [83]. Release of organic acids has also been detected in soil, which lowers the pH, resulting in increased microbial activity and greater P solubilization. For example, when soil pH decreases from

9.0 to 7.0, $Ca_3(PO_4)_2$ forms more soluble entities by chelating Ca^{2+} ions. According to Gaur [84] who has carried out extensive work on P solubilization and field application, use of co-inoculants such as *Azotobacter, Bacillus* and *Pseudomonas* with Mussoorie rock phosphate (MRF) could make P availability equivalent to 50 kg of P_2O_5 applied in the form of single superphosphate (SSP).

Gyaneshwar et al. [83, 85] reported three bacterial isolates from pigeon pea (*Cajanus cajan*) rhizosphere which were able to secrete about 50 mM gluconic acid. The bacterial isolate, identified as *Enterobacter absureae*, was able to solubilize "P" when grown in the presence of glucose as C source and ammonia or nitrate as N source. Gluconic acid secretion in *Acinetobacter caloaceticus, Erwinia herbicola* and *Pseudomonas cepacia* is dependent on PQQ dependent GDH (EC 1.1.99.17). This enzyme is differentially regulated in these bacteria by the "C" source. According to these workers, GDH of *Enterobacter absuriae* is increased approximately 5-fold under P starvation conditions and the increase is dependent on protein synthesis. In its activity and function, GDH is similar to type B enzyme of *Acinetobacter caloaceticus*. Confirmation of the role of GDH in P solubilization was tested by developing mutants deficient in GDH activity; these failed to release phosphate from alkaline soils [86]. In a continuing study of *Enterobacter absureae*, Vyas et al. [87] studied the growth and yield of *Vigna radiata* (mung bean) in pot experiments and under field conditions in Vadodora area. Increased grain yield of 155% was obtained by bacterial seed treatment; a commercial PSB preparation resulted in 111% increase in grain yield. In a related study, Gyaneshwar et al. [88], were successful in cloning the mineral phosphate solubilizing (mps) genes from the unicellular cyanobacterium *Synechocystis* PCC 6803 into *E. coli*; the transformants solubilized rock phosphate. Based on the premise that secretion of organic acids is sufficient to confer the P-solubilizing property, Srivastava et al. [89] attempted metabolic engineering for the development of P-solubilizing strains of *Pseudomonas fluorescens* which is a broad-host range PGPR; to achieve this, incorporation of the PPC gene of *Anacystis nidulans* (*Synechococcus* 7942) was carried out. While non-transformed strains secreted low levels of organic acids under in vitro conditions or solubilized P in the medium, growth of the transformants resulted in a change of pH from 7.0 to below 5.0 with resultant release of 300–400 μM phosphate in the medium; HPLC analysis showed liberation of gluconic and citric acids as the main components.

Improvement of P solubilization and siderophore production in *Pseudomonas* isolate EM85, recovered from maize rhizosphere, was studied by Pal et al. [74]. Transposon mutagenesis resulted in recovery of 8 isolates with improved (2–3 fold) phosphate solubilization and release of catechol type siderophore. In an extensive screening of over 1000 wheat rhizobacteria from low fertility soils, approx. 80% isolates could solubilize phosphorus in the plate assay (Gaur et al., unpublished).

7.6 Field Evaluation

Extensive field evaluation of various rhizobacteria as monoculture and as co-cultures with rhizobia, phosphobacteria and others has been carried out in India.

This work has been extensively reviewed [9, 10, 22]. The salient information is presented here.

Rhizobacteria from green gram [*Vigna radiata* (L.) Wilczek and *Vigna glabrescens* (n=42)] were recovered from rhizosphere, rhizoplane and endorhizosphere by Gupta et al. [8,90]. Two isolates when tested in combination with *Bradyrhizobium* spp. (*Vigna*) strains Cog 15 and S 24 influenced shoot biomass, N content and grain yield of the plant. Dileep et al. [91] have reported improved growth and yield of bhindi (*Abelmoschus esculentus* L.), paddy (*Oryza sativa* L.) and peanut (*Arachis hypogea* L.) by *Pseudomonas* FPO 4 and FPC 32 recovered from the rhizoplane of paddy and chilli, respectively; both the isolates sustained populations of over 10^4 cfu for up to 50 d suggesting strong rhizosphere colonization. Employing two strains belonging to the genus *Proteus*, Barthakur and Bezbaruah [92] found effective colonization of roots of gram (*Cicer arietinum*), bean (*Phaseolus radiatus*) and mung bean (*Phaseolus mungo*) with the resultant improvement of root and shoot biomass.

Chickpea (*Cicer arietinum* L.) is a major pulse crop of India but is severely affected on account of either the disease or salt since it is cultivated in fairly large areas which are salt affected. Saxena and Rewari [93] have reported that the availability of Zn^{2+} at 5 ppm provided protection to the plants under saline conditions (salinity level, 4.34 and 8.3 dSm^{-1}); the protection was traced to a reduced Na^+:K^+ ratio in shoot biomass. Saxena and Rewari [94] performed a field experiment on five chickpea cultivars (Pusa 209, 212, 240, 261, 312) and eight strains of *Rhizobium* spp. (G-257-3; GS-24-84; F/75, IC/76, IC 94, KG 31, KG 46, P-114-3) at IARI Crop Research Centre in a sandy loam soil (pH 7.5). The yield response was dependent on the cultivar-*Rhizobium* spp. interaction. Maximum grain yield (660 kg ha^{-1}) was obtained with Pusa 240 (cultivar) and F-75 (*Rhizobium*); three other combinations (Pusa 240-IC 76; Pusa 240-KG 31; Pusa 312-KG 31) were also significantly better than control.

Malik et al. [95] isolated a thermotolerant Gram-positive bacterium from the rhizosphere of wheat, *Kurthia* sp., that could enhance grain yield of rapeseed (*Brassica campestris* v. *toria* cv. T-9) in a pot experiment. Further improvement in yield was possible by the use of a consortium of rhizobacteria comprising *Klebsiella planticola*+*Bacillus subtilis* and *Proteus vulgaris*+*Bacillus subtilis.* Kaushik et al. [96] isolated *Azosprillium brasilense* strains from wheat endorhizosphere that could grow at sub-optimum temperature. Improved growth of wheat cultivars HD 2285 and WH 547 in pot experiments with a 25–27% increase in grain yield was observed in HD 2285 with efficient colonization of the endorhizosphere by two such isolates.

In view of the fact that rhizobia, in general, tend to be poor competitors, several reports of co-inoculation with rhizobacteria exist in the Indian literature [10]. Dileep Kumar et al. [97] studied FLPs recovered from Indian soils and strains S 97 and S 510 representing Swedish soils for co-inoculation with three strains belonging to *Rhizobium leguminosarum* biovar. *viciae* (R 304, R 313, R 361-27). Co-inoculation of FLPs and rhizobia improved growth of pea (*Pisum sativum* L. cv. Capella) in terms of root and shoot biomass.

Pal et al. [98] carried out detailed analysis of plant growth promoting FLPs from pea rhizosphere on yield and nutrient uptake of peanut cultivar JL 24 in a

Table 8. Effect of PGPR on the growth, yield and nutrient uptake in peanut cultivar GG2 in the rainy season of 1999 under field conditions* [98]

Isolate	Pod yield (kg ha^{-1})	Plant biomass (g $plant^{-1}$)	N content of shoot (%)	P content of shoot (%)
Control	1872	17.91	2.155	0.192
PGPR 1	2350	24.48	2.370	0.227
PGPR 2	2320	27.28	2.368	0.221
PGPR 3	2170	21.48	2.290	0.201
PGPR 4	2315	24.48	2.371	0.225
PGPR 5	2157	19.06	2.305	0.223
PGPR 6	2175	20.94	2.225	0.242
PGPR 7	2045	20.55	2.217	0.238
PGPR 8	1955	18.48	2.283	0.245
PGPR 9	1945	19.52	2.590	0.218
CD (0.05)	258	4.72	0.212	0.028

black calcareous soil with normal doses of fertilizers (20 kg N; 40 kg P_2O_5 ha^{-1}). Four isolates (PGPR 1, PGPR 2, PGPR 4, PGPR 9) significantly improved nodule number in a pot trial in two consecutive seasons (Table 8). There was a general increase of 30 to 60% in plant biomass with all isolates except for PGPR 9. In the field trials of two years, higher pod yield was recorded after maturity, i.e., 110 d. Isolates PGPR 1 and PGPR 2 significantly enhanced plant biomass, nodule dry mass and pod yield (20 to 30%) compared to control. There was an increase in N content of plants (5–9%) and kernels (6–12%) after field inoculation with, particularly, PGPR 1 and PGPR 2. This was also reflected in better solubilization and availability of phosphorus to plants.

In view of the limited availability of phosphorus, considerable field experimentation has been carried out with PSB. Employing preparations of *Arthrobacter awamori*, *Bacillus polymyxa* and *Pseudomonas striata*, Gaind et al. [9] have reported improved yields of chickpea, rice, soybean and wheat (Table 9). Since *Pseudomonas striata* is a preferred organism, Singh and Tilak [99] have reported increased P uptake and gain yield in a range of crops with this bacterium (Table 10).

Recent information with non-symbiotic nitrogen fixers shows that the application of *Azotobacter* and *Azospirillum* can improve yields of both annual and perennial grasses (Table 11) [100]. With the current changes and emphasis towards ap-

Table 9. Effect of phosphorus solubilizing bacteria on grain yield [9]

Treatment	Yield (kg ha^{-1})			
	Wheat	Rice	Soybean	Chickpea
Control	3440	2490	1050	2370
Pseudomonas striata	3730	2540	1786	2460
Bacillus polymyxa	3660	2580	1730	2920
Arthrobacter awamori	3620	2560	1724	2780

Table 10. Influence of application of *Pseudomonas striata* as P solubilizer on performance of various crops [99]

Crop	% increase in P uptake	% increase in grain yield
Green gram	31.4	–
Chickpea	27.3	25.0
Sorghum	18.3	12.6
Sorghum	20.4	24.7
Canola	13.0	–
Rice	14.9	8.7
Chickpea	30.99	–
Rice	–	16.7

Table 11. Effect of non-symbiotic N-fixers on yield of some cultivated non-legume fodder crops and range grasses [100]

Crops	Green forage yield (q ha^{-1})		
	Control	*Azotobacter*	*Azospirillum*
Annual			
Jowar	412	470	455
Maize	388	433	428
Oat	425	490	465
Perennial			
Napier bajra hybrid	1368	1460	1539
Guinea grass	792	855	905
Pasture grass (Buffel grass)	287	325	308

plication of FYM and use of microbial consortia, it is relevant to test the potentiality of new inoculants in combination with various other inputs. Fluorescent pseudomonad GRP_3 was found to perform well with *Azotobacter*, PSB and FYM in a field experiment conducted at agricultural station, Sagar in MP (Table 12). This field trial also confirmed the stability of the isolate recovered from the rhizoplane of soybean at Pantnagar on a different crop and agroecological zone (Johri et al. unpublished).

7.7 Technological Developments

From the traditional farming systems based on FYM and cow dung, modern agriculture has seen intensification of technological inputs to boost production. However, this has been at a cost; from a fertilizer consumption of 69,000 tonnes in1951, the level increased to approximately 17 million tonnes in 1998–1999. The consumption of technical grade chemical pesticides between 1948–1949 to 1990–1991 increased from 160 tonnes to 75,000 tonnes, followed by a decline to about 49,150 tonnes in 1998–1999 on account of the environmental concerns and adoption of IPM strategies [101]. Based on the input:output ratio of the nutrients,

Table 12. Influence of fluorescent pseudomonad GRP_3 as coinoculant with other rhizobacteria and organics on wheat yield [a]

Treatments	Mortality (% root rot)	Tillers/m²	Grain yield (q/ha)
FYM 0+No culture	14.54	394.0	32.86
FYM 5 t+No culture	14.24	493.3	34.18
FYM 0+GRP_3	0.04	642.6	37.72
FYM 5 t+GRP_3	5.46	736.0	42.58
FYM 5 t+*Azotobacter*	14.20	652.0	34.94
FYM 5 t+PSB	9.33	700.0	35.63
FYM 0+GRP_3+PSB	5.91	848.0	41.19
FYM 5 t+GRP_3+PSB	6.02	808.0	42.58
FYM 5 t+PSB+*Azotobacter*	9.01	798.6	37.40
SEM±	1.29	58.89	2.06
CD at 5%	3.78	172.73	6.08

[a] Based on a trial conducted at agricultural research station at Sagar, MP (Johri et al. unpublished).

Indian soils exhibit a negative balance of about 10 million tonnes of N, P and K. Therefore, there is a necessity to apply appropriate management practices, taking into account the chemical and biological inputs in an integrated manner.

The organic carbon status of the soils is low in India whereas phosphorus availability is generally medium to low; even potash levels in some areas are lower than the plant demand. While average nutrient consumption in the country is around 89 kg ha^{-1}, there are many areas where this does not reach even 50 kg and in smaller pockets, organics alone provide the necessary nutrient resource to the farmers [101]. Use of organics in the form of bioinoculants (biofertilizers, biopesticides), composts, green manure, crop residues, animal manures etc., has not only picked up momentum but organic agriculture is slowly making inroads with farmers at large. As per the present estimates, about 650 million tonnes of rural compost and 16 million tonnes of urban compost require more effective and intensive utilization since current efficacy relates to application of only one-third of the available product [101]. In addition, biofertilizers are to be given greater impetus because there is a tremendous gap in the supply and demand of various kinds of products. The total estimated demand for the biofertilizers based on various microorganisms is as follows (calculated based on area under different crops and the biofertilizer dose [2]):

Type of biofertilizer	Requirement (tonnes)
Azospirillum	4,82,000
Azotobacter	1,62,610
Azolla	20,380
Blue green algae	2,67,720
Phosphate solubilizers	2,75,500
Rhizobium	35,730

As against the projected demand, the production scenario of biofertilizers in various sectors in the country is as follows:

Sector	Production (tonnes)
Central Government	375
State Agricultural Universities	245
State Agricultural Departments	665
Agro Industries Corporations	1,130
Private Sector	195

To achieve the twin objectives of maintaining the quality and integrating the bioinoculant use in sustainably managed production systems, the following factors require critical appraisal and application to boost their utilization and production [2].

- The production sector (Govt./Private) should have adequate R & D infrastructure for conservation and characterization of germplasm employing both traditional and modern tools.
- Scale up fermentation facilities and downstream processing infrastructure must meet international standards to maintain batch-to-batch quality.
- Quality assessment should follow use of immunological and DNA fingerprinting protocols. This should be applied to random samples collected from the market.
- Legal provisions for registration of the product and subsequent marketing system are necessary along with a redressal cell for farmers.
- Efforts towards improved formulations with greater shelf-life are necessary to target larger audiences.

7.7.1
Newer Developments: Role of rDNA Techniques

The role of recombinant techniques has been appreciated by researchers involved in bioinoculant investigations, however, no GMO has been released for biofertilizer and biopesticides on a large scale. The necessity to manipulate likely target genes was however felt for some time. Based on a Brainstorming Session in May 2000 on Transgenic Plant Growth Promoting Rhizobacteria, a National Network Programme has recently been initiated by the Department of Biotechnology, Government of India on cyanobacteria, *Azosprillum*, *Azotobacter*, *Pseudomonas* and other systems to develop more efficient bioinoculants for problem soils and other stressed environmental situations (Sinha, personal communication). The choice of organisms includes those with growth promotory activities and others exhibiting both biofertilizer and biocontrol potentiality, e.g., *Pseudomonas*.

7.7.2
Manipulation of Genes

Pseudomonas species have been used as both biofertilizers and biocontrol agents since in several isolates the two properties can be found together. Among these, mineral phosphate solubilization is a commonly reported trait. Krishnaraj et al. [102] used random mutagenesis and Tn5 insertions into the genome of *Pseudomonas* sp. Psd 201 and recovered four MPS and two delayed solubilization (MPSd) mutants. Tn5 insertion yielded four Psd 201::Tn5 derivatives that were MPS. The pleiotrophy shown by these mutants suggested that lesions may have occurred in the regulatory *mps* loci, resulting in altered expression and solubilization capacity.

Saha et al. [103] have described the plasmid borne determinants of colony morphology, pigmentation, antibiotic resistance and antibiosis in *Pseudomonas* species antagonistic to bacterial blight of cotton caused by *Xanthomonas axonopodis* pv. *malvacearum*. Restriction sites on the plasmid were present for *Bam*HI and *Eco*RI. Curing resulted in loss of these characters whereas transformation with plasmid DNA restored these properties, confirming their plasmid borne nature.

7.7.3
Dissection of Biocontrol System

The significance of biotechnological developments in plant protection in the country was highlighted in a symposium held at BHU in February 2000 [104]. Major themes included genetic engineering and plant protection, biochemical and molecular approaches to plant defense and biotechnology and biocontrol: new options. The discussions by Indian researchers brought to light the necessity to employ molecular tools in disease diagnosis, epidemiology and control involving use of recombinant rhizobacteria such as *Bacillus, Pseudomonas* and others. Detailed analysis of signal molecules such as PR-proteins was highlighted in the context of bacterial, fungal and viral diseases and attempts at integration of various practices were suggested.

While reviewing the status of bacterial blight disease of rice caused by *Xanthomonas oryzae* pv. *oryzae*, Gnanamanickam et al. [105] have laid considerable emphasis on biological control agents, particularly the antagonistic rhizobacteria. Considerable disease suppression was achieved by foliar spray of *Pseudomonas putida* strain V14i, a bacterial strain also effective against the sheath blight pathogen, *Rhizoctonia solani*. The authors have reported a direct correlation between the endophytic survival of *P. putida* in rice tissues and the extent of disease suppression. According to these workers, a future scenario in biological control must take into account all three major rice pathogens, viz., *X. oryzae* pv. *oryzae*, *Magnoporthe grisea* and *Rhizoctonia solani*. Among these pathogens, rice blast caused by *M. grisea* (anamorph: *Pyricularia grisea*) has been analyzed for genetic diversity, fertility and virulence and a global atlas of *M. grisea* and rice blast database have been prepared [106]. This group and others [107] have subsequently analyzed genetic variability in *X. oryzae* pv. *oryzae* employing RAPD-PCR and IS 112-based PCR. Based on the use of seven primers, simple, specific and reproducible fingerprints were obtained which were helpful in differentiating the bacterial isolates. This was also true for primers PJEL1 and PJEL2, a component of insertion sequence IS-1112 based PCR. Availability of such rapid tools is a help in proper selection of efficient rhizobacteria against this important disease.

7.7.4 *Bt in Control of Plant Disease*

Bacilllus thuringiensis has been used extensively in control of mosquitoes and as a source of insect resistance genes in various crops. While this unique bacterium has also been studied for control of nematode and soil borne plant diseases, the mechanism of such actions is not known. Amer et al. [108] have reported that Bt suppressed the growth of *Pythium ultimum* and *Fusarium oxysporum* f. sp. *lycoperisici* by lysing the mycelium. Scanning electron microscopy showed polar, external attachment of the bacterial cells to hyphae of *P. ultimum*, resulting in deformation and lysis. Cells attached randomly on the hyphae of *F. oxysporum* f. sp. *lycopersici* at some points but bacterial cells also entered the hyphae and caused lysis. Thus, physical attachment/adherence appeared as an essential step in control of fungal diseases by Bt. In an alternate approach, Naik and Vedamurthy [109] have used red rot (*Colletotrichum falcatum* Went) toxin in MS medium for selection of sugarcane lines (*Saccharum officinarum* L.) resistant to the fungal toxin; growth of var. CoC 671 was completely inhibited at 0.5% toxin level in the basal medium.

Variability in soil fusaria, responsible for severe pathogenicity of plant roots, has been difficult to diagnose because of the available morphological characteristics. Chakrabarti et al. [110] have, however, described the use of the restriction enzyme (*Eco*RI) based pattern of the nuclear ribosomal DNA (5.8S and 25S regions) in *Fusarium oxysporum* f. sp. *ciceris* to study the existing polymorphism; based on molecular analysis, these authors suggested that of the four types of races prevalent in India, races 1 and 4 are same.

7.7.5 *Risk Potential of Biological Agents*

Selection of effective rhizobacterial biocontrol agents is often based on short-term in vitro assays followed by field evaluation. In the case of fungal pathogens that grow vegetatively through the mycelial phase and reproductive phase comprising asexual and sexual structures, such short-term assessment can become counterproductive. In an interesting report of the likely risk potential of *Bacillus subtilis*, effective in mycelial inhibition and sclerotium formation in *Sclerotium rolfsii*, Prithviraj and Singh [111] found regrowth of hyphae on prolonged incubation of 10 to 12 d. This fluffy growth was different from the normal vegetative spread and the thin hyphae produced hymenia with fertile basidia and basidiospores. Depletion of sugar in the medium checked the antagonistic activity and formation of sexual structures. The frequency of formation of sexual structures in *S. rolfsii* by *B. subtilis* was of the order 20 to 30%, suggesting that this biocontrol bacterium had a direct role in the genetic recombination of the pathogen which may pose a danger of creating wide genetic variability with unpredictable results. Brasier [112] had earlier reported induction of spores in *Phytophthora infestants* by the potential biocontrol agent, *Trichoderma viride.*

7.7.6
Genetic Manipulation of PGPR

Attempts have been made to improve the performance of rhizobacteria through mutation selection and gene cloning strategies. Anith et al. [75] examined the molecular basis of antifungal toxin production in *Pseudomonas* strain EM85 by selecting a defective mutant AN 21. Preparation of a genomic library in cosmid vector pLAFR1 followed by complementation analysis resulted in recovery of a cosmid clone pANF 17 which complemented the defective character in mutant AN 21. The size of the complementing DNA fragment was 23.5 kb, and its chromosomal origin was confirmed by Southern blot; sub-clones derived from *Eco*RI fragmentation, however, failed in the complementation test [76]. Extraction of the toxin and its in situ detection confirmed the stable nature of the complementing cosmid. In continued investigations on mutants of the isolate EM85, Pal et al. [74] reported that Sid^- and HCN^- mutants failed to inhibit *Rhizoctonia solani* under in vitro conditions although in the soil microcosm these mutants suppressed damping off disease by 52%. This was not possible with mutants deficient in production of fluorescent pigment (Flu^-) and antifungal antibiotic (Afa^-). The proliferation of wild type and *lac*Z marked isolates showed nearly identical proliferation in the rhizosphere of cotton. Kaushik et al. [113] selected *Tn*5:*lac*Z mutants of *Azospirillum brasilense* that were isogenic to wild type but capable of growth at suboptimal temperatures. Mutants MC48 and MA3 fixed nitrogen and produced IAA in an isogenic manner to the wild type strains, CDJA and A40. The colonization potential of mutants for wheat was adversely affected when they were coinoculated with the respective wild type in a ratio of 1:1.

Sharma et al. [114] have developed a simple method to assess nodule occupancy of *Bradyrhizobium* sp., utilizing *gus*A gene as a marker. A construct of broad host range *Bradyrhizobium* sp., Br9038U (from *Cajanus cajan*) was prepared by random *Tn*5 mutagenesis employing plasmid pCAM111 loaded with mini *Tn*5 containing *gus*A gene, expressed from a constitutive promoter. This strain formed characteristic brown-colored nodules on the host plants, viz., pigeonpea, cluster bean, mung bean and *Acacia*. The nodule sap (bacteroid suspension) in GUS assay buffer containing X-Glc A developed a blue color in the wells of an ELISA microtitre plate within 2–3 h; such a color change was not detected in sap from nodules formed by the parent or native strains. Based on this technique, the occupancy of GUS marked *Bradyrhizobium* sp., on pigeonpea, cluster bean, mung bean and *Acacia* was found to vary between 37.1 to 41.2% in unsterilized soil. This group [115] has also reported successful characterization of nodulation genes in strains of *Rhizobium ciceri* by Southern hybridization using nod genes of *R. leguminosarum* bv. *viciae* as probes. In addition, these workers analyzed the nod region of *R. ciceri* strain HS-1 using a combination of restriction enzymes based on which a tentative restriction map of the nod region was prepared.

Garg et al. [116] have described high efficiency transformation of *Rhizobium leguminosarum* by electroporation employing a 15.1 kb plasmid, pMP154 (Cm^r) containing *nod*ABC-*lac*Z fusion. A transformation efficiency of 10^8 transformants μg^{-1} DNA was achieved. The virtue of the system developed by these workers lies in its wider applicability in contrast to conjugation which, in rhizobia, is limited to species with special plasmids carrying a gene transfer function.

7.7.7
Monitoring with Molecular Markers

Besides assessment of the quality of the bioinoculant product at the production center based on molecular fingerprinting and immunological tools, monitoring of the introduced biofertilizer and biocontrol agents in the natural environment of root, leaf and soil poses problems. The commonly used markers have included Lux, LacZ, GusA and XylE, on the basis of which the tagged bacteria could be detected based on unique colored products after cell growth on specific media. However, the gene encoding green fluorescent protein (*gfp*) from jelly fish *Aequorea victoria* is currently a preferred biomarker since it does not require any substrate or additional cofactor for fluorescence. Saha et al. [117] have used gfp to monitor populations of cowpea *Rhizobium* sp. transformants recovered from root nodules of *Dalbergia melanoxylon* in low cost carriers such as paddy husk, sawdust and groundnut shells. Paddy husk served as a good carrier and the population structure was comparable with commonly used lignite although the bacterial population declined from 50 d onwards; groundnut and sawdust were relatively inferior. *Rhizobium* transformants carrying the reporter gene, *gfp*, were stable up to 154 d.

LacZ as a stable chromogenic marker was used by Garg et al. [118] to monitor the colonization of *Azospirillum lipoferum* ICM-1001 on pearl millet (*Pennisetum americana* L.) roots. Microscopic observations showed that *Azospirillum* remained localized on the outer surface of root; cells were not detected intra- or intercellularly. In an associated study, Garg et al. [119] screened isolates belonging to *Azospirillum, Herbaspirillum* and *Spirillum* for production of calcoflor binding EPS. Based on *Tn*5 mutagenesis of *A. lipoferum* ICM 1001, six EPS mutants were recovered among a pool of 1800 clones, which failed to fluoresce in the presence of calcoflor (Cal^-). Cal^- mutants lost the ability to flocculate, grew poorly in a malate medium and produced 67 to 96% less exopolysaccharide. Such mutants were defective in colonization of pearl millet roots, confirming a direct role of EPS in the spread of this isolate on seed (spermosphere) and root.

Pal et al. [74] employed the *Tn*5::*lac*Z marker to monitor the rhizobacterium *Pseudomonas glumae* EM85, an effective biocontrol agent against *Rhizoctonia solani* from cotton. Introduction of *lac*Z into the chromosome resulted in the development of isogenic, deficient and overproducing mutants for antifungal antibiotics, siderophore, fluorescent pigment, HCN and other properties selected based on the Lac^+ phenotype. Assessment of cotton root colonization revealed that mutants deficient in these properties built up higher populations than the wild type. The inactivation or overproduction of antifungals did not impair establishment, population dynamics and ecological fitness of *P. glumae* EM85. Sankarasubramanian and Kaushik [120] had introduced *gus* and *lux* genes in cyanobacteria, *Synechocystis* sp. and *Anabaena cylindricae* to assess the stability of these markers in soil. Chromosomal integration resulted in higher stability compared to plasmid borne characters; the former were stable up to 60 d in soil microcosms prepared from a paddy field soil.

7.7.8
Future Scenario

In view of their unique position as bioinoculants, the genus *Pseudomonas* with its PGPR properties and biocontrol action has attracted most attention from not only the commercial angle but also from the view point of functional genomics [28]. The availability of the complete genome sequence of the genus *Pseudomonas* is seen as an immense help in these developments. There is considerable progress in dissecting the rhizosphere competence genes, role of signal mechanism in cell density dependent regulation of especially antifungal molecules and other secondary metabolites and a host of other developments that should see future bioinoculants with greater specificity, less variability, and stability under much wider abiotic and biotic conditions. In *Ggt*, the causal agent for take-all disease of wheat, involvement of the antifungal metabolite, Phl is now established beyond doubt. In maize [121] too, a larger proportion of the root surface *Pseudomonas* population (~15%) harbored Phl genes compared to non-rhizosphere soil (<0.65%). However, in our analysis of approximately 200 FLP isolates from the rhizosphere of wheat based on colony hybridization and a PCR based assay, only 6 were positive for Phl (Gaur et al., unpublished). Whether it is a consequence of the absence of Ggt or a result of different soil type, is under investigation. In addition to Phl, other molecules involved in biocontrol action include pyoluteorin, pyrrolnitrin and phenazines. Neilsen et al. [122] have characterized a cell-free cyclic lipodepsipeptide, viscosinamide, from *Pseudomonas* which can control *Pythium ultimum* in soil microcosms. However, it is being argued that rhizosphere competence, i.e., competition in the rhizosphere environment and root colonization are crucial to the development and sustenance of an efficacious bioinoculant. To understand these issues, use of gfp and other bioluminescent techniques has become necessary. In situ monitoring based on CLSM has shown the formation of microcolonies on the roots of crop plants [123, 124]. In terms of improvement of biocontrol efficacy, greater attention is focused towards the change in timing of the production of antifungals such as Phl so that immediate protection to plants can be provided. Besides modulation of gene expression at the transcriptional level, the role for QS, i.e., cell density dependent control of gene expression in *Pseudomonas aureofaciens* and the discovery of *N*-acylhomoserine lactones in *Pseudomonas fluorescens* F 113 have opened up new vistas of metabolite production and control [28]. In addition, a second level of control for secondary metabolite secretion operates at the post-transcriptional level through the global regulators GacS/GacA, the environmental sensor kinase and response regulator [125]. According to Walsh et al. [28], the following strategies based on application of molecular tools are likely to result in improved biocontrol strains of *Pseudomonas.*

1 Modification of post-transcriptional regulation for increased production of antifungal metabolites.
2 Regulation of antifungal metabolite biosynthetic genes based on exudates and rhizosphere inducible promoters.
3 Improvement of rhizosphere competence by introduction of additional genes involved in plant root colonization.
4 Introduction of gene clusters for production of additional fungal metabolites.
5 Modification of gene dosage of antifungal biosynthetic genes to further improve the levels of antifungal metabolites.
6 Development of consortia of microbial inoculants for increased and broad spectrum antifungal action.

These molecular approaches are likely to yield isolates with more efficient antifungal action, better root colonization and sustenance in the rhizosphere. The distinction between biofertilizer and biocontrol action is slowly falling apart since many rhizobacteria exhibit both properties. There are now instances where microorganisms such as rhizobia show good biocontrol action in addition to nitrogen fixation under symbiotic conditions. The future emphasis shall see reprogramming of the relation of genes at both transcriptional and post-transcriptional levels in order to achieve greater and better coordination of the production of antifungals involved in disease control so that improved plant health is achieved through targeted inoculants.

8 Interactions of Rhizobacteria with other Microorganisms

Rhizobacteria and other microorganisms occupy the rhizosphere through competition and interactions that have far reaching influence on the stability and sustenance of not only the indigenous useful forms but also have a direct bearing on the introduced inoculants.

Interactive studies involving the use of *Azotobacter, Azospirillum* and phosphate solubilizers show that the free-living bacteria usually maintain the combined beneficial influence which is reflected in improved plant biomass and productivity [10]. Two other major systems of interest, viz., nitrogen-fixing rhizobia and phosphorous mobilizing AM, have been studied in detail. Saxena and Tilak [10] have summarized the findings of Indian researchers in the context of AM fungi and agriculturally beneficial soil microorganisms that take into account the rhizobacteria. A general conclusion is that the synergistic behavior of *Azospirillum, Azotobacter, Acetobacter diazotorophicus* and rhizobia result in improved plant productivity in the presence/absence of other inorganic/organic inputs. There are reports that show a positive and rather close association of rhizobacteria with AM fungi that help fungal spread and colonization, the so-called MHB. The advantage of introducing a rhizobacterium, which is also MHB, takes care of several associated beneficial properties that are linked to the fungal symbiont, viz., stress tolerance, P-mobilization, metal tolerance etc.

9 Future of Rhizobacteria as New Inoculant

The future of rhizobacteria as emerging inoculants is dependent on appropriate shelf-life, quality products and predictable field performance.

9.1 Problems of Shelf-Life and Field Performance

Considerable effort in the country has been made by various groups to search for new and more effective PGPRs employing the routine approach of dilution plating of rhizosphere soil and/or root macerates on to general/selective media, followed by functional analysis as under (Fig. 1).

The above approach for selection of PGPRs including FLPs has met with success and a fairly large gene pool of effective bacteria is available in the country. However, except for a few groups, these selections are usually based on screening of rather small gene pools whereby there is every likelihood of missing out on bacteria with the most desired traits. Nautiyal [29] has focused on the utility of a raw soil-based assay for screening of a large gene pool of native rhizobacteria with effective root colonization potential. Subsequent potentiality is assessed through a two-stage, plant-based microcosm assay. This group working at NBRI has been successful in the transfer of the sand-live soil assay method based technology to MBI International, USA and to Dhampur Sugar Mills in UP. Thus, more concerted efforts are necessary in future to develop a new inoculant for sustainable soil and plant health.

9.2 Acceptability by the End User

Research on PGPR in India and elsewhere has not met with a great deal of success on the technological and product formulation front. This is related to not only the inherent variability of PGPR and inconsistencies associated with the influence of plant and soil type on its performance but also the lack of acceptable products in the market and the associated quality assurance. According to the information available on the USDA Web Site [28], no more than ten *Pseudomonas*-based commercial products are available on account of inconsistent field behavior. Among the more widely available commercial formulations, use of dry powdered products has been preferred globally on account of ease in transport. Most formulations in India are dry charcoal, lignite-based preparations. However, this leads to drying of cells and consequent loss of cell number and quality on storage. Therefore, there is an urgent need to develop alternate formulations and inoculant delivery systems which provide better nutrient environment with appropriate moisture availability for not only cell proliferation but also for release of desired secondary metabolites. This could mean applying some form of amendments to already available preparations. Several researchers in India have utilized FYM along with charcoal or similar inert carrier material for this purpose.

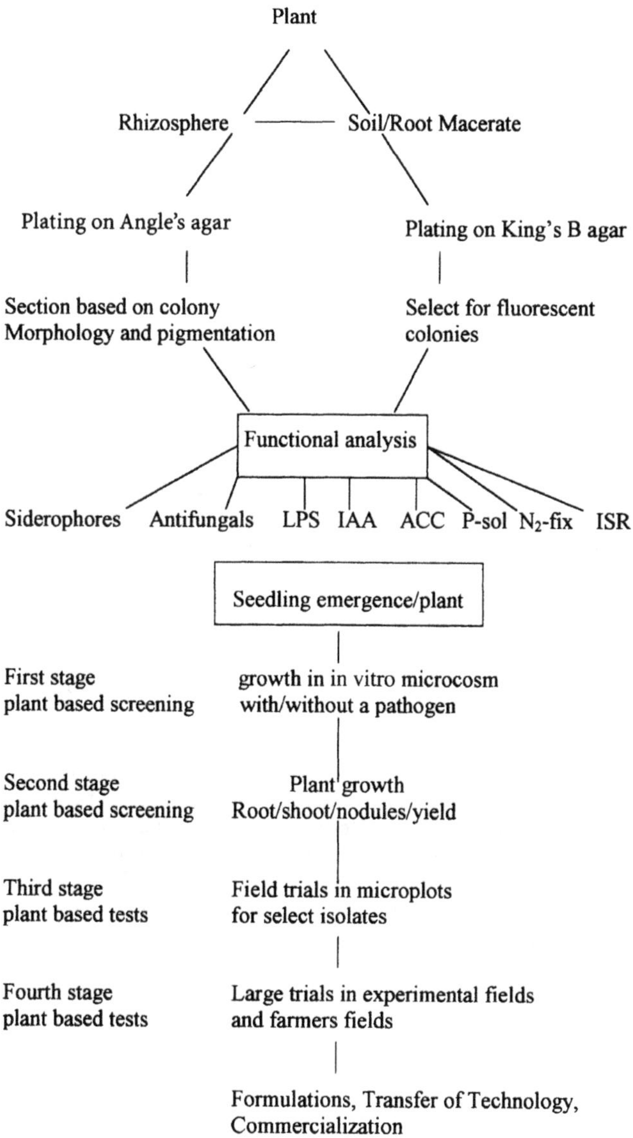

Fig. 1. Approach for selection of PGPRs including FLPs

9.3 Influence on Non-Target Organisms

The assessment of PGPR-based preparations on non-target organisms is an area where little, if any, work has been done by Indian researchers. The reasons are three fold – firstly, that many PGPR preparations tested in India have been used for plant growth promotion activity, presuming that the biocontrol action is not inherent in such an exercise; secondly, that a biofertilizer is not required to be registered on the lines of pesticides/insecticides; thirdly, the guidelines issued by the government for safety assessment of genetically modified organisms (GMOs) do not apply to naturally developed strains. However, in doing so, a point has been left out that many PGPR, such as FLPs and others including bacilli, exhibit several properties in one strain and therefore what is being considered as growth promotory, is also endowed with antifungal metabolites, siderophores, HCN or lytic enzymes. Since sufficiently large bacterial populations are released in the root zone through seed coating and root dipping protocols, the levels of such molecules could be considerable to hit the non-target organisms. Furthermore, most plants harbor the natural symbiont, AM, as a beneficial partner in this ecosystem which could be a target of the antifungals released by PGPR. During the last few years Barea and his group in Spain have made this assessment for not only *Pseudomonas* but also with *Azospirillum* and *Trichoderma* [126, 127]. They have concluded that the preparations tested did not influence AM symbiosis. However, there could be some influence of the antifungals on the culturable fungal population [128]; interestingly this influence was not as large as that noted with repeated growth of cucumber in the same soil. Considering the fact that culturable methods reveal no more than 1 to 5% of the existing microbial diversity, Lottman et al. [129] used a culture-independent approach, viz., DGGE to assess the effect of inoculation of transgenic potatoes with a *P. putida* strain and concluded that the age of the plant was mainly responsible for the change in the community structure and not the bacterial inoculant. These new reports are a reflection of our growing concern for critical assessment of both abiotic and biotic processes in the rhizosphere before releasing a new rhizobacterial inoculant. Luckily, the use of newer tools of microbial community analysis and soil processes based on enzymes permit the choice of appropriate markers that can be used with ease and could be coupled to application of in vitro methodologies to monitor in-house the localization in the root zone before releasing an inoculant for large-scale field trials and commercialization. The time has come when any new process/product shall have to meet the global standards and therefore it would be appropriate to search for an industrial partner when an initiative on PGPR research is taken up as a long-term goal.

9.4 Community Analysis, Population Dynamics and Integrated Management

Until now, most rhizobacterial research in the country has revolved around the recovery of rhizobacteria from various plant rhizospheres and their subsequent evaluation for plant growth promotion and soil processes. In doing so, it is diffi-

Table 13. DBT supported INM initiative operated in the country (1999–2002)

S. No.	Institution/PI	Cropping system/ organisms	Major work
A	Molecular markers, fingerprinting and monitoring		
	SPIC, Chennai (Dr. W. Hopper)	*Acetobacter, Azospirillum, Bradyrhizobium, Rhizobium*, phosphobacteria	DNA fingerprinting, tagging and monitoring
	Madurai Kamraj University, Madurai (Dr. S. Gunasekaran)	Fluorescent pseudomonads	DNA fingerprinting, tagging and in situ analysis
B	On-farm evaluation/validation-cum-demonstration trials of biofertilizer use in rice/rice system		
	Vivekananda Institute of Biotechnology, Nimpith, West Bengal (Dr. S.K. Das)	Rice Legume	*Azolla* and Blue Green Algae, *Azospirillum* and *Azotobacter* *Rhizobium*
	MS Swaminathan Research Foundation, Pondicherry, TN (Dr. H. D. Subashini)	Rice Groundnut	*Azospirillum*+PSB *Rhizobium*
	Council of Science & Technology, Lucknow (Dr. C. P. Dwivedi)	Rice Wheat Soybean	Blue Green Algae *Azosprillium*+PSB *Bradyrhizobium*
C	Soil nutrient dynamics, microbial dynamics, yield with biofertilizers for main and intercrops employing complementation/supplementation/substitution of chemical fertilizers		
	Indian Agricultural Research Institute, New Delhi (Dr. B. D.Kaushik, Dr. Anil Saxena, Dr. Sunil Pabbi)	Rice Wheat Mung bean	*Azolla*+Blue Green Algae *Azolla., Az* & PSB *Bradyrhizobium* and PSB
	P. D. Krishi Vishwavidyalya, Akola, Maharashtra (Dr. S. R. Potdukhe)	Pearl millet/Chickpea Mungbean/ wheat	PSB+Organic matter+*Az* PSB+*Rhizobium*+OM
	Rajasthan Agricultural University, Udaipur (Dr. S. C. Bhandari)	Sorghum Chickpea Winter maize	*Az*+PSB+VAM+OM *Bradyrhizobium*+PSB+ VAM+OM *Az* & PSB+VAM+OM
	Tata Energy Research Institute, New Delhi (Dr. Alok Adholeya)	Wheat-Mungbean in Poplar system	*Rhizobium*+VAM
	G. B. Pant University of Agriculture and Technology, Pantnagar (Dr. B. N. Johri)	Sugarcane-based cropping system	*Az*+*Azosprillium*+ *Pseudomonas*+OM

Table 13 (continued)

S. No.	Institution/PI	Cropping system/ organisms	Major work
D	Selection, characterization and evaluation of biofertilizer strains and validation in nursery and field		
	Bhavnagar University, Bhavnagar, Gujarat (Dr. H. H. Patel)	Arid salinity prone ecosystem	Phosphorus solubilizers
	Assam Agricultural University, Jorhat (Dr. N. C. Talukdar)	Rice-Legume (*Phaseolus*)	*Azospirillum- Rhizobium*
E	Selection of sugarcane genotypes for high BF response		
	Indian Institute of Sugarcane Research, Lucknow (Dr. Archana Suman)	Sugarcane	Native and constructed strains of *Acetobacter diazotrophicus*
F	Rhizospheric and endophytic diazotrophs and their role in N, P, Fe nutrition		
	Banaras Hindu University (Dr. Ashok Kumar)	Rice	Characterization of endophytes, fingerprinting and efficiency

Source: Department of Biotechnology, GOI (Sinha, Personal communication).

cult to assess their in situ role in overall microbial community dynamics and nutrient dynamics. To bridge this gap, the Department of Biotechnology, Government of India had initiated a National Network Programme in 1999 on the Role of Biofertilizers in Integrated Nutrient Management – A Biotechnological Approach, with inputs from microbiologists, molecular biologists, agronomists, soil scientists and extension workers. A total of 17 research centres were involved in this programme (Table 13) to cover various cropping systems and agroecological zones to assess the potentiality of biofertilizers as a component of the current agricultural practices and the likely savings this could bring on the use of chemical fertilizers. The microbial population and community structure were assessed coupled to soil health in order to look at any subtle changes such as improved biomass and soil aggregation to follow the residual influence of the available inputs on the succeeding crops. This programme is in its final phase and it is expected that an appropriate package of practices will be generated that can be recommended for farmers to use. This programme has also resulted in the development of protocols for quality assessment of bioinoculants and their monitoring based on molecular fingerprinting tools which would help in the production of quality products in the coming years. The ultimate aim of this scientific initiative is to devise packages of practice for farmers based on the integration of chemical fertilizers, biofertilizers, and organics.

10 Conclusion

It has been recognized that the rhizosphere of plant species is a unique ecosystem which harbors a multitude of microorganisms. Some of these, the rhizobacteria, have developed a close relationship with the root zone based on the availability of nutrients and other yet, unknown benefits. Fluorescent pseudomonads comprise one such group that has been studied extensively by various groups in India and elsewhere in the world since various strains can provide direct or indirect benefits and help improve soil and plant health. In addition, *Azotobacter, Azosprillium, Enterobacter* and others too have been studied in various crop systems with a view to utilize them individually or in association with other rhizobacteria for improved performance of various crops in the country. Research in the broad sphere of rhizosphere biotechnology with emphasis on new bioinoculants has received support from the Department of Biotechnology and Indian Council of Agricultural Research, Government of India and appropriate technologies have been developed. However, the general acceptance of such technologies with the farmers is still beset with skepticism since the shelf-life and performance of the products exhibit inconsistencies. This is in spite of proven data on the inputs that such biologicals can provide, viz., nitrogen, phosphorous, iron etc., besides creating an environment non-conducive to proliferation of phytopathogens. Currently, several laboratories have embarked on systematic characterization of potentially effective rhizobacteria of indigenous origin from the rhizosphere by employing the screening of large bacterial gene pools based on both traditional and molecular tools with extended assessment in laboratory and field. In addition, there is an effort to integrate such applications with integrated management practices and organics to reduce fertilizer use and move towards greener technologies. On the anvil are programmes whereby the development of new bioinoculants based on the incorporation of specific genes is underway so that future bioinoculants will have greater predictability, stability and efficacy in a given set of ecological environments. Also, efforts are underway to improve upon the formulation and packaging of new products based on comprehensive quality assurance protocols so that the acceptability of the products improves. These approaches coupled to continued financial support are likely to result in better utilization of rhizobacteria as future inoculants.

Acknowledgement. The writing of this article has been possible with the active support from a large number of rhizobacterial researchers in the country who provided published and unpublished data. However, authors are specially thankful to Dr. K. V. B. R. Tilak, Dr. A. K. Saxena, Dr. H. C. Dube, Dr. S. C. Nautiyal, Dr. B. S. Dileep Kumar, Dr. K. K. Pal, Dr. B. Venkateswarlu and Dr. A. K. Tripathi for scientific support. BNJ would like to thank the Department of Biotechnology, Government of India for support provided for the work cited in this chapter.

11 References

1. Srivastava JP, Mukhopadhyay M (1997) Sustainable intensification of rice-wheat cropping systems in India. A Report of the World Bank, Washington, DC
2. Kannaiyan S (2000) Biofertilizers: Key factor in organic farming, The Hindu Survey of Indian Agriculture, p 165
3. Lynch JM (1990) The Rhizosphere. Wiley Interscience, Chichester
4. Dube HC, Yeole RD (1999) In: Bagyaraj DJ, Verma A, Khanna KK, Kehri HK (eds), Modern approaches and innovations in soil management. Rastogi Publications, Meerut, p 33
5. Sinha R, Goel R, Johri BN (2001) Molecular markers in rhizosphere microbiology. In: Maheshwari DK, Dubey RC (eds), Innovative approaches in microbiology. Bishen Singh Mahendra Pal Singh, Dehradun, p 255
6. Bowen GD and Rovira AD (1991) In: Waisel Y, Eschel A, Kafkafi U. (eds), Plant Roots. Marcel Dekker Inc, New York, p 349
7. Glick BR (1995) Can J Microbiol 41:109
8. Gupta A, Saxena AK, Gopal M, Tilak KVBR (1998) Microbiol Res 153:113
9. Gaind S, Mathur RS, Tilak KVBR (2000) In: Yadav AK, Singh MR, Rauchaudhary S (eds), Recent advances in biofertilizer technology. Society for Promotion and Utilization of Resources and Technology, New Delhi, India, p 190
10. Saxena AK, Tilak KVBR (1998) In: Varma A (ed), Microbes: for health, wealth and sustainable environment. Malhotra Publishing House, New Delhi, p 25
11. Rao VR (1998) Methane emission from rice based cropping system. Project Report, ICAR AP Cess Fund Project, Central Rice Research Institute, Cuttack
12. James EK, Gyaneshwar P, Barraquio WL, Mathan N, Ladha JK (2000) In: Ladha JK, Reddy PM (eds), The quest for nitrogen fixation in rice. International Rice Research Institute, Los Banos, Philippines, p 119
13. Tripathi AK, Verma SC, Ron EZ (2002) Res Microbiol 153:579
14. Barraquio WL, Segubre EM, Gonzalez MS, Verma SC, James EK, Ladha JK, Tripathi AK (2000) In: Ladha JK, Reddy PM (eds), The quest for nitrogen fixation in rice. International Rice Research Institute, Los Banos, Philippines, p 93
15. Kanungo PK, Panda D, Adhya TK, Ramakrishanan B, Rao VR (1997) J Sci Food Agric 73:485
16. Kumaraswamy S, Ramakrishnan B, Satpathy SN, Rath AK, Misra S, Rao VR, Sethunathan N (1997) Plant Soil 191:241
17. Gopalaswamy G, Kannaiyan S, O'Callaghan J, Davey MR, Cocking EC (2000) Proc Royal Soc. Lond B 267:103
18. Nautiyal CS (1999) FEMS Microbiol Lett 170:265
19. Johri BN, Rao ChVS, Goel R (1997) In: Dadarwal KR (ed), Biotechnological approaches in soil microorganisms for sustainable crop production. Scientific Publishers, Jodhpur, p 193
20. Tilak KVBR, Singh G, Mukherji KG (1999) In: Mukherji KG et al. (eds), Biotechnological approaches in biocontrol of plant pathogens, Kluwer Academic/Plenum Publishers, New York, p 115
21. Dileep Kumar BS (1998) World J Microbiol Biotechnol 14:735
22. Saxena AK, Pal KK, Tilak KVBR (2000) In: Upadhyay RK, Mukherji KG, Chamola BP (eds), Biocontrol potential and its exploitation in sustainable agriculture. Vol. 1. Crop diseases, weeds and nematodes. Kluwer Academic/Plenum Publishers, New York, p 25
23. Gunasekaran P (2001) Role of plant growth promoting rhizobacteria in integrated nutrient management. Project Report, Department of Biotechnology, New Delhi
24. Hopper W (2001) Development of rapid molecular methods for evaluation of microorganisms. Project Report, Department of Biotechnology, New Delhi
25. Rao ChVS, Johri BN (1999) Indian J Microbiol 39:31
26. Rao ChVS, Sachan IP, Johri BN (1999) Indian J Microbiol 39:23
27. Rainey PB (1999) Environ Microbiol 1:248

28. Walsh, UF, Morrissey JP, O'Gara F (2001) Curr Opi Biotechnol. 12:289
29. Nautiyal CS (2000) Proc Natl Acad Sci India 70 B:107
30. van Loon L, Bakker PAHM, Pieterse CMJ (1998) Ann Rev Phytopahol 36:453
31. Srivastava R, Johri BN, Sharma A (1999) Indian J Microbiol 39:205
32. Smith LM, Tola E, de Boer P, O'Gara F (1999) Environ Microbiol 1:495
33. Schnider-Keel U, Seematter A, Maurhofer M, Blumer C, Duffy B, Gigot-Bonnefoy C, Reimmann C, Notz R, Defago G, Haas D, Keel C (2000) J Bacteriol 182:1215
34. Venkateswarlu B (1999) Final Technical Project Report. Indian Council of Agricultural Research, New Delhi
35. Roesti D, Mishra P, Gaur R, Jeet K, Shani N, Rossi P, Johri BN, Aragno M (2001) Microbial community analysis of wheat rhizosphere in low fertility soils of the Indo-Gangetic plain region of Western UP. Proceedings of 9th International Symposium on Microbial Ecology, Amsterdam, p. 270 (abstract)
36. Chandra D (2002) MSc Thesis, GBPUA & T, Pantnagar
37. Kaushik BD (2001) Integrated role of BGA in phosphorus dynamics in rice-wheat-mung bean cropping system. Project Report, Department of Biotechnology, New Delhi
38. Johri BN (2001) Microbial dynamics and nutrient budgeting in sugarcane based cropping system. Project Report, Department of Biotechnology, New Delhi
39. Adholeya A (2001) Integrated nutrient management in poplar-*Eucalyptus* based sustainable agroforestry system. Project Report, Department of Biotechnology, New Delhi
40. Talukdar NC (2001) Development of biofertilizers for use as a component of integrated nutrient management in rice based cropping system. Project Report, Department of Biotechnology, New Delhi
41. Rao AV, Tarafdar JC, Sharma SK, Praveen Kumar, Aggrawal RK (1995) J Arid Environ 31:237
42. Maier RM, Soberon-Chavez G (2000) Appl Microbiol Biotechnol 54:625
43. Stanghellini ME, Rasmussen SL, Kim DH, Rorabaugh PA (1996) Plant Dis 80:422
44. Loper JE, Henkels MD (1999) Appl Environ Microbiol 65:5357
45. Meyer JM (2000) Arch Microbiol 174:135
46. Poole K, Neshat S, Krebes K, Heinrichs D (1993) J Bacteriol 175:4597
47. Castignetti D (1997) Curr Microbiol 174:135
48. Bitter W, Marugg JD, de Wager LA, Tommassen J, Weisbeck PJ (1999) Mol Microbiol 5:647
49. Koster M, van der Vossenberg J, Leong J, Weisbeck PJ (1993) Mol Microbiol 8:591
50. Sauer M, Hantke K, Braun V (1990) Mol Microbiol 4:427
51. Munsch P, Geoffroy VA, Alatossava T, Meyer JE (2000) Appl Environ Microbiol 66:4834
52. Dave BP, Dubey HC (2000) Ind Phytopathol 53:97
53. Yeole RD, Dave BP, Dube HC (2001) Ind J Expt Biol 39:464
54. Shah SS, Karkhamis V, Desai AJ (1992) Curr Microbiol 25:347
55. Jadhav RS, Desai AJ (1992) Curr Microbiol 24:137
56. Jadhav RS, Desai AJ (1994) FEMS Microbiol Lett 115:185
57. Pandya S, Desai A (1998) Curr Microbiol 36:36
58. Pandey A, Sharma E, Joshi M, Palni LMS (1998) Microbiol Res 154:259
59. Saikia N, Bezbaruah B (1995) Ind J Expt Biol 33:571
60. Dileep C, Dileep Kumar BS, Dube HC (1998) Indian J Expt Biol 36:429
61. Dave BP, Dube HC (1999) Pl Dis Res 14:139
62. Sharma A (1999) MSc Thesis, GBPUA & T, Pantnagar
63. Cook RJ, Thomashow LS, Weller DM, Fujimoto D, Mazzola M, Banger G, Kim DS (1995) Proc Natl Acad Sci USA 92:4171
64. O'Sullivan DJ, O'Gara F (1992) Microbiol Rev 56: 662
65. Dileep Kumar BS, Bezbaruah B (1996) Indian J Microbiol 36:45
66. Dileep Kumar BS, Bezbaruah B (1997) Indian J Expt Biol 35:289
67. Dileep Kumar BS, Bezbaruah B (1999) J Plant Crops 27:9
68. Dileep Kumar BS (1996) Indian J Exp Biol 34:794
69. Dileep Kumar BS (1999) Biol Fertil Soils 29:87
70. Dileep C, Dileep Kumar BS (2000) Indian J Expt Biol 38:1245

71. Tripathi M (1999) Ph D Thesis, GBPUA & T, Pantnagar
72. Pandey A, Palni LMS, Hebbar KP (2001) Microbial Res 156:191
73. Pal KK, Tilak KVBR, Saxena AK, Dey R, Singh CS (2001) Microbiol Res 156:209
74. Pal KK, Tilak KVBR, Saxena AK, Dey R, Singh CS (2000) Microbiol Res 155:233
75. Anith KN, Tilak KVBR, Khanuja SPS, Saxena AK (1998) World J Microbiol Biotechnol 14:939
76. Anith KN, Tilak KVBR, Khanuja SPS (1999) Curr Sci 77:671
77. Yeole RD, Dube HC (2000) J Mycol Pl Pathol 30:335
78. Kishore KG, Pande S, Podile AR, Rao JN (2000) Seminar on Biological Control and Plant Growth Promoting Rhizobacteria (PGPR) for Sustainable Agriculture. April 3–4, University of Hyderabad, Hyderabad
79. Vasudevan P, Gnanamanickam SS (2000) Seminar on Biological Control and Plant Growth Promoting Rhizobacteria (PGPR) for Sustainable Agriculture. April 3–4, University of Hyderabad, Hyderabad
80. Shah S, Li J, Moffatt BA, Glick BR (1998) Can J Microbiol 44:833
81. Leinhos V, Vace O (1994) Microbiol Res 49:31
82. Narisan V, Thakkar J, Patel HH (1999) Indian J Microbiol 34:113
83. Gyaneshwar P, Naresh Kumar G, Parekh LJ (1999) World J Microbiol Biotechnol 14:669
84. Gaur AC (1990) Phosphate solubilizing microorganisms as biofertilizers. Omega Scientific Publishers, New Delhi, p 176
85. Gyaneshwar P, Naresh Kumar G, Parekh LJ (1998b) Curr Sci 74:1097
86. Gyaneshwar P, Parekh LJ, Archana G, Poole PS, Collins MD, Hutson RA, Naresh Kumar G (1999). FEMS Microbiol Lett 171:223
87. Vyas D H, Anthony SM, Archana G, Vyas PR, Patel TC, Mevada KD, Patel JJ, Nafde SD, Naresh Kumar G (2000) International conference on managing natural resources for sustainable agricultural production in the 21st Century. Feb. 14–18, New Delhi, p 652
88. Gyaneshwar P, Naresh Kumar G, Parekh L J (1998) In: Deshmukh AM (ed), Biofertilizers and Biopesticides. Technoscience Publications, Japan, p112
89. Srivastava S, Kulkarni A, Yadav N, Archana G, Naresh Kumar G (2000) International conference on managing natural resources for sustainable agricultural production in the 21st Century. Feb 14–18, New Delhi, p 650
90. Gupta A, Saxena AK, Gopal M, Tilak KVBR (1998) J Sci Ind Res 57:726
91. Dileep C, Dileep Kumar BS, Dube HC (1998) Indian J Expt Biol 36:399
92. Barthakur M, Bezbaruah B (1999) Indian J Expt Biol 37:919
93. Saxena AK, Rewari RB (1991) World J Microbiol Biotechnol 7:202
94. Saxena AK, Rewari RB (1992) Biol Fertil Soils 13:31
95. Mali KRDS, Kavimandan SK, Tilak KVBR (1999) Indian J Expt Biol 37:92
96. Kaushik R, Saxena AK, Tilak KVBR (2000) World J Microbiol Biotechnol 16:567
97. Dileep Kumar BS, Berggren I, Mirtensson AM (2001) Plant Soil 229:25
98. Pal KK, Dey R, Bhatt DM, Chauhan SM (2000) Proceedings Fifth International PGPR Workshop. 29 Oct – 3 Nov 2000, Cordoba, Argentina. Auburn University Web Site (www.argentina/pdfmanuscripts/kloepper.pdf)
99. Singh G, Tilak KVBR (2001) Fert News 46:33
100. Biswas BC, Das S, Kalwe SP (2001) Fert News 46:15
101. Motsara MR (2000) Plant nutrients: Role of micronutrients. The Hindu Survey of Indian Agriculture, p 175
102. Krishnaraj PU, Sadasivam KV, Khanuja SPS (1999) Curr Sci 76:1032
103. Saha S, Singh RP, Verma JP, Jayaraman J (2000) Curr Sci 79:1384
104. Arora DK, Upadhyay RS (2000) Curr Sci 79:1049
105. Gnanamanickam SS, Priyadarshini VB, Narayanan NN, Vasudevan P, Savitha S (1999) Curr Sci 77:1435
106. Babujee L, Gnanamanickam SS (2000) Curr Sci 78:248
107. Gupta VS, Rajebhosale MD, Sodhi M, Singh S, Gnanamanickam SS, Dhaliwal HS, Ranjekar PK (2001) Curr Sci 80:1043
108. Amer GA, Aggarwal R, Singh DV, Srivastava K (1997) Curr Sci 73:284

109. Naik GR, Vedamurthy AB (1997) Curr Sci 73:367
110. Chakraborti A, Mukherjee PK, Sherkane PD, Bhagwat AS, Murthy NBK (2001) Curr Sci 80:571
111. Prithviraj B, Singh UP (1997) Curr Sci 73:643
112. Brasier CM (1971) Nature New Biol 231:283
113. Kaushik R, Saxena AK, Tilak KVBR (2000) World J Micriobiol Biotechnol 16:567
114. Sharma PK, Kamboj DV, Rustagi N, Dogra RC (1997) Ind J Microbiol 37:91
115. Srivastava P, Kamboj DV, Sharma PK, Dogra RC (1997) Ind J Microbiol 37:139
116. Garg B, Dogra RC, Sharma PK (1999) Appl Environ Microbiol 65:2802
117. Saha AK, Deshpande MV, Kapadnis BP (2001) Curr Sci 80:669
118. Garg R, Kamboj DV, Sharma PK, Kundu BS (1995) Ind J Microbiol 35:121
119. Garg RK, Sharma PK, Kundu BS (1996) Ind J Microbiol 36:193
120. Sankarasubramanian S, Kaushik BD (2001) World J Microbiol Biotechnol 17:535
121. Picard C, DiCello F, Ventura M, Fani R, Guckert A (2000) Appl Environ Microbiol 66:948
122. Nielson TH, Christopersen C, Anthonie V, Sorensen J (1999) J Appl Microbiol 86:80
123. Normander B, Henderiksen NB, Nybroe B (1999) Appl Environ Microbiol 65:4646
124. Bloemberg GV, Wijfies AH, Lamers GE, Stuurman N, Lugtenberg BJ (2000) Mol Plant Microbe Interact 13:1170
125. Haas D, Bluemer C, Keel C (2000) Curr Opin Biotechnol 11:290
126. Barea JM, Andrade G, Bianciotto V, Dowling D, Lohrke S, Bonfante P, O'Gara F, Azcon-Aguilar C (1998) Appl Environ Microbiol 64:2304
127. Vazquez MM, Cesar S, Azcon R, Barea JM (2000) Appl Soil Ecol 15:261
128. Girlanda M, Perotto S, Moenne-Loccoz Y, Bergero R, Lazzari A, Defago G, Bonfante P, Luppi AM (2000) Appl Environ Microbiol 67:1851
129. Lottman J, Heuer H, de Vries J, Mahn A, During K, Wackernagel W, Smalla K, Berg G (2000) FEMS Microbiol Ecol 3:41

Received: May 2002

Adv Biochem Engin/Biotechnol (2003) 84: 91–121
DOI 10.1007/b11037

Plant Molecular Biology and Biotechnology Research in the Post-Recombinant DNA Era

Akhilesh K. Tyagi · Jitendra P. Khurana

Department of Plant Molecular Biology, University of Delhi South Campus, New Delhi 110021, India. *E-mail: akhilesh@genomeindia.org*

After the beginning of the recombinant DNA era in the mid-1970s, researchers in India started to make use of the new technology to understand the structure of plant genes and regulation of their expression. The outcome started to appear in print in early the 1980s and genes for histones, tubulin, photosynthetic membrane proteins, phototransduction components, organelles and those regulated differentially by developmental and extrinsic signals were sequenced and characterized. Some genes of biotechnological importance like those encoding an interesting seed protein and the enzyme glyoxalase were also isolated. While work on the characterization of genome structure and organization was started quite early, it remained largely focused on the identification of DNA markers and genetic variability. In this context, the work on mustard, rice and wheat is worth mentioning. In the year 2000, India became a member of the international consortium to sequence entire rice genome. Several laboratories have also given attention to regulated expression of plastid and nuclear genes as well as to isolate target-specific promoters or design promoters with improved potential. Simultaneously, transgenic systems for crops like mustard, rice, wheat, cotton, legumes and several vegetables have been established. More recently, genes of agronomic importance like those for insect resistance, abiotic stress tolerance, nutritional improvement and male sterility, isolated in India or abroad, have been utilized for raising transgenics for crop improvement. Some of these transgenics have already shown their potential in containment facility or limited field trials conducted under the stipulated guidelines. Plant molecular biology and biotechnology are thus clearly poised to make an impact on research in basic biology and agriculture in the near future.

Keywords. Agriculture, Crop improvement, DNA markers, Gene function, Genomics, Organelle genome, Plant transgenics, Regulatory elements

1 **Introduction** 93

2 **Organelle Genome and its Function** 94

3 **Gene Organization, Structure, Expression and Function** 97

4 **Genome-Wide Analysis** 101

4.1 Framework Maps and Marker-Assisted Breeding 101
4.2 Fingerprinting and Genetic Diversity 105
4.3 Genome Sequencing 108

© Springer-Verlag Berlin Heidelberg 2003

5 **Transgenics** 110

6 **Perspectives** 114

7 **References** 116

Abbreviations

A Adenine
ABRE Abscisic acid responsive element
AFLP Amplified fragment length polymorphism
AGL9 Agamous-like factor 9
AP1 Apetala 1
bar Bialaphos resistance gene
bZIP Basic leucine-zipper transcription factor(s)
BAC Bacterial artificial chromosome
C Cytosine
CaMV35S Cauliflower mosaic virus 35S mRNA
CIM Composite interval mapping
cM CentiMorgan
CMS Cytoplasmic male sterile
codA Choline oxidase gene
COP1 Constitutively photomorphogenic 1 gene
cry Crystalline protein gene of *Bacillus thuringiensis*
DBT Department of Biotechnology
EST Expressed sequence tag
FISH Fluorescence in situ hybridization
G Guanine
gfp Green fluorescent protein gene
GPC Grain protein content
GTP Guanosine triphosphate
gus β-Glucuronidase gene
ICAR Indian Council of Agricultural Research
IRGSP International Rice Genome Sequencing Project
ISSR Inter simple sequence repeats
kb Kilobase
kDa KiloDalton
MAS Marker assisted selection
NIL Near isogenic line
nptII Neomycin phosphotransferase II gene
OsMADS1 Rice MADS-box gene similar to *AP1/AGL9*
PAC P1 derived artificial chromosome
PCNA Proliferating cell nuclear antigen
PCR Polymerase chain reaction
PHS Pre-harvest sprouting
QTL Quantitative trait loci

R0	Transgenic plants from gene delivery
R1	Progeny of R0 transgenic plants
RAPD	Randomly amplified polymorphic deoxyribonucleic acid
RFLP	Restriction fragment length polymorphism
RGA	Resistance gene analogues
RIL	Recombinant inbred line
SAMPL	Selective amplification of microsatellite polymorphic loci
SCAR	Sequence characterized amplified region
SMA	Single marker analysis
SSR	Simple sequence repeats
STMS	Sequence tagged microsatellite site
STS	Sequence tagged site
T	Thymine
WA	Wild abortive

1 Introduction

The need to understand the structure and function of plants and application of new knowledge to improve crop plants is more relevant today to India than ever before if an environmental homeostasis is to be allowed to succeed in maintaining the balance between human beings and nature. To feed the increasing population of over one billion with dwindling land reserves, water, and other natural resources, Indian farmers have to opt for producing more with fewer resources. This requires a scientifically accurate approach to conserve and improve our resources. Thus, scientific knowledge is required to develop technologies which help grow crops in saline or drought-prone areas, protect crops from pathogens or insects, and improve nutritional value and yield. "Who will feed India in another 20 or 30 years?" is again being asked by international experts like Dr. Lester Brown of the World Watch Institute. The record production of food grain over 200 million tons in the last years is a breakthrough and testimony to architects of the "Green Revolution" including scientists and farmers. The 21st century needs an "Evergreen Revolution" to cope with the trends of last decade when the growth of the population has been more than the rate of growth in food grain production [1]. It is estimated that the demand for food grains, vegetables and fruits in India will increase to 260, 194 and 106 million tons in 2030, as compared to 195, 91 and 52 million tons, respectively, in the year 2000.

The advent of recombinant DNA technology in 1970s provided an immense potential for manipulating genes, precisely and across the incompatibility barriers. Plant science has moved from genes to genomics and genes can be inserted into plant genomes with relatively high efficiency. Indian plant scientists also started to use recombinant DNA technology to answer questions emanating from their scientific investigations. The establishment of the Department of Biotechnology (DBT) in 1980s and keen interest of the Indian Council of Agricultural Research (ICAR) to incorporate new technology into its crop-based programmes gave a push to the research in the area of plant molecular biology and biotechnology, and its impact started to become evident by the mid-1990s. Major initiatives taken in India include establishment of the International Center for Genetic Engineering and Biotechnology, New Delhi, the National Research Centre for Plant Biotechnology, New Delhi, M S Swaminathan Research Foundation, Chennai, the National Centre for Plant Genome Research, New Delhi, Centres for Plant Molecular Biology at Calcutta, Coimbatore, Delhi, Hyderabad, Lucknow, Madurai, net-work programs on insect resistance, bioprospecting, molecular taxonomy and marker-assisted breeding by DBT and mission-mode projects of ICAR. Other agencies under the Ministry of Science and Technology have also given due emphasis to biotechnology research. Contributions of the Rockefeller Foundation and the World Bank for National Agricultural Technology Project have been significant in improving infrastructure, research quality and human resource development.

The aim of the present article is to provide a review of Indian plant molecular biology and biotechnology research in the post-recombinant DNA era. The emphasis has been placed on research related to the use of recombinant DNA technology for gene discovery, gene expression and function, transgenics and genome-wide approaches in plants. The article, therefore, does not include interesting work on plant-tissue culture, protein biology, enzymology or biochemistry and pathogens or pests. Furthermore, the authors realized during the course of compiling the literature that citations to all publications would not be possible if emphasis is to be given to highlight the recent progress and significant achievements. It is hoped that readers of this article will understand such a limitation and can benefit by consulting earlier efforts in this direction [2–5].

2 Organelle Genome and its Function

Plant chloroplasts and mitochondria contain their own genome and perform vital cellular functions. Work on the characterization of plastid genomes of *indica* rice, mothbean and poplar was initiated in early 1990s and several photosynthesis-related genes have been cloned [6–8]. Chloroplast tRNA genes have been investigated in cucumber [9]. These studies also revealed the typical circular structure of plastid genomes in having a large single copy region and a small single copy region separated by inverted repeats. While efforts are ongoing to sequence the entire chloroplast genome of poplar, cloned genes are being used to investigate genome architecture and gene expression. By using gene-specific probes for *psbA* and *psbD*, the somatic hybrids involving *Trachystoma ballii+B. juncea* were analyzed and a novel intergenomic recombinant plastome of a cytoplasmic male sterile (CMS) line of *Brassica juncea* was identified [10]. Interestingly, chloroplast-related functions of the CMS line remained stable over generations.

Significant insights have been gained into the components involved in the maintenance of the plastid genome. A pea chloroplast-based system could replicate DNA molecules containing pea chloroplast origin of replication sequences. The presence of *OriA* and *OriB* sequences resulted in the mode of replication (θ) similar to in vivo conditions, but the presence of either of these could show replicative intermediates having sigma structures [11]. Topoisomerases are known to take care of topological constraints of DNA replication. One 69 kDa topo I enzyme has been characterized from pea chloroplasts, which does not require Mg^{++} and removes both positive and negative supercoils [12]. In addition, a plastid DNA polymerase accessory activity in the form of a 43 kDa protein was identified. The corresponding cDNA has been characterized, which revealed the presence of hydroxyproline-rich protein motifs. The protein is glycosylated and the deglycosylated protein retained DNA and chloroplast DNA polymerase binding activity but it failed to stimulate DNA polymerase activity [13, 14]. Recently, a 70 kDa DNA polymerase from pea chloroplasts has been purified and shown to have high processivity and moderate fidelity [15]. DNA helicases can also assist DNA replication. Plastidic DNA helicases have been purified and characterized [16–18]. One of them, chloroplast DNA helicase II, is stimulated by fork-like replication structures, which makes it a candidate for a DNA replicative helicase. Other chloroplast nucleic acid-related proteins identified include a single-strand DNA-specific endonuclease [19]. Work on ribonucleoproteins from pea chloroplasts was also initiated [20]. Recently, a cDNA and its genomic counterpart for a plastid-specific ribonucleoprotein from pea has been characterized and found to be stimulated by light [21].

The expression of chloroplast genes has been analyzed in a legume (*Vigna aconitifolia*), rice (*Oryza sativa*) and a tree (*Populus deltoides*) by using homologous or heterologous probes since chloroplast genes show a high degree of sequence conservation [22–25]. The investigated genes either produced single transcript (*psaA/B, psbA, psbE*) or multiple transcripts (*rbcL, psbC-D, psbB*) reflecting on the organization of genes in operons and extensive transcript processing. Interestingly, *rbcL* produced single transcript in rice [23], but two transcripts in *V. aconitifolia* [22].

An interaction of the developmental program of a plant vis-à-vis light stimulation of chloroplast gene expression has been established in *V. aconitifolia* as well as *indica* rice [22, 24]. In rice, temporal development between 5–8 days after germination resulted in 10-, 2.3-, 7.0- and 8.0-fold increases in transcript levels for *psbA, psbD, psaA-B* and *rbcL*, which could be further enhanced by 5-, 11.4-, 6.6- and 7.8-fold, respectively, on exposure to light during this period, thereby increasing the steady-state transcript levels ~25–60-fold during chloroplast biogenesis [24]. Several photomorphogenic mutants of *Arabidopsis thaliana* have also been investigated for plastid gene expression [26, 27] and found to express plastid genes in the dark, reflecting on the repressive role of the native gene product(s). A variation in the developmental programme of seedlings has also been observed in sorghum, rice, wheat and barley, and depending on the extent of development of the leaf in dark, the plastid gene transcript levels show variation in their accumulation, reflecting on the role of native developmental regulators in plastid gene expression (Tyagi and coworkers, unpublished).

Two types of RNA polymerases transcribe plastid genes. To determine the RNA polymerase involved in light regulation, tagetitoxin – an inhibitor of plastid-encoded RNA polymerase – was used and it blocked the light-dependent accumulation of mRNA of photosynthesis-related chloroplast DNA-encoded genes in rice. No influence of tagetitoxin was observed on 16S rRNA, which is apparently transcribed by a nuclear-encoded RNA polymerase transported into the chloroplast [28]. The light signal involved in the expression of *psbA*, *psaA* and *rbcL* was found to be perceived by phytochrome [29]. Use of agonists and antagonists of their action identified other important components of the signal transduction chain involved in plastid gene expression with rice seedlings as a system. The major components include G-proteins, calcium, calmodulin and the phosphorylation status [28, 29]. Dephosphorylated status was correlated with plastid gene expression and chloroplast development. Work is in progress to identify proteins and their genes influencing the transduction of signal to chloroplast transcription machinery.

Spatial control of plastid gene expression was also investigated in leaf and root of rice. The high level expression was found to occur in leaves and in the case of *psbB* and *psbD-C* operons qualitative changes in transcripts were also observed [23]. Within the leaf, a gradient of gene expression was observed from leaf base onwards and this was also influenced by calcium as well as protein phosphatases/kinases [30].

Attempts have also been made to transform plastids to investigate the basic mechanisms of plastid gene expression and for the production of commercially/pharmaceutically important biomolecules. Plastid transformation constructs have been prepared by employing components of the rice plastid genome and evaluated for their function [28]. Production of transplastomic rice, however, requires more effort, as the plastids in embryogenic calli of rice are very small as well as being less developed, thereby limiting their amenability for plastome transformation. The rice *psbA* transcriptional elements have been shown to work in tobacco plastids for expressing *cry1Ia5*, a gene encoding insecticidal protein, up to 3% of soluble protein, thereby making transplastomic tobacco resistant to insect larvae [31]. The expression of interferon gamma as a fusion protein with beta-glucuronidase/His-tag and cellulolytic alkali-thermostable xylanase at ~6% level has also been successful in tobacco chloroplasts (Reddy and coworkers, unpublished). Since both the products retain their biological activity and purification is simplified, the use of transplastomics as bioreactors would become a reality.

The mitochondrial genome has attracted the attention of workers in particular due to its possible role in male sterility. The mitochondrial genome in plants is highly heterogeneous and it is known to exist as multiple sub-genomic molecules as also like a plasmid. The presence of mitochondrial plasmids has been detected in three male sterile lines but not in fertile maintainer lines of sorghum, although their correlation with male sterility has not been established [32]. The rice mitochondrial genome has been shown to contain five sub-genomes of 117, 130, 95, 70 and 50 kb by pulse-field gel electrophoresis and a physical map of the 95 kb sub-genome was developed [33]. The organization of the *cob2* gene for apocytochrome *b* was found to vary in fertile and wild-abortive (WA)-type male sterile cytoplasm of rice. This mitochondrial gene is present as a functional copy (*cob1*) and as a pseudogene (*cob2*) in fertile IR36, but *cob2* was absent in the WA-type [34]. Such a recombination of genome can influence male sterility/fertility of lines but the nature and mechanism of the influence are not completely known. The utility of RAPD analysis for mitochondrial DNA of such cytoplasms, which could not be distinguished by RFLP and Southern hybridization, has also been demonstrated [35]. The uniqueness of mitochondrial DNA fragments can also be used to investigate transfer of cytoplasmic male sterility in cybrids by protoplast fusion as demonstrated for *Brassica* [36, 37] and rice [38].

Thus, the organelle genomes of crops analyzed and lessons learnt from them are being put to use for basic research as well as crop improvement.

3 Gene Organization, Structure, Expression and Function

Indian scientists started to analyze genes from crops of interest in the early 1980s. From a collection of *BamHI* fragments of rice (IR20) genome cloned in pBR322, a fragment of 6.64 kb was selected and genes for H2A and H2B histones on one end and for H4 on the other end were localized [39]. These were found to be transcribed in a bidirectional fashion from opposite strands. A genomic library of *Vigna radiata* in the phage Charon 4A was made and used to isolate clones containing a repetitive DNA and β-tubulin gene using a heterologous probe from chicken [40]. Interestingly, a 300 bp sequence repeat of *V. radiata* was found to express and be present in $\sim 10^5$ copies per haploid genome [41]. Similarly, the gene for β-tubulin in *V. radiata* has been sequenced [42]. By the late 1980s, the potential of recombinant DNA technology attracted the attention of several groups in India and efforts were made to isolate genes of importance for nutrition, oil quality, photosynthesis, and stress tolerance. Their structure and expression were investigated in detail and regulatory components analyzed. While some useful genes have been engineered to improve crops (Sect. 5), others are of interest to understand novel plant processes and gene expression, which continues to be a frontier area of investigation [43].

Non-availability of nutritionally balanced food to a large section of population in India is a major problem. Engineering of proteins with balanced amino acid composition and vitamins in staple food crops can help eliminate malnutrition. An albumin protein with an amino acid composition comparable to the World Health Organization's recommended value for highly nutritious proteins has been identified. Its gene was isolated in the form of a 1.2 kb cDNA capable of encoding an 304-amino acid polypeptide synthesized specifically in seed during early embryogenesis and the mid-maturation phase [44]. The 2S albumin protein gene has also been characterized from *Brassica* species and its promoter analyzed for seed-specific elements [45]. Additionally, rice genes homologous to seed storage protein, kafirin [46], chickpea genes homologous to pea legumin [47, 48], barley B_1-hordein homologue [49], and α-gliadin homologue of wheat [50] have been isolated. In certain cases, expression studies have revealed seed-specific expression of genes. For manipulation of amino acid levels, genes of key enzymes in their biosynthetic pathways would be useful. A cDNA for threonine deaminase, involved in the biosynthesis of the essential amino acid isoleucine, was cloned from chickpea [51]. This enzyme is targeted to the plastids and expressed at higher levels in flowers. Another category of biomolecules represented by glucans and present in the cell walls of the endosperm of certain members of Poaceae can be of utility for a mixed-diet programme and help modulate glucose and cholesterol levels. With the objective of introducing such a biomolecule in staple food, a gene for the reversibly glycosylated polypeptide, associated with glucan biosynthesis, of *indica* rice has been characterized [52]. Furthermore, analysis of its expression in millets like *Eleucine* and *Echinochloa*, as well as rice, revealed the temporal control of gene expression after fertilization and a low expression in rice, indicating a need to improve the expression level.

Indians consume a large amount of lipids as cooking oil. Metabolic engineering of lipid biosynthesis could provide a way to avoid undesirable and enhance specialty components. With such an objective in view, genes for stearoyl-acyl carrier protein desaturase [53], fatty acid elongation [54], plastidial omega-3 desaturase [55, 56], and oleoyl-ACP desaturase for chain length determination in oleic acid biosynthesis [57] have been cloned from *Brassica juncea*. As expected, these genes express mainly in the seed and their promoters have been found to contain conserved seed-specific *cis*-acting elements.

Abiotic and biotic stresses result in a significant loss of crop yield due to variable environmental and geographic conditions in India. It is, therefore, expected that genes for components involved in stress phenomenon will attract the attention of scientists. Considering the importance of polyamines in stress phenomenon, the expression of arginine decarboxylase was analyzed in salt-sensitive (M-1–48) and -tolerant (Pokkali) cultivars of rice [58]. A PCR-amplified probe from oat showed a 20-fold mRNA accumulation in Pokkali and an only 7-fold accumulation in M-1–48 after salt stress, thereby indicating a differential expression pattern of the gene in sensitive and tolerant varieties. Another abscisic acid and salt/mannitol inducible gene from the rice cultivar Pokkali has been characterized and shown to be similar to the *Asr* genes of other species [59]. Similarly, a developmentally-regulated novel *S*-adenosyl-L-methionine synthetase gene of rice cultivar Pusa Basmati 1 was found to be induced by salt, drought, abscisic acid, injury, osmotic- or heat-shock but not by cold stress [60]. The search for stress-inducible genes in rice continues and several expressed sequence tags (ESTs) from a normalized cDNA library from drought-stressed seedlings of rice cultivar Nagina 22 have been sequenced. Their in silico analysis has revealed several stress-inducible homologue genes [61]. In a detailed investigation on abiotic stress responses in rice cultivar IR54, transcript levels of several genes encoding glycolytic and fermentation enzymes were shown to change, albeit to variable extents, when subjected to cold, desiccation, salt, high temperature and oxygen deprivation [62]. A gene for glyoxalase I has been isolated from a dicot, *Brassica juncea*, and shown to be upregulated by sodium chloride, mannitol and zinc chloride [63]. Another stress-responsive gene from peanut has been characterized and found to be homologous to the flavonol 3-*O*-glucosyltransferase responsible for anthocyanin biosynthesis [64]. It is expected that several of these genes will turn out to be important for crop improvement.

Transcripts homologous to the wheat gene *EmBP1* (bZIP class factor) capable of recognizing the abscisic acid-responsive element (ABRE) were found to accumulate at higher levels in the roots of salt-treated Pokkali rice [65]. The analysis of ABRE-binding factor interactions in rice nuclear extracts revealed that spermidine, proline and GTP enhance DNA binding, which may be one of the ways to modulate the activity of the transcription factor. South-Western blot analysis also showed two polypeptides of 22 and 28 kDa, binding to ABRE in equal amounts, with or without salt-stress, indicating a modulation of DNA binding of pre-existing polypeptides under stress. Analysis of more transcription factors involved in the stress-inducible expression of genes is highly desirable to develop a precise and efficient strategy for manipulating stress tolerance.

Plant genes involved in resistance against pathogens or insects have also been studied. A gene for the 34 kDa basic protein involved in the systemic resistance of susceptible plants against plant viruses has been cloned from *Clerodendron aculeatum* and shown to have homology to ribosome inactivating proteins [66]. The recombinant product was found to be capable of inhibiting in vitro protein synthesis in eukaryotic translation systems but not in prokaryotic translation systems. A gene for cowpea lectin has also been characterized with a possible utility in engineering insect resistance [67].

The developmental programme of plants is greatly influenced by light and hormones from seed germination to flowering and seed-set [26]. Several labs in India have attempted to understand the mechanism of photoperception, signal transduction and gene expression in plants. The expression of phytochrome 1 in maize was found to be down-regulated by light, as opposed to up-regulation of nitrate reductase mRNA under similar conditions, and involved signal transduction components like PI cycle intermediates or protein kinase C [68, 69]. The gene for calmodulin was cloned from *Arabidopsis* and characterized for its regulated expression [70]. The gene for another calcium-binding protein, calnexin, was cloned from pea and shown to be constitutively expressed [71]. It has, however, been shown to be associated with the microsomal membrane and is known to perform its function as a molecular chaperone. The activity of several light-regulated genes is suppressed during skotomorphogenic development. The gene for a key repressor *COP1* (*constitutively photomorphogenic 1*) has been sequenced and characterized from *indica* rice and shown to be similar to its homologous gene from *Arabidopsis* [72]. In contrast to the dicot system, its transcripts were shown to be regulated by developmental cues and light. Another gene belonging to *Aux/IAA* class of transcription factors has also been characterized from rice [73]. The expression of gene is stimulated by auxin(s) and down-regulated by light. It remains to be seen whether the effect of light is direct or via its influence on endogenous auxin levels. Induction of haustoria in *Cuscuta* is influenced by cytokinins and associated genes have also been characterized [74, 75]. Other interesting plant responses include the touch response of *Mimosa pudica*. A gene for a chromophore-binding apyrase has been cloned from *M. pudica* [76]. It has been suggested that apyrase may also play some role in the light responses of *M. pudica*.

Since light is known to be an important factor for chloroplast biogenesis, investigations on nuclear genes for chloroplastic proteins have been performed. Three genes (*psbO*, *psbP*, *psbQ*) for 33-, 23- and 16-kDa polypeptides involved in oxygen evolution were characterized from *Arabidopsis* and shown to be light-regulated in an organ-specific manner [77–79]. Their expression in photomorphogenic mutants further revealed the nature of the genetic loci involved in keeping their expression repressed in dark [27]. A gene for the Ftsz protein involved in chloroplast division has been isolated from pea and shown to be light-regulated and young leaf-specific [80]. The involvement of this protein in division mechanisms has been demonstrated by complementation studies in an *E. coli ftsz* mutant. The above studies identified various components of light-regulated processes in plants. The knowledge about interaction of such components can help understand plant development better.

Investigations on flower development have shown that three sets of genes (A, B and C) are involved in floral organ formation [81]. The expression of the *AGAMOUS* gene was analyzed during *in vitro* flowering from cotyledons of ground nut and found to be associated with flower bud initiation [82]. Expression of *OsMADS1*, a rice MADS-box gene belonging to AP1/AGL9, was investigated in rice. It starts to express in spikelet meristems and the expression is limited to lemma and palea with weak expression in the carpel [83]. In transgenic plants, ectopic expression of *OsMADS1* causes stunted panicles and outer rudimentary glumes are transformed to lemma/palea-like organs in spikelets. Ectopic expression of *SUPERMAN* in rice at high levels proved detrimental to the plants [84]; this transcription factor is involved in the maintenance of the boundary between stamens and carpels of dicots. But, plants with a low level of expression survived and produced flowers with carpel expansion and destabilized stamen-carpel whorls, reflecting on its conserved function in monocots. Efforts are in progress to characterize the *dyad* gene as the *Arabidopsis* mutant in *dyad* shows that megaspore mother cell meiosis is compromised as reflected by an arrest at the end of meiosis 1 in ovules [85]. Such investigations would help understand the molecular basis of female reproductive development.

Plant cells also require several enzymes for their DNA-related processes. These include topoisomerases, PCNA (proliferating cell nuclear antigen) and helicases. A gene for topoisomerase I, defined by its activity of nicking and sealing one of the two strands and a change in the linking number of the supercoiled DNA by a step of one, has been characterized from pea and its product localized in the nucleus [86]. The DNA relaxation activity of the enzyme is repressed by PCNA from pea, the gene for which has been characterized and expressed in bacteria [87]. As expected, this PCNA also showed stimulation of DNA polymerase activity. The gene and promoter for pea topisomerase II, defined by activity on both strands of DNA and a change in the linking number of supercoiled DNA by steps of two, have also been characterized and the expression of the gene was found to be more in proliferative tissues, and was sensitive to light and hormone application [88]. A comparative analysis of the gene structure revealed that the deduced product is a nuclear topisomerase II. The helicases catalyze ATP-dependent unwinding of duplex DNA to generate a single-stranded DNA template to allow various DNA-related transactions. The cDNA for plant helicase has been characterized from pea [89] and the deduced sequence shows homology to elF-4A. The product is located in the nucleus as well as the cytosol where it influences the activity of topoisomerase I and translation, respectively. Another DNA helicase of 65 kDa has been shown to be a homologue of human DNA helicase [90]. The protein is localized in the nucleolus, gets activated by protein kinases and shows DNA as well as RNA duplex unwinding activities. Further investigations would be required to confirm its suggested role in rDNA transcription and pre-rRNA processing.

4
Genome-Wide Analysis

Plant genome research has made great strides in recent years. The whole genome sequence of *Arabidopsis* is now available for functional analysis and the "draft sequence" of the rice genome has also been published very recently. The questions regarding genetic relatedness among individuals, phylogenetic relationships, mapping of quantitative trait loci, and analysis of genomes and their organization are being addressed simultaneously using a variety of molecular tools. These approaches are particularly useful for understanding plant genomes that are rather too large to sequence at the present time. Several laboratories in India are engaged in research on the above aspects with emphasis on one or the other crop species, including cereals, legumes, oil-seed crops and some tree species, and their accomplishments are highlighted below.

4.1
Framework Maps and Marker-Assisted Breeding

Molecular markers are becoming invaluable in developing detailed genetic and physical chromosome maps of plant systems. These molecular markers, usually DNA-based, also have profound application in plant breeding as they can be linked to traits of economic interest, can be scored at any developmental stage and are relatively free from any influence of the environment. A rather extensive programme has been launched in India for the development and use of molecular makers for genetic mapping and marker-assisted selection (MAS) in wheat. The majority of the economically useful traits are quantitative in nature and controlled by multiple genes and are thus referred to as quantitative trait loci (QTL). The QTL analysis has, therefore, been initiated for at least four growth parameters (early growth habit, plant height, days to heading, and days to maturity) and two leaf characters (leaf color and waxiness), utilizing the International Triticeae Mapping Initiative Population (Gupta and coworkers, personal communication).

The problem of pre-harvest sprouting (PHS), particularly in amber kernels, is quite common in major wheat growing regions of the world, including India. In contrast, the red kernels are known to be tolerant and thus are a good source for identifying the markers associated with this trait. The improvement in grain protein content (GPC) and its composition in bread wheat is also a difficult proposition but remains a major concern of plant breeders. Recently, QTL/genes for PHS tolerance [91] and GPC [92, 93] have been tagged using STMS and STS markers. These markers were also assigned to specific chromosome arms (2DL, 6BS and 7DL) using nullisomic-tetrasomic lines and ditelocentric stocks. In addition, marker validation for two STMS markers (*WMC415* and *WMC41*) has been successfully conducted using near isogenic lines for high GPC. Subsequently, physical mapping has been undertaken using additional markers located on these arms, along with the above three markers identified. This integrated physical map thus prepared harbors 27 maker loci on 2DL, 42 marker loci on 6BS, and 54 marker loci on 7DL [94]. Such integrated physical maps should be useful for study of relative gene density across the chromosome arms as well as for map-based cloning etc. These PCR-based STMS and STS markers were also useful for detecting polymorphism between two parents and between the two bulks of recombinant inbred lines (RILs) representing low and high grain weight [95]. At least one STMS marker, *Xwnc333*, present on chromosome 1, could be associated with grain weight. Likewise, using inter simple sequence repeat (ISSR) markers, three markers associated with small seed size and four linked to large seed size were identified [96]. Using deletion lines, the three small seed size associated QTLs could be assigned to chromosomes 6BL, 2DL and 1DS.

In continuation of these studies, QTL interval mapping for PHS tolerance and GPC, for 110 and 100 RILs, respectively, has been conducted. This has lead to the identification of five QTLs spread over four chromosomes (1D, 2D, 4B and 6A) for PHS tolerance by composite interval mapping (CIM). On the other hand, QTL mapping for GPC was conducted by CIM as well as by single marker analysis (SMA) and interval mapping (IM). For this purpose, a framework genetic map was prepared, which had 173 loci involving 171 SSR primer pairs. At least 13 QTLs, spread over eight different chromosomes have been detected by all the three methods employed. Among these, 7 QTLs were identified by CIM, and two of these 7 QTLS were identified by all three methods. The map is being saturated further using SAMPL and AFLP markers (P. K. Gupta and coworkers, personal communication).

The QTLs linked to GPC have also been identified by subjecting the RILs to ISSR and RAPD analysis [97]. The RILs were phenotyped for GPC at two diverse agroclimatic locations. Three of the markers were found to be associated with GPC at both the locations, whereas two markers were found to be specific to Pune and four markers were specific to Ludhiana. This study shows that GPC is greatly influenced by the agroclimatic conditions. Several QTLs have also been identified for kernel hardness in wheat and this trait found to be interdependent on various related components like kernel weight and protein content [98].

A molecular genetic map of wheat using SSRs has been prepared by a collaborative effort under the auspices of the International Wheat Microsatellite Mapping Network. As part of this consortium, ~400 SSR primer pairs were developed; 58 out of the 176 primer pairs tested were found to be polymorphic between the parents of the mapping population used. Using this population and the framework map, 66 microsatellite loci, distributed on 20 chromosomes, have been mapped [99]. These 66 microsatellite loci add to the already existing 384 microsatellite loci mapped earlier in bread wheat.

Wheat rust continues to be a recurrent problem in areas which predominantly cultivate wheat. There is thus a continuous struggle to develop rust-resistant cultivars against the ever-evolving pathogen races. It is indeed necessary to do pyramiding of major rust-resistant genes but it will be time-consuming to achieve this objective by conventional breeding procedures. Attempts are thus being made to identify molecular markers linked to the rust-resistant genes. In one such study, an STS (sequence-tagged-site) marker linked to *Lr28*, a wheat leaf rust-resistance gene, has been identified [100]. It was accomplished by RAPD analysis of NILs of *Lr28* in eight varietal backgrounds. Of the 80 random primers tested, one RAPD marker distinguished the NILs and the donor parent from the susceptible recurrent parent. The STS marker thus developed was further confirmed by subjecting the F3 families to bulk segregant analysis. In a similar study, an STMS marker *gwm368* of chromosome 4BS has been shown to be tightly linked to *Aegilops triuncialis* rust-resistance gene, *LrTr*, transferred to wheat [101]. This in fact is just the beginning of the task to identify many more molecular markers for additional leaf rust-resistance genes, which will eventually be useful for gene pyramiding.

In rice, one aim of marker-assisted breeding is to develop resistance against gall midge (*Orseolia oryzae*), a major insect pest of rice. In most cases, the resistance to gall midge biotypes is governed by single dominant gene [102]. An effort has thus been mounted at the national level to screen rice germplasm for new sources for resistance genes effective against one or more biotypes of the pest. To map the gall midge resistance genes in rice, the RFLP approach was adopted to study *Gm2*, a dominant gene, conferring resistance to some biotypes of gall midge [103]. The mapping population of 40 $F_{5\text{-}6}$ recombinant inbred lines was derived from a cross between "Phalguna" (resistant variety) and "ARC 6650" (a susceptible land race). The *Gm2* gene was assigned to chromosome 4, close to the markers RG329 and RG476. For validation of RFLP-based mapping, RAPD/bulk segregant analysis was also carried out. The sequence information of two phenotype-specific RAPD fragments tightly linked to gall midge biotype 1 resistance was utilized to design specific primers. These primers were shown to amplify specifically 1.7 and 0.6 kb fragments in the susceptible and resistance plants, respectively [104]. Another gall midge resistance gene, *Gm4t*, which is non-allelic to *Gm2* and is known to confer resistance against insect biotypes 1, 2, 3 and 4, has also been tagged using RAPD analysis in combination with bulked-segregant analysis on an F3 population derived from a cross between two *indica* rice varieties, "Abhaya", resistant parent, and "Tulsi", susceptible parent [105]. Using F3 mapping populations, derived from two *indica* parents, "Abhaya" and "Shyamala", somehow the *Gm4t* gene could not be mapped due to lack of useful polymorphism. However, an alternative strategy of employing the F2 population derived from a cross between "Nipponbare", a *japonica* variety, and "Kasalath", an *indica* variety, clicked and *Gm4t* was mapped to chromosome 8, between markers R1813 and S1633 B [106]. While efforts are being made by Mohan and coworkers to clone the gall midge resistance gene by adopting a map-based approach, in parallel, they have used degenerate oligos, based upon conserved motifs of previously isolated disease resistance genes from other plant species, to amplify analogous sequences from the resistant variety "Phalguna". The putative resistance gene analogues (RGAs) were cloned, sequenced and mapped [107]. Of the 14 clones analyzed in detail, a major cluster of 8 RGAs was assigned to chromosome 11. Several of these putative RGAs were found to be tightly linked to known disease resistance genes. Among these, RGA7, which mapped to chromosome 4, showed cosegregation with a few markers previously shown to be tightly linked to the *Gm2* gene [107, 108]. Recently, using RAPD analysis, another gene, *GM-6(t)*, which confers resistance against four biotypes of Asian rice gall midge, has also been mapped on chromosome 4, located about 16 cM from the *Gm-2* resistance gene [109]. Further work, however, is necessary to establish unequivocally the presence of such genes and their role in conferring disease resistance against gall midge.

Other ongoing significant efforts on marker-assisted breeding and gene pyramiding include resistance to blast, blight, sheath rot, drought and salinity, as well as QTL analysis for traits like yield [110–113]. The bacterial blight (BB) of rice is caused by *Xanthomonas oryzae*. In a recent study, rice genotypes have been surveyed for BB resistance genes using STMS and STS markers. The *Xa5* and *Xa21* genes could be identified in eight and two lines, respectively, whereas the *Xa13* gene could not be detected among the genotypes tested [114]. Attempts are also being made to introgress these traits. Using marker assisted selection, three BB resistance genes, *Xa5*, *Xa13* and *Xa21* have been pyramided into cultivar PR106 of rice [115], which is widely grown in the Indian state of Punjab. These genes in combination have provided a high level of resistance to several races of pathogen causing disease in rice.

Brassica juncea, commonly known as Indian mustard, is a major annual oil-yielding crop for India. For molecular mapping and character tagging, cultivar "Varuna" and an exotic collection BEC144 were selected based upon their morphological and molecular differences and subjected to RFLP analysis [116, 117]. A high degree of RFLP was detected between the two varieties and the majority of the genomic DNA probes used revealed duplicate loci. At least 17 significant marker-quantitative trait associations could be established in respect of 6 quantitative traits, i.e., primary branches/plant, secondary branches/plant, days to flowering, plant height, siliques per secondary branch and seeds per silique. One marker in fact showed a tight linkage with the yellow seed coat color locus. In addition, a linkage relationship among the markers was established and 25 of the markers could be arranged in nine linkage groups, covering a of total of 243.3 cM [117]. This study has been extended to map the locus conferring resistance to white rust disease caused by the fungus *Albugo candida* [118]. For this purpose, a set of 94 RILs derived from a cross between BEC144 (resistant accession) and "Varuna" (susceptible isolate) were screened using 186 decamer primers and 11 RAPD markers identified through bulk segregant analysis. Five of these markers showed linkage with the white rust resistance locus, two of which were linked in coupling and repulsion phase on either side of the locus. These two markers could be potentially useful in mustard breeding programmes for transfer of the resistance gene. In another study on seed coat color, a dominant AFLP marker has been converted into a simple codominant SCAR (sequence characterized amplified region) that can easily distinguish between yellow- and brown-seeded *B. juncea* [119]. This marker could be used for developing yellow seed brassicas as well as for the genes controlling seed coat color. Recently, a high-density genetic linkage map of *B. juncea* (2n=36) has been constructed with the help of 996 AFLP and 33 RFLP markers using an F1-derived doubled haploid population of 123 individuals raised by crossing a variety "Varuna" and a canola line "Heera" (Pental and coworkers, personal communication). This map will be useful for dissection and transfer of agronomically important traits to cultivated Indian varieties of brassicas.

4.2 Fingerprinting and Genetic Diversity

The importance of understanding genome organization and evolution in higher plants is being increasingly realized. Attempts are thus being made to learn the physical organization of genes and repeated sequences, which in fact constitute a rather large chunk of DNA in higher plants. These repeat elements show extensive differences in sequence motifs and abundance even between closely related species. Analysis of repeat DNA is thus important for evolutionary, genetic and taxonomic studies. The study of phylogenetic relationships and genetic relatedness also depends heavily on an accurate assessment of genetic diversity. A series of techniques and molecular markers are being deployed to address some of these questions, which are otherwise difficult to resolve by conventional methods.

The genus *Oryza* represents one of the agronomically important cereal crops and is the principal staple food in developing countries, including India. Many of the wild species under this genus provide a rich source of genes that may confer resistance to biotic and abiotic stresses. Genetic diversity in the rice germplasm, including wild rice species and cultivated *indica* and *japonica* rice varieties, has been analyzed using various molecular markers [120–122]. To detect intra- and inter-specific variability, oligonucleotide probes specific for simple repetitive DNA sequences such as $(GATA)_4$, $(GGAT)_4$, $(GAA)_6$, $(CAC)_5$ and $(GACA)_4$, were used. All these probes generated a level of polymorphism that could easily distinguish several rice genotypes [121]. Some of these oligonucleotide probes were also successfully used recently to monitor changes in DNA variability in successive generations of tissue culture-derived somaclonal variants of an elite rice cultivar, Indrayani [123]. Subsequently, several hypervariable DNA sequences were used as probes to generate DNA fingerprints in rice. Among several probes tested, R18.1, a bovine genomic clone with poly(GT) stretches, the bacteriophage M13 repeat probes, and PV47, a human minisatellite probe, generated fingerprints peculiar for all the rice cultivars tested, with probe R18.1 showing the highest level of polymorphism [122]. In both these studies, the DNA fingerprint generated was cultivar specific, without any variability among the plants of one cultivar. The wild rice germplasm has also been subjected to genetic variability studies at specific loci using *Rm122* and *knotted-1* homeobox locus specific markers [124, 125]. The potential of ISSRs (inter simple sequence repeats), the regions that lie within the microsatellite repeats and offer great advantage in determining intragenomic and intergenomic diversity, has also been exploited [126]. The ISSR analysis revealed that the genus *Oryza* might have evolved following a polyphyletic pathway. The ISSR marker based DNA profiles also generated typical fingerprints for various cultivars, species and genera. Genetic variation in nine aromatic and four non-aromatic rice varieties using RAPD analysis has been studied [127]. Using 26 random primers, 177 PCR products were generated, 98 of which were polymorphic. In a recent study, identification and classification of aromatic rice genotypes by RAPD profiling has been done using 58 random primers. Cluster analysis using 314 polymorphic bands generated could demarcate the short- and long-grain genotypes into two distinct groups [128]. Essentially a similar study has been conducted for identifying genotypes of basmati rice available in genetic resource collections, using random primers, and RAPD profiles developed that could easily discriminate the various accessions [129]. In a rather more extensive study, SSR and ISSR markers have been developed to reveal genetic relationships and distinguish the traditional and evolved basmati and non-basmati rice varieties [130]. Some of these molecular genetic markers used in the above studies could thus be easily utilized not only for diagnostic fingerprints of various rice genotypes and their evolutionary relationship, but also in aiding introgression of genes from the wild rice genotypes in the breeding programmes of cultivated varieties.

As mentioned earlier, *Brassica juncea* is the oilseed brassica of choice for the Indian subcontinent. To improve the yield, heterosis breeding can be done and hybrids between the genotypes of different geographic origins can be expected to be more productive. To characterize these different heterotic groups, it is imperative to study the genome organization, phylogeny and polymorphism in the genus *Brassica*. RAPD analysis to examine genetic relationship and inherent variations among 12 Indian and 11 exotic *B. juncea* genotypes has been performed [131]. The Indian genotypes showed a low level of variability, whereas considerable polymorphism was detected among the exotic species. Based upon the pair-wise comparison of amplification products, a dendrogram was constructed, which clustered the genotypes in two groups. This cluster analysis may be useful for selection of parents for heterosis breeding. In another study, AFLP markers were utilized to assess the genetic diversity among 21 natural and 9 synthetic brassicas (some developed recently) and several other lines originating from various parts of the world. Based upon the cluster analysis of the 1251 fragments amplified (using 21 primer pairs), the 30 *B. juncea* lines could be grouped into three distinct clusters, where the recently formed synthetics formed a separate cluster [132]. Four of the primer pairs were unique in that they identified all the genotypes tested, and thus could be used for routine varietal identification.

A large variety of grain legumes, including pigeonpea (*Cajanus cajan*), is commonly grown in India. To understand the relationship between *C. cajan* and its wild relatives, RFLP/RAPD analysis has been carried out [133]. However, further work is required to develop consensus regarding the genomic relationships in *C. cajan*. The phylogenetic relationship and rDNA variation among several *Cajanus* species and *C. cajan* accessions has been examined recently using heterologous probes [134]. Three rDNA repeat classes were identified among the eight species examined. The dendrogram developed on the basis of RFLP analysis revealed the degree of relationship between different species.

There are major mangrove formations along the Indian coast. However, a steady decline has been registered in mangrove cover in India due to inhabitation. Phylogenetic analysis of several mangrove species, including *Acanthus*, *Avicennia* and *Excoecaria*, has been carried out using molecular markers [135–137]. In the more recent study on *E. agallocha*, genomic DNA of 84 individuals from seven mangrove populations was subjected to RAPD and RFLP analysis. The results of this study indicate that considerable intrapopulation and interpopulation genetic variation exists in this mangrove genus. It thus appears that the morphological uniformity observed across the species cannot be ascribed to the lack of genetic variation.

The genus *Populus* consists of more than 30 species and has potential use not only in agroforestry but also in basic forest tree biology research. It is thus only natural that efforts are being made to develop superior clones by conventional breeding programmes. The repetitive DNA elements can be useful tools in the evaluation of hybrids. An investigation primarily to study the organization of repetitive DNA sequences in the *Populus* genome has been carried out [138]. A 145-bp tandem repeat family, accounting for ca 1.5% of the *Populus* genome, has been characterized, but it does not show any homology to tandem repeats already known from plants. This repeat family was present in 13 of the 14 *Populus* species examined, indicating the ancient origin of this family. Female plants of several dioecious angiosperms including papaya are commercially valuable. A PCR-based assay for seedling sex diagnosis has been developed for early sexing of papaya [139]. This is essentially based on a male-specific RAPD fragment that has been cloned and can be used as a male-specific SCAR marker.

Genetic diversity has also been assessed in neem, an evergreen multipurpose tree with medicinal properties and a wide distribution range in India. RAPD profiles were generated for 34 accessions of neem using 200 random primers, with 49 primers giving reproducible amplification products. The dendrogram constructed using these data suggests that neem may have a narrow genetic base [140]. In a parallel study, another set of 37 neem accessions, representing different ecogeographic regions within India, and 4 exotic lines from Thailand, were analyzed by AFLP markers [141], and several rare and accession-specific bands identified. The phenetic dendrogram generated could easily demarcate the Indian genotypes from the four Thai lines. However, the cluster analysis indicated that the neem germplasm within India constitutes a broad genetic base, which is at variance to the conclusion drawn by the earlier study. The relative efficiency of AFLP and SAMPL markers has also been compared for their applicability in genetic diversity studies in neem [142], and SAMPL markers were found to be particularly useful in resolving differences between closely related accessions. The inter- and intra-specific variation in *Withania somnifera* (35 accessions) and *W. coagulans* (5 accessions), the species with important anticancerous properties, has also been examined by AFLP analysis [143]. The dendrogram generated classified these species into two major clusters and *W. somnifera* further subdivided into Kashmiri, Nagori and an intermediate group. These AFLP markers can thus be used for detecting genetic variation both at the species as well as the intraspecific level.

The distribution of 18S-26S and 5S rRNA gene families on the chromosomes of 19 of the 24 species/subspecies belonging to three species complexes of the genus *Vicia* by FISH (fluorescence in situ hybridization) has been studied. These rDNA gene families were physically mapped on the basis of chromosome length, the arm-length ratio and the relative position of the hybridization signals [144]. Essentially, a similar study was performed on two alkaloidal plants, *Papaver somniferum* and *Hyoscyamus niger* and the distribution of 18S-26S and 5S rDNA families examined by FISH [145]. The distribution of these two rDNA families at specific places on the chromosomes of the species investigated should be useful in identification of the chromosomes as well as the relatedness of species.

To study genetic diversity in wheat, several different approaches have been adopted. A set of 55 elite and exotic wheat genotypes (originated in 29 countries) has been subjected to molecular analysis using SSR, SAMPL (selective amplification of microsatellite polymorphic loci) and AFLP markers [146–148]. The binary data for these markers have been utilized for the study of marker trait associations involving data scored on 14 phenotypic traits in diverse wheat genotypes investigated. In these studies, a total of 351 molecular markers (SSR, 131; AFLP, 166; SAMPL, 54) were identified, and practically each of these markers showed significant association with at least one of the 14 traits examined. Some of these markers, after validation studies, could be of potential use for MAS in wheat breeding.

4.3 Genome Sequencing

Keeping in view the importance of rice as the largest component of food grain production in India, DBT and ICAR took an initiative in the year 2000, and supported a venture that will mark a beginning in India for high throughput genome sequencing. India became a formal member of the International Rice Genome Sequencing Project (IRGSP) and claimed a region spanning 56.9 to 109.3 cM on the long arm of chromosome 11 for high throughput genome sequencing and gene discovery (Fig. 1). The scientific programme is being implemented at the Department of Plant Molecular Biology, University of Delhi South Campus, and the National Research Centre for Plant Biotechnology, Indian Agricultural Research Institute, New Delhi. Indian groups started with the generation of a physical map for their region by screening PAC and BAC libraries and generated a map covering about 90% of the region. Some of this work has contributed to the development of an integrated physical and genetic map of the rice genome [149]. About 4.0 Mb of the sequence have already been released to GenBank by Indian laboratories, out of the 6.0 Mb of the sequence assembled at various levels by May 2002. Using an array of gene prediction programmes, the annotation of some of the PAC/BAC clones has helped in the identification of genes, which may define disease resistance, receptor kinases, hormone-inducible components, insect resistance, and several quantitative trait loci (QTL). Details are available on the website (www.genomeindia.org) and links provided therein.

The rice genome sequence could serve as a standard to assess variability in rice and its wild relatives and help allele mining. New genes with potential application can be discovered and validated. Functional genomics of rice, including gene tagging, micro-array analysis and regulation biology would get a boost. Availability of markers would pave the way for precision breeding. Analysis of QTLs for traits like yield per se will also become possible. India can contribute significantly to new knowledge and improve rice production by making use of the information emerging world-wide on rice genomics.

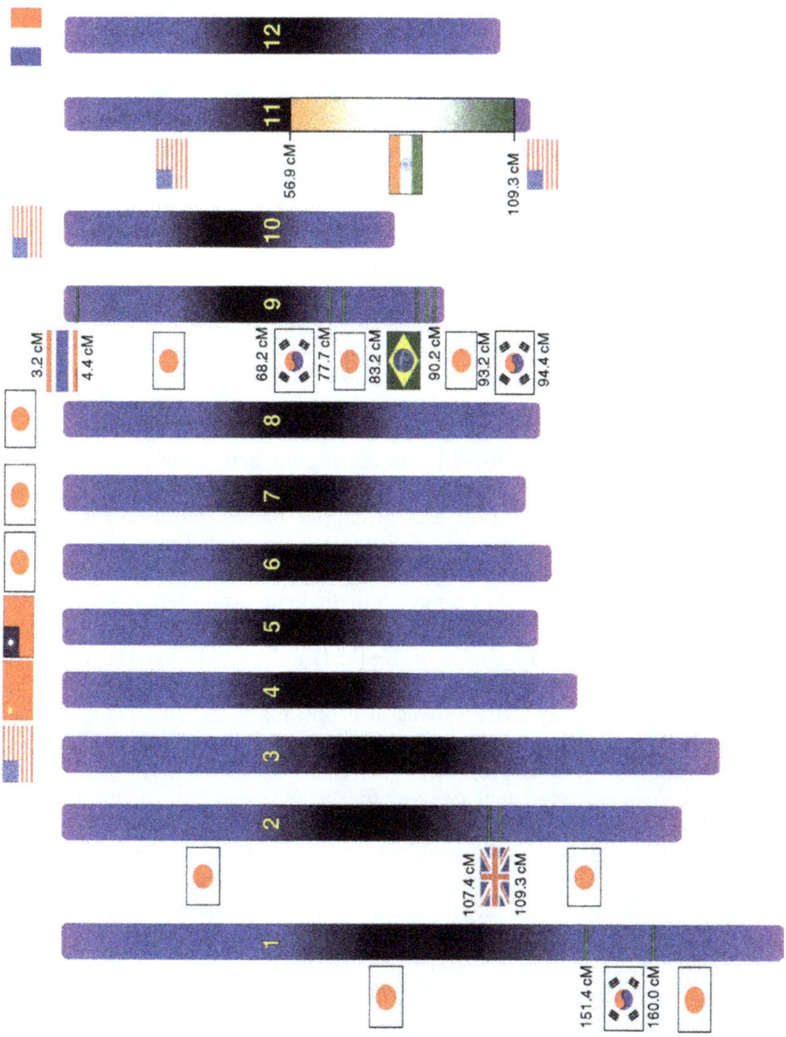

Fig. 1. Sharing of rice chromosomes for high throughput sequencing by member countries in the International Rice Genome Sequencing Project

5 Transgenics

The development of reliable methods for genetic transformation of higher plants with desirable genes was achieved in the 1980s. These included methods based on *Agrobacterium,* electroporation and biolistics. However, due to variable responses of different plant species and low frequency of genetic transformation, the search for novel methods still continues. In the 1990s, several transgenic crops were tested in the open field and covered a large cultivated area world-wide, particularly in USA. The importance of transgenic crops in providing means of utilizing vast reservoir of genes, across sexual compatibility barriers and for novel purposes, has been emphasized [150, 151]. The efforts of Indian laboratories on optimization of genetic transformation and production of transgenics in crop plants as well as to understand gene function are described below.

During transformation, success largely depends on efficient gene delivery and high frequency of regeneration from transformed cells. The efficiency of gene delivery can often be monitored by measuring the activity of a reporter gene. Such transient expression of the delivered gene can also help evaluate the potential of regulatory sequences and other parameters required for efficient delivery. Electroporation-mediated delivery into rice protoplasts was optimized by monitoring the transient expression of the β-glucuronidase (*gus*) gene and used to analyze the activity of photosynthetic gene promoters [152, 153]. As a matter of fact, electroporation-mediated gene delivery was also possible in intact cells of germinating seed embryos of rice [154] and could be used to analyze the activity of soybean heat-shock gene (*Gmhsp* 17.5-E) promoter in rice [155]. Biolistic-mediated gene delivery was also optimized for rice [156], *Albizzia lebbeck* [157] and millets like *Eleucine coracana* and *Echinochloa crusgalli,* where the efficiency of various promoters was also evaluated [158]. The utility of *Agrobacterium*-mediated gene delivery for dicots is well known, but it has also been shown to work in the case of monocots like wheat [159]. Other novel methods of gene delivery evaluated by transient gene expression include cellular permeabilization in *Brassica* [160] and wheat [161]. Polyethylene glycol-mediated gene delivery in *Arabidopsis thaliana* protoplasts was used to evaluate promoter activity and the homologous *ubiquitin* promoter was found to be the most effective followed by the *CaMV35S* promoter [162].

A "designers' promoter" was made by analysis of a large number of plant promoters and found to be effective, several times better than *CaMV35S* promoter, in various plants tested [163]. Similarly, the context of the initiator codon was optimized and found to result in enhanced reporter gene (*gus, gfp*) activity, largely due to post-transcriptional events like protein stability as evident after transient expression [164]. The promoter of *PetH* encoding ferredoxin-NADP$^+$-oxidoreductase and other photosynthetic promoters of *PsaF* from spinach were found to be active in green tissues of transgenic rice or leaf protoplasts of rice in a transient expression assay, respectively [153]. The promoter of *PsaF* was also evaluated in tobacco and the involvement of heterotrimeric G-proteins, phospholipids, calcium and phosphorylation status was found to be important for its light-dependent activity [165]. For cloning 5′ flanking regions representing promoters of target genes, a new PCR-based approach has been developed which allows directional genome walking from a defined point within the gene [166].

Basic mechanisms involved in *Agrobacterium*-mediated gene transfer have also been investigated. For example, rice scutella and scutellum-derived calli could induce *virE* gene expression as well as T-strand generation but leaf segments failed to do so [167]. By introducing a plant selectable marker gene (*nptII*) at an appropriate position, it was shown that DNA transfer events starting at the left border and moving away from the T-DNA region also occur, thereby emphasizing the need for optimal analysis of transgenic events [168]. A series of versatile binary vectors has also been developed for genetic transformation of rice and analyzed by reporter gene expression [169].

Early attempts to transform plants in India included oncogenic/non-oncogenic strains of *Agrobacterium* or physical means of gene delivery. Thus, transformation and regeneration of *Brassica oleracea* by *Agrobacterium* was reported [170]. Protoplast-based genetic transformation of *B. oleracea* as well as *Agrobacterium*-mediated transformation of *B. campestris* at high frequency were reported subsequently [171, 172]. Indian oilseed mustard (*B. juncea*) has also been transformed using herbicide resistance gene (*bar*) and homozygous lines produced to evaluate their utility [173]. Another member of the Cruciferae family, *Arabidopsis thaliana*, and an additional oilseed crop, safflower (*Carthamus tinctorius*), have also been transformed by *Agrobacterium* [174, 175]. Transformation of legumes by *A. tumefaciens* resulted in transgenic calli in the case of *Vigna mungo* [176] or plantlets in *V. unguiculata* [177]. Use of cotyledonary node explants and super binary vector (pTOK233) helped produce transgenic plants in *V. radiata* [178], which has otherwise proved to be a recalcitrant grain legume. Cotyledonary explants were also used to transform peanut (*Arachis hypogaea*) at high frequency with *Agrobacterium tumefaciens* [179, 180]. Inheritance of the transgene was investigated in the T1 generation and molecular evidence provided for gene integration [179]. Other interesting work on *Agrobacterium*-mediated genetic transformation has been carried out on *Cajanus cajan* [181], *Capsicum annuum* [182], *Santalum album* [183], *Camellia sinensis* [184] and Indian cotton [185]. In the case of cotton, progeny analysis revealed Mendelian inheritance of the transgene.

Due to a lack of clarity about the utility of *Agrobacterium* for monocot transformation, first attempts to transform cereal crops involved physical methods of gene transfer. Biolistics and improved culture systems were used to transform two *indica* cultivars of rice, namely IR64 and Karnal Local [186]. Rice variety IR36 was transformed by electroporation of protoplasts [187]. Both groups used the herbicide resistance gene (*bar*) as a selectable marker. *Agrobacterium*-mediated transformation of *indica* rice variety Pusa Basmati 1 was achieved at high frequency by using super-binary vector (pTOK233) and the transgene was shown to be inherited up to the R2 generation [188]. Simultaneously, *Agrobacterium*-mediated transformation of IR64 and Karnal Local was achieved by using pTOK233 as well as pCAMBIA 1301, albeit with the supervirulent strain AGL1 [189]. Work on rice variety Pusa Basmati 1 transformation with super-virulent binary vector was confirmed and the utility of non-supervirulent vectors proved by another group [190]. While these studies used mature seed-derived calli, callus from coleoptiles of a *japonica* cultivar was also used for transformation with *A. tumefaciens* LBA4404 (pTOK233) [191].

Genetic transformation of three Indian wheat cultivars, namely CPAN 3004, Sonalika and UP2338, was achieved by biolistic method [192]. The inheritance of a selectable marker gene (*bar*) and a reporter gene (*gus*) was investigated in R1 progeny and it was found that *bar* under the control of *ubiquitin* promoter can provide resistance to herbicide levels capable of killing weeds in field.

To improve nutritive value, diploid potato genotype A16 was transformed with a non-allergenic and nutritionally balanced protein gene *AmA1* under the control of a tuber-specific or constitutive promoter [193]. This resulted in 35–45% more protein and 4–8-fold increases in certain amino acids besides increases in growth and production of tubers in a restricted experimental plot. The protein was localized in the cytoplasm as well as the vacuole of transgenic tubers. It is proposed that other crops can be improved for nutritional value in a similar manner. Another category of transgenics of interest for human health was produced in tomato to express cholera toxin B subunit (non-pathogenic) with a view to evaluate its potential as a candidate antigen for vaccine development [194]. The protein was expressed in fruits and pentamerized as demonstrated by its size and capacity to bind the GM1-ganglioside receptor. The immune potential of the antigen expressing in tomato is being evaluated in model systems.

A large area in India is not amenable to optimal cultivation due to the prevalence of abiotic stresses like drought, salinity, and submergence. Therefore, engineering plants for abiotic stress tolerance remains a priority area of research. Glyoxalase I gene from *Brassica juncea* was overexpressed in tobacco under the control of *CaMV35S* promoter and found to confer tolerance to methylglyoxal and high salt [63]. The gene for choline oxidase (*codA*) from *Arthrobacter globiformis* can convert choline into glycinebetaine and hydrogen peroxide, which can serve as a protectant against osmotic stress and a signal for stress response, respectively. The *codA gene* with *RbcS* transit peptide coding region for chloroplast targeting under the control of the *CaMV35S* promoter was expressed in transgenic *Brassica juncea* [195]. The gene was found to confer salt tolerance during germination and seedling growth. The acquired potential to convert choline into glycinebetaine made mustard tolerant to exogenously supplied choline [196]. This construct was also used to transform an elite *indica* rice variety Pusa Basmati 1 [197]. Biochemical, molecular and genetic analysis of transgenics was performed in R0 and R1 generations. Salt stress to R1 seedlings followed by a recovery period revealed that in some cases more than 50% of transgenic plants survived and set seed whereas wild-type plants failed to recover after exposure to salt (Fig. 2). The bacterial gene for mannitol-1-phosphodehydrogenase (*mtlD*) was introduced for overexpression in eggplant [198] and found to impart tolerance to osmotic stresses in the T0 and T1 generations. Osmotin gene overexpression in tobacco imparted drought and salinity resistance to transgenic tobacco [199]. Calcium has been recognized as an important signal for cellular responses including abiotic stress. A novel calcium-binding protein gene from *Entamoeba histolytica*, namely *EhCaBP*, has been found to confer salt tolerance to transgenic tobacco during seed germination and seedling growth [200].

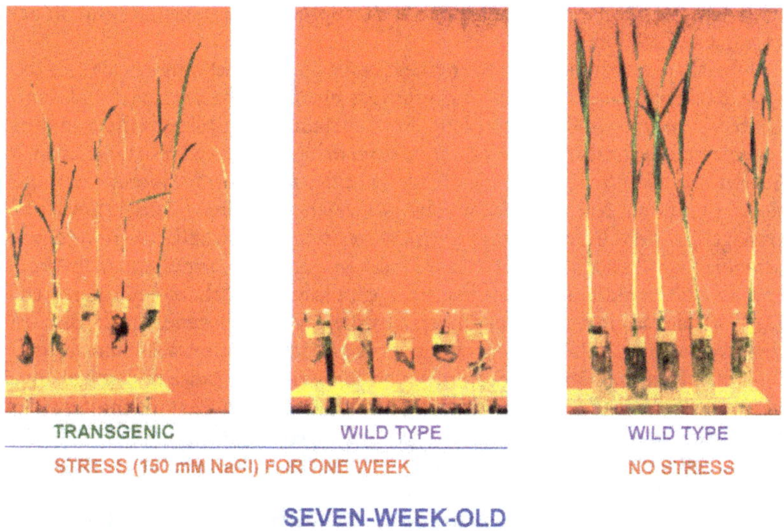

Fig. 2. Transgenics of rice variety Pusa Basmati 1 expressing *codA* survive on 150 mM NaCl but wild-type fail to recover. Compare growth with wild-type seedlings without stress

For imparting insect resistance to crop plants, δ-endotoxin genes of *Bacillus thuringiensis* have been deployed over 50 million hectares cultivated area worldwide and have been considered of great value also for India in combination with practices capable of avoiding the development of resistance in insects [201]. Among the first efforts on crop plants, a synthetic *cry1Ab* gene was transferred to eggplants and a significant level of resistance to the insect (*Leucinodes orbonalis*) was observed in eggplant fruits [202]. Recently, work has been extended to tomato [203] and potato [204]. Another truncated gene for *cryIAc* was introduced into chickpea (*Cicer arietinum*) by biolistics and feeding assays indicated its efficacy in inhibiting the development of *Heliothis armigera* larvae [205]. A new gene for *cry1Ia5* was cloned and found to be as effective as synthetic *cry1Ab* or *cry1Ac* in transgenic tobacco plants against *Heliothis armigera* [206].

The yellow stem borer (*Scirpophaga incertulas*) of rice is known to cause massive yield loss. A reconstructed *cry1Ac* gene under the control of maize *ubiquitin* promoter was used to transform *indica* cultivar IR64 by biolistics [207]. The expression level was found to vary between 0.02 to 0.025% of total soluble protein and was stable up to the T2 generation. The transgenic plants were shown to be toxic to insect larvae. Another synthetic *cry1Ac* gene codon optimized for rice, with similar regulatory control, was used to transform *indica* rice breeding lines (IR-64, Pusa Basmati 1 and Karnal Local) by biolistics as well as the *Agrobacterium* approach [208]. Selected lines of Pusa Basmati 1 and IR-64 with expression level above 0.1% of total soluble protein were tested against yellow stem borer and shown to cause 100% mortality of larvae in cut stem as well as in vegetative whole plant assays. But, assays at the reproductive stage did result in occasional infestation.

Efforts are also in progress to engineer resistance against fungi. A gene for oxalate decarboxylase from *Collybia velutipes* has been characterized and engineered in tobacco and tomato [209]. Since certain fungi utilize oxalic acid during infestation, these transgenics were shown to have improved resistance against *Sclerotinia sclerotiorum*. A maize *C2* gene under *ubiquitin* promoter was engineered in Taipei 309 rice with a view to influence flavonoid biosynthesis and transgenics were found to have an increase in blast resistance in preliminary screening of R1 plants [210]. Increased resistance against *Cercospora arachidicola* was also observed in peanut transgenics expressing the tobacco chitinase gene [211]. Viral resistance can be engineered by the coat protein gene, which has been deployed against Indian peanut clump virus and introduced via *Agrobacterium*-mediated transformation of peanut, but evaluation of resistance remains to done [179]. Transgenics for tomato showing resistance against Physalis mottle virus have been produced by *Agrobacterium*-mediated transformation [212]. The potato virus Y coat protein gene has been cloned and evaluated for its efficacy in transgenic tobacco [213].

The increase in yield potential has been obtained by heterosis for several years. But, hybrid technology requires male-sterile lines and restoration of fertility in the progeny. Male sterile lines of *Brassica juncea* have been obtained by using *barnase* gene expression in tapetum [214]. To avoid leakiness in barnase expression, incorporation of a spacer fragment between the barnase gene and a selectable marker cassette was found to be useful. Towards the aim of fertility restoration, the *barstar* gene was also introduced into *B. juncea* and several crosses were made with male sterile lines. Out of 30 cross-combinations between 3 male sterile lines and 14 *barstar* lines, one combination was found to be effective in restoration of male fertility in F1 progeny [215]. Such a trait can be diversified and should prove useful for heterosis breeding.

6 Perspectives

The research work in the area of plant molecular biology and biotechnology in India has come out of its lag phase and shown significant progress in the last five years as reflected in the text earlier. Interaction of scientists within the country and internationally to work on a common theme is appreciable and rewarding. Several net-work or mission-mode projects are in progress to analyze genes involved in abiotic and biotic stress responses, with the eventual aim of their engineering to improve crop plants. New genes and promoters for targeted expression are being analyzed. These include novel organ-specific and inducible genes of rice [216], inositol synthase genes from rice and wild rice, glyoxalase II and ion transport protein genes, lipid biosynthesis pathway genes, heat-shock protein and annoxia-related genes, post-harvest related genes, and genes from viruses. Several transgenics are ready for limited field trails as per established guidelines [217] and others are being produced by introducing various genes for pathogen resistance, pest resistance and abiotic stress tolerance. Efforts are also ongoing to transform fruit crops like banana, grape and papaya as well as to develop metabolic engineering for vitamins and herbal products. Finger-printing of several cultivated and wild flora, including medicinal plants, coffee and spices, is being carried out and efforts are on to generate and extend DNA marker-based genetic maps of crops like mustard, chickpea and wheat. Genome level sequencing of rice should pave the way for better understanding of cereal genomes and applications for functional genomics and precision breeding. Pyramiding of genes by a DNA marker-based approach is being attempted in rice and wheat as mentioned earlier. It is expected that molecular approaches would find a wider-application in breeding and the role of transgenics will increase with the development of nation-wide infrastructure and extension of training to more scientists.

Although involvement of the private sector is currently limited, it is improving as the corporate sector takes established technologies and products like hybrid seeds to consumers and looks forward to incorporate new tools of crop improvement. While several in the seed industry have moved for imported transgenic seeds of vegetables, cotton, oil crops, maize, soybean with traits like insect resistance, herbicide resistance, disease resistance and male sterility for heterosis breeding; others are developing partnership with public sector research laboratories for crop improvement by biotechnological approaches. The recent clearance given to insect-resistant transgenic cotton for commercialization should pave the way for other improved crops in the near future. Other private groups have established facilities to provide commercial services in gene sequencing and fingerprinting. India is making its foray into bioinformatics by making use of the nation-wide net-work established by the Department of Biotechnology and the interest generated among the private enterprises. A glimpse of the developments can be obtained from the Annual Report of the Department of Biotechnology (www.dbtindia.nic.in) or from reports prepared for assessing plant molecular biology and agricultural biotechnology in India [218, 219].

To sustain this activity and make it grow to logical ends in terms of new intellectual property and applications, a planned approach with higher in-puts in terms of human and financial resources would be absolutely essential. Complementation of resources and expertise would pay more dividends but requires better managing skills. As new biotechnological applications would emerge from basic research, it should be supported with full vigor, particularly if India wants to avoid long lag phases and incubation periods in emerging frontier areas. Notwithstanding the relatively high cost for performance and debates on various ethical and environmental issues, scientists as well as consumers in India are poised to develop and take advantage of molecular plant biotechnology for sustainable food security [220].

Acknowledgement. Authors are thankful to Dr. R. P. Sharma, IARI, and Professor S. K. Sopory, ICGEB, for critical reading of the manuscript. However, the responsibility for final version rests with the authors. We also thank Ms. Monika Anand for assistance during the survey of the literature.

7 References

1. Khush GS (1999) Genome 42:646
2. Pental D (1998) Plant Mol Biol Rep 16:93
3. Ghose TK, Bisaria VS (2000) Advances in Biochem Engg/Biotech 69:87
4. Sopory SK, Maheshwari SC (2001) Curr Sci 80:270
5. Raghuram N (2002) Trends Plant Sci 7:92
6. Kapoor S, Maheshwari SC, Tyagi AK (1991) In: Singhal GS, Ramasarma T (eds) Trends in Bioenergetics and Biotechnological Processes. Today & Tomorrow's Printers & Publishers, New Delhi, p 65
7. Kelkar NY, Tyagi AK, Maheshwari SC (1992) J Plant Biochem Biotech 1:1
8. Trivedi PK, Nath P, Sane PV (1994) J Plant Biochem Biotechnol 3:97
9. Hande S, Sreenagesh MV, Jayabaskaran C (1996) Ind J Biochem Biophys 33:448
10. Baldev A, Gaikwad K, Kirti PB, Mohapatra T, Prakash S, Chopra VL (1998) Mol Gen Genet 260:357
11. Reddy MK, Choudhury NR, Kumar D, Mukherjee SK, Tewari KK (1994) Eur J Biochem 220:933
12. Mukherjee SK, Reddy MK, Kumar D, Tewari KK (1994) J Biol Chem 269:3793
13. Chen W, Gaikwad A, Mukherjee SK, Roychoudhury NR, Kumar D, Tewari KK (1996) Nucl Acids Res 24:3953
14. Gaikwad A, Tewari KK, Kumar D, Chen W, Mukherjee SK (1999) Nucl Acids Res 27:3120
15. Gaikwad A, Hop DV, Mukherjee SK (2002) Mol Gen Genet 267:45
16. Tuteja N, Phan TN, Tewari KK (1996) Eur J Biochem 238:54
17. Tuteja N, Phan TN (1998) Biochem Biophys Res Commun 244:861
18. Tuteja N, Phan TN (1998) Plant Physiol 118:1029
19. Kumar D, Mukherjee S, Reddy MK, Tewari KK (1995) J Exp Bot 46:767
20. Subbaiah CC, Tewari KK (1993) Eur J Biochem 211:171
21. Reddy MK, Nair S, Singh BN, Mudgil Y, Tewari KK, Sopory SK (2001) Gene 263:179
22. Kelkar NY, Maheshwari SC, Tyagi AK (1993) Plant Sci 88:55
23. Kapoor S, Maheshwari SC, Tyagi AK (1993) Plant Cell Physiol 34:943
24. Kapoor S, Maheshwari SC, Tyagi AK (1994) Curr Genet 25:362
25. Dixit R, Trivedi PK, Nath P, Sane PV (1999) Curr Genet 36:165
26. Khurana JP, Kochhar A, Tyagi AK (1998) Crit Rev Plant Sci 17:465
27. Khurana JP, Tyagi AK, Khurana P, Kochhar A, Jain PK, Raychaudhuri A, Chawla R, Bharti A, Laxmi A, Dasgupta U (2001) In: Sopory SK, Oelmuellar R, Maheshwari SC (eds), Signal Transduction in Plants: Current Advances. Kluwer Academic/Plenum, New York, p 25
28. Tyagi AK, Khurana JP, Sharma AK, Mohanty A, Dhingra A, Raghuvanshi S, Gaur T (2001) In: Sopory SK, Oelmuellar R, Maheshwari SC (eds), Signal Transduction in Plants: Current Advances. Kluwer Academic/Plenum, New York, p 297
29. Grover M, Dhingra A, Sharma AK, Maheshwari SC, Tyagi AK (1999) Physiol Plant 105:701
30. Grover M, Sharma AK, Dhingra A, Maheshwari SC, Tyagi AK (1998) Plant Sci 137:185
31. Reddy VS, Leelavathi S, Selvapandian A, Raman R, Giovanni F, Shukla V, Bhatnagar RK (2002) Mol Breed 9:259
32. Sane AP, Jaiswal P, Nath P, Dabholkar AR, Sane PV (1994) J Plant Biochem Biotech 3:121
33. Senthilkumar P, Narayanan KK (1999) J Biosci 24:215
34. Narayanan KK, Senthilkumar P, Sridhar VV, Thomas G, Thomas J (1995) Theor Appl Genet 90:1087
35. Sane AP, Seth P, Ranade SA, Nath P, Sane PV (1997) Theor Appl Genet 95:1098
36. Mukhopadhayay A, Arumugam N, Pradhan AK, Murthy HN, Yadav BS, Sodhi YS, Pental D (1995) Theor Appl Genet 89:19
37. Arumugam N, Mukhopadhayay A, Gupta V, Sodhi YS, Verma JK, Pental D, Pradhan AK (2000) Theor Appl Genet 100:1043

38. Bhattacharjee B, Sane AP, Gupta HS (1999) Mol Breed 5:319
39. Thomas G, Padayatty JD (1983) Nature 306:82
40. Sen S, Roy P (1986) Biochem Biophys Res Commun 137:788
41. Roy P, Bhattacharya N, Biswas BB (1988) Gene 73:57
42. Raha, D, Sen K, Biswas BB (1987) Plant Mol Biol 9:565
43. Tyagi AK (2001) Curr Sci 80:161
44. Raina A, Datta A (1992) Proc Natl Acad Sci USA 89:11774
45. Dasgupta S, Dasgupta J, Mandal RK (1993) Gene 133:301
46. Joshi KS, Varma VR, Gupta VS, Ranjekar PK (1992) J Plant Biochem Biotech 1:33
47. Kansal R, Ramkumar TP, Koundal KR (1992) J Plant Biochem Biotech 1:69
48. Mandaokar AD, Koundal KR (2000) J Plant Biochem Biotech 9:7
49. Tyagi A, Santha IM, Mehta SL (1993) J Plant Biochem Biotech 2:125
50. Venkateswari J, Johari RP, Mehta SL (1996) J Plant Biochem Biotech 5:23
51. John SJ, Srivastava V, Guha-Mukherjee S (1995) Plant Physiol 107:1023
52. Gupta P, Raghuvanshi S, Tyagi AK (2000) J Plant Biochem Biotech 9:99
53. Vageeshbabu HS, Venkateswari J, Kirti PB, Chopra VL (1996) J Plant Biochem Biotech 5:51
54. Venkateswari J, Kanrar S, Kirti PB, Malathi VG, Chopra VL (1999) J Plant Biochem Biotech 8:53
55. Garg R, Kiran MD, Santha IM, Lodha ML, Mehta SL (2000) J Plant Biochem Biotech 9:13
56. Garg R, Santha IM, Lodha ML, Mehta SL (2001) J Plant Biochem Biotech 10:13
57. Mandal MNA, Santha IM, Lodha ML, Mehta SL (2002) J Plant Biochem Biotech 11:27
58. Chattopadhyay MK, Gupta S, Sengupta DN, Ghosh B (1997) Plant Mol Biol 34:477
59. Vaidyanathan R, Kuruvilla S, Thomas G (1999) Plant Sci 140:21
60. Mukhopadhyay A, Sharma S, Tyagi AK (2001) J Plant Biochem Biotech 10:25
61. Reddy AR, Ramakrishna W, Sekhar AC, Ithal N, Babu PR, Bonaldo MF, Soares MB, Bennetzen JL (2002) Genome 45:204
62. Minhas D, Grover A (1999) Plant Sci 146:41
63. Veena, Reddy VS, Sopory SK (1999) Plant J 17:385
64. Gopalakrishna R, Kumar G, Krishna Prasad BT, Mathew MK, Udaya Kumar M (2001) Biochem Biophys Res Commun 285:574
65. Gupta S, Chattopadhyay MK, Chatterjee P, Ghosh B, Sengupta DN (1998) Plant Mol Biol 37:629
66. Kumar D, Verma HN, Tuteja N, Tewari KK (1997) Plant Mol Biol 33:745
67. Datta S, Kansal R, Koundal KR (2000) J Plant Biochem Biotech 9:67
68. Raghuram N, Sopory SK (1995) Plant Mol Biol 29:25
69. Chandok MR, Sopory SK (1998) J Biol Chem 273:19235
70. Chandra A, Upadhyaya KC (1993) Cell Mol Biol Res 39:509
71. Ehtesham NZ, Phan TN, Gaikwad A, Sopory SK, Tuteja N (1999) DNA Cell Biol 18:865
72. Raghuvanshi S, Kelkar A, Khurana JP, Tyagi AK (2001) DNA Res 8:73
73. Thakur JK, Tyagi AK, Khurana JP (2001) DNA Res 8:193
74. Subramaniam K, Mahadevan S (1994) Gene 149:375
75. Subramaniam K, Ranie J, Srinivasa BR, Sinha AM, Mahadevan S (1994) Gene 141:207
76. Ghosh R, Biswas S, Roy S (1998) Eur J Biochem 258:1009
77. Kochhar A, Khurana JP, Tyagi AK (1996) DNA Res 3:277
78. Jain PK, Kochhar A, Khurana JP, Tyagi AK (1998) DNA Res 5:221
79. Grover M, Gaur, T, Kochhar A, Maheshwari SC, Tyagi AK (1999) DNA Res 6:173
80. Gaikwad A, Babbarwal V, Pant V, Mukherjee SK (2000) Mol Gen Genet 263:213
81. Maheshwai SC, Sopory SK, Bhalla-Sarin N, Khurana JP (2001) Phytomorphology, Golden Jubilee Issue, p 117
82. Kumar TA, Reddy GM (1997) J Plant Biochem Biotech 6:81
83. Prasad K, Sriram P, Kumar CS, Kushalappa K, Vijayraghavan U (2001) Dev Genes Evol 211:281
84. Nandi AK, Kushalappa K, Prasad K, Vijayraghavan U (2000) Curr Biol 10:215
85. Siddiqui I, Ganesh G, Grossniklaus U, Subbiah V (2000) Development 127:197
86. Reddy MK, Nair S, Tewari KK (1998) Plant Mol Biol 37:773

87. Hop DV, Gaikwad A, Yadav BS, Reddy MK, Sopory S, Mukherjee SK (1999) Plant J 19:153
88. Reddy MK, Nair S, Tewari KK, Mudgil Y, Yadav BS, Sopory SK (1999) Plant Mol Biol 41:125
89. Pham XH, Reddy MK, Ehtesham NZ, Matta B, Tuteja N (2000) Plant J 24:219
90. Tuteja N, Beven AF, Shaw PJ, Tuteja R (2001) Plant J 25:9
91. Roy JK, Prasad M, Varshney RK, Balyan HS, Blake TK, Dhaliwal HS, Singh H, Edwards KJ, Gupta PK (1999) Theor Appl Genet 99:336
92. Prasad M, Varshney RK, Kumar A, Balyan HS, Sharma PC, Edwards KJ, Singh H, Dhaliwal HS, Roy JK, Gupta PK (1999) Theor Appl Genet 99:341
93. Harjit-Singh, Prasad M, Varshney RK, Roy JK, Balyan HS, Dhaliwal HS, Gupta PK (2001) Plant Breed 120:273
94. Varshney RK, Prasad M, Roy JK, Roder MS, Balyan HS, Gupta PK (2001) Cer Res Commun 29:33
95. Varshney RK, Prasad M, Roy JK, Kumar N, Singh H, Dhaliwal HS, Balyan HS, Gupta PK (2000) Theor Appl Genet 100:1290
96. Ammiraju JSS, Dholakia BB, Santra DK, Singh H, Lagu MD, Tamhanker SA, Dhaliwal HS, Rao VS, Gupta VS, Ranjekar PK (2001) Theor Appl Genet 102:726
97. Dholakia BB, Ammiraju JSS, Santra DK, Singh H, Katti MV, Lagu MD, Tamhankar SA, Rao VS, Gupta VS, Dhaliwal HS, Ranjekar RK (2001) Biochem Genet 39:325
98. Galande AA, Tiwari R, Ammiraju JSS, Santra DK, Lagu MD, Rao VS, Gupta VS, Misra BK, Nagarajan S, Ranjekar PK (2001) Theor Appl Cenet 103:601
99. Gupta PK, Balyan HS, Edwards KJ, Isaac P, Korzun V, Roder M, Gautier MF, Joudrier P, Schlatter AR, Dubcovsky J, De la Pena RC, Khairallah M, Penner G, Hayden MJ, Sharp P, Keller B, Wang RCC, Hardouin JP, Jack P, Leroy P (2002) Theor Appl Genet 105:413
100. Naik S, Gill KS, Rao VSP, Gupta VS, Tamhankar SA, Pujar S, Gill BS, Ranjekar PK (1998) Theor Appl Genet 97:535
101. Aghaee-Sarbarzeh M, Singh H, Dhaliwal HS (2001) Plant Breed 120:259
102. Mohan M, Nair S, Bhagwat A, Krishna TG, Yano M, Bhatia CR, Sasaki T (1997) Mol Breed 3:87
103. Mohan M, Nair S, Bentur JS, Rao UP, Bennett J (1994) Theor Appl Genet 87:782
104. Nair S, Bentur JS, Rao UP, Mohan M (1995) Theor Appl Genet 91:68
105. Nair S, Kumar A, Srivastava MN, Mohan M (1996) Theor Appl Genet 92:660
106. Mohan M, Sathyanarayanan PV, Kumar A, Srivastava MN, Nair S (1997) Theor Appl Genet 95:777
107. Mago R, Nair S, Mohan M (1999) Theor Appl Genet 99:50
108. Rajyashri KR, Nair S, Ohmido N, Fukui K, Kurata N, Sasaki T, Mohan M (1998) Theor Appl Genet 97:507
109. Katiyar SK, Tan Y, Huang B, Chandel G, Xu Y, Zhang Y, Xie Z, Bennett J (2001) Theor Appl Genet 103:953
110. Abstracts (2001) 8th National rice biotechnological network meeting. Aurangabad, India
111. Srinivasachary, Hittalmani S, Kumar KG, Shashidhar HE, Vaishali MG (2002) Curr Sci 82:133
112. Prasad SR, Bagali PG, Hittalmani S, Shashidhar HE (2000) Curr Sci 78:162
113. Naqvi NI, Bonman JM, Mackill DJ, Nelson RJ, Chattoo BB (1995) Mol Breed 1:341
114. Davierwala AP, Reddy APK, Lagu MD, Ranjekar PK, Gupta VS (2001) Biochem Genet 39:261
115. Singh S, Sidhu JS, Huang N, Vikal Y, Li Z, Brar DS, Dhaliwal HS, Khush GS (2001) Theor Appl Genet 102:1011
116. Sharma A, Mohapatra T, Sharma RP (1994) J Plant Biochem Biotech 3:85
117. Upadhyay A, Mohapatra T, Pai RA, Sharma RP (1996) J Plant Biochem Biotech 5:17
118. Mukherjee AK, Mohapatra T, Varshney A, Sharma R, Sharma RP (2001) Plant Breed 120:483
119. Negi MS, Devic M, Delseny M, Lakshmikumaran M (2000) Theor Appl Genet 101:146
120. Gupta VS, Ramakrishna W, Rawat SR, Ranjekar PK (1994) Biochem Genet 32:1
121. Ramakishana W, Lagu MD, Gupta VS, Ranjekar PK (1994) Theor Appl Genet 88:402

122. Ramakrishna W, Chowdari KV, Lagu MD, Gupta VS, Ranjekar PK (1995) Theor Appl Genet 90:1000
123. Chowdari KV, Ramakrishna W, Tamhankar SA, Hendre RR, Gupta VS, Sahasrabudhe NA, Ranjekar (1998) Plant Cell Rep 18:55
124. Deshpande AD, Ramakrishna W, Mulay GP, Gupta VS, Ranjekar PK (1998) Theor Appl Genet 97:135
125. Ramkrishna W, Davierwala AP, Gupta VS, Ranjekar PK (1998) Biochem Genet 36:323
126. Joshi SP, Gupta VS, Aggarwal RK, Ranjekar PK, Brar DS (2000) Theor Appl Genet 100: 1311
127. Baishya S, Sachdev A, Johari RP, Mehta SL (2000) J Plant Biochem Biotech 9:23
128. Ray Choudhury P, Kohli S, Srinivasan K, Mohapatra T, Sharma RP (2001) Euphytica 118:243
129. Verma SK, Khanna V, Singh N (1999) Electrophoresis 20:1786
130. Nagaraju J, Kathirvel M, Kumar RR, Siddiq EA, Hasnain SE (2002) Proc Natl Acad Sci USA 99:5836
131. Jain A, Bhatia S, Banga SS, Prakash S, Lakshmikumaran M (1994) Theor Appl Genet 88:116
132. Srivastava A, Gupta V, Pental D, Pradhan AK (2001) Theor Appl Genet 102:193
133. Ratnaparkhe MB, Gupta VS, Ranjekar PK (1995) Theor Appl Genet 91:893
134. Parani M, Lakshmi M, Senthilkumar P, Parida A (2000) Curr Sci 78:1235
135. Lakshmi M, Rajalakshmi S, Parani M, Anuratha CS, Parida A (1997) Theor Appl Genet 97:1121
136. Lakshmi M, Parani M, Ram N, Parida A (2000) Genome 43:110
137. Parani M, Lakshmi M, Elango S, Ram N, Anuratha CS, Parida A (1997) Genome 40:487
138. Rajagopal J, Das S, Khurana DK, Srivastava PS, Lakshmikumaran M (1999) Genome 42:909
139. Parasnis AS, Gupta VS, Tamhankar SA, Ranjekar PK (2000) Mol Breed 6:337
140. Farooqui N, Ranade SA, Sane PV (1998) Bioch Mol Bio Int 45:931
141. Singh A, Negi MS, Rajagopal J, Bhatia S, Tomar UK, Srivastava PS, Lakshmikumaran M (1999) Theor Appl Genet 24:272
142. Singh A, Chaudhury A, Srivastava PS, Lakshmikumaran M (2002) Plant Sci 162:17
143. Negi S, Singh A, Lakshmikumaran M (2000) Genome 43:975
144. Raina SN, Mukai Y, Kawaguchi K, Geol S, Jain A (2001) Theor Appl Genet 103:839
145. Lavania UC (1998) Hereditas 128:273
146. Prasad M, Varshney RK, Roy JK, Balyan HS, Gupta PK (2000) Theor Appl Genet 100:584
147. Roy JK, Balyan HS, Prasad M, Gupta PK (2002) Theor Appl Genet 104:465
148. Roy JK, Lakshmikumaran MS, Balyan HS, Gupta PK (2004) Biochem Genet 42 (in press)
149. Chen M, Presting G, Barbazuk WB, Goicoechea JL, Blackmon B, Fang G, Kim H, Frisch D, Yu Y, Sun S, Higingbottom S, Phimphilai J, Phimphilai D, Thurmond S, Gaudette B, Li P, Liu J, Hatfield J, Main D, Farrar K, Henderson C, Barnett L, Costa R, Williams B, Walser S, Atkins M, Hall C, Budiman MA, Tomkins JP, Luo M, Bancroft I, Salse J, Regad F, Mohapatra T, Singh NK, Tyagi AK, Soderlund C, Dean RA, Wing RA (2002) Plant Cell 14:537
150. Sharma AK, Mohanty A, Singh Y, Tyagi AK (1999) Curr Sci 77:524
151. Tyagi AK, Mohanty A (2000) Plant Sci 158:1
152. Chaudhary A, Maheshwari SC, Tyagi AK (1993) Physiol Plant 89:842
153. Mohanty A, Grover M, Chaudhury A, Rizwan-ul-Haq Q, Sharma AK, Maheshwari SC, Tyagi AK (2000) Ind J Biochem Biophys 37:447
154. Chaudhary A, Maheshwari SC, Tyagi AK (1995) Plant Cell Rep 14:215
155. Chowdhary CN, Sood N, Maheshwari N, Tyagi AK (1996) Proc Nat Acad Sci, India 66:1
156. Minhas D, Bajaj S, Grover A, Rajam MV (1996) Curr Sci 71:1005
157. Khurana J, Khurana P (2000) J Plant Biochem Biotech 9:31
158. Gupta P, Raghuvanshi S, Tyagi AK (2001) Plant Biotech 18:275
159. Mahalakshmi A, Khurana P (1995) J Plant Biochem Biotech 4:55
160. Parihar DS, Maheshwari SC, Khurana P (1994) J Plant Biochem Biotech 3:107

161. Mahalakshmi A, Chugh A, Khurana P (2000) Plant Biotech 17:235
162. Gandhi R, Maheshwari SC, Khurana P (1999) In Vitro Cell Dev Biol-Plant 35:232
163. Sawant S, Singh PK, Madanala R, Tuli R (2001) Theor Appl Genet 102:635
164. Sawant S, Kiran K, Singh PK, Tuli R (2001) Plant Physiol 126:1630
165. Chandok MR, Sopory SK, Oelmueller R (2001) Mol Gen Genet 264:819
166. Reddy MK, Nair S, Sopory SK (2002) Anal Biochem 306:154
167. Vijayachandra K, Palanichelvam K, Veluthambi K (1995) Plant Mol Biol 29:125
168. Ramanathan V, Veluthambi K (1995) Plant Mol Biol 28:1149
169. Katiyar-Agrawal S, Kapoor A, Grover A (2002) Curr Sci 82:873
170. Srivastava V, Reddy AS, Guha-Mukherjee S (1988) Plant Cell Rep 7:504
171. Mukhopadhyay A, Topfer R, Pradhan A, Sodhi YS, Steinbiss HH, Schell J, Pental D (1991) Plant Cell Rep 10:375
172. Mukhopadhyay A, Arumugam N, Nandkumar PBA, Pradhan AK, Gupta V, Pental D (1992) Plant Cell Rep 11:506
173. Mehra S, Pareek A, Bandyopadhyay P, Sharma P, Burma PK, Pental D (2000) Curr Sci 78:1358
174. Gandhi R, Maheshwari SC, Khurana JP, Khurana P (2001) In Vitro Cell Dev Biol-Plant 37:629
175. Rao KS, Rohini VK (1999) Plant Biotech 16:201
176. Karthikeyan AS, Sarma KS, Veluthambi K (1996) Plant Cell Rep 15:328
177. Muthukumar B, Mariamma M, Veluthambi K, Gnanam A (1996) Plant Cell Rep 15:980
178. Jaiwal PK, Kumari R, Ignacimuthu S, Potrykus I, Sautter C (2001) Plant Sci 161:239
179. Sharma KK, Anjaiah V (2000) Plant Sci 159:7
180. Venkatachalam P, Geetha N, Khandelwal A, Shaila MS, Lakshmi Sita G (2000) Curr Sci 78:1130
181. Geetha N, Venkatachalam P, Lakshmi Sita G (1999) Plant Biotech 16:213
182. Manoharan M, SreeVidya CS, Lakshmi Sita G (1998) Plant Sci 131:77
183. Shiri V, Rao KS (1998) Plant Sci 131:53
184. Mondal TK, Bhattacharya A, Ahuja PS, Chand PK (2001) Plant Cell Rep 20:712
185. Satyavathi VV, Prasad V, Gita Lakshmi B, Lakshmi Sita G (2002) Plant Sci 162:215
186. Khanna H, Raina SK, Srinivasulu, Kumar K (1997) J Plant Biochem Biotech 6:75
187. Bhattacharjee B, An G, Gupta HS (1997) J Plant Biochem Biotech 6:69
188. Mohanty A, Sarma NP, Tyagi AK (1999) Plant Sci 147:127
189. Khanna H, Raina SK (1999) Aust J Plant Physiol 26:311
190. Kumria R, Waie B, Rajam MV (2001) Plant Cell, Tissue, Organ Culture 67:63
191. Kant T, Kothari SL, Hodges HK, Hodges TK (2001) J Plant Biochem Biotech 10:121
192. Gopalakrishnan S, Garg GK, Singh DT, Singh NK (2000) Curr Sci 79:1094
193. Chakraborty S, Chakraborty N, Datta A (2000) Proc Natl Acad Sci USA 97:3724
194. Jani D, Meena LS, Rizwan-ul-Haq QM, Singh Y, Sharma AK, Tyagi AK (2002) Transgenic Res 11:447
195. Prasad KVSK, Sharmila P, Kumar PA, Pardha Saradhi P (1999) Mol Breed 6:489
196. Prasad KVSK, Sharmila P, Pardha Saradhi P (2000) Plant Sci 159:233
197. Mohanty A, Kathuria H, Ferjani A, Sakamoto A, Mohanty P, Murata N, Tyagi AK (2002) Theor Appl Genet 106: 51
198. Prabhavati V, Yadav JS, Kumar PA, Rajam MV (2002) Mol Breeding 19: 137
199. Barthakur S, Babu V, Bansal KC (2001) J Plant Biochem Biotech 10:31
200. Pandey GK, Reddy VS, Reddy MK, Deswal R, Bhattacharya A, Sopory SK (2002) Plant Sci 162:41
201. Tuli R, Bhatia CR, Singh PK, Chaturvedi R (2000) Curr Sci 79:163
202. Kumar PA, Mandaokar A, Sreenivasu K, Chakrabarti SK, Bisaria S, Sharma SR, Kaur S, Sharma RP (1998) Mol Breed 4:33
203. Mandaokar A, Goyal RK, Shukla A, Bhalla R, Chaurasia A, Reddy VS, Altosaar I, Sharma RP, Kumar PA (2000) Crop Protection 19:307
204. Chakrabarti SK, Mandaokar A, Pattanayak D, Shukla A, Naik PS, Sharma RP, Kumar PA (2000) Potato Res 42:227

205. Kar S, Basu D, Das S, Ramakrishnan NA, Mukherjee P, Nayak P, Sen SK (1998) Transgenic Res 6:177
206. Selvapandiyan A, Reddy VS, Kumar PA, Tewari KK, Bhatnagar RK (1998) Mol Breed 4:473
207. Nayak P, Basu D, Das S, Basu A, Ghosh D, Ramakrishnan NA, Ghosh M, Sen S (1997) Proc Natl Acad Sci USA 94:2111
208. Khanna HK, Raina SK (2002) Transgenic Res 11:411
209. Kesarwani M, Azam M, Natarajan K, Mehta A, Datta A (2000) J Biol Chem 275:7230
210. Gandikota M, de Kochko A, Chen L, Ithal N, Fauquet C, Reddy AR (2001) Mol Breed 7:73
211. Rohini VK, Rao KS (2001) Plant Sci 160:889
212. SreeVidya CS, Manoharan M, Lakshmi Sita G (2000) J Plant Physiol 156:105
213. Ghosh SB, Nagi LHS, Ganapathi TR, Khurana SMP, Bapat VA (2002) Curr Sci 82:855
214. Jagannath A, Bandyopadhyay P, Arumugam N, Gupta V, Burma PK, Pental D (2001) Mol Breed 8:11
215. Jagannath A, Arumugam N, Gupta V, Pradhan A, Burma PK, Pental D (2002) Curr Sci 82:46
216. Tyagi AK, Khurana JP, Sharma AK, Mohanty A, Dhingra A, Raghuvanshi S, Mukhopadhyay A, Gupta V, Anand S, Kathuria H, Bhushan S, Thakur J, Kumar D (2003) In: Proceedings of International Rice Genetics Symposium IV 2000 (in press)
217. Ghosh PK (2000) In: RIs Biotechnology and Development Review. Research and Information system for the Non-aligned and Other Developing Countries, New Delhi, p 39
218. Pursuit and promotion of science – the Indian experience (2001) Indian National Science Academy, New Delhi
219. Alam G (2001) The acquisition of agricultural biotechnology in India: a study of rice, cotton and spices. UNCTAD and UNDP, New Delhi
220. Swaminathan MS (2000) J Plant Biol 27:203–211

Received: May 2002

Adv Biochem Engin/Biotechnol (2003) 84: 123 – 141
DOI 10.1007/b11038

Drug Targets in Malaria Parasites

G. Padmanaban

Department of Biochemistry, Indian Institute of Science Bangalore 560 012, India
E-mail: geepee@biochem.iisc.ernet.in

Malaria ranks with tuberculosis and AIDS in terms of its ability to destroy human health. In India there are at least two million cases seen annually. Although mortality may not be as high as it is in Africa, the trauma due to morbidity and debility and loss of productive man hours are colossal. Since resistance to chloroquine and antifolates is spreading rapidly, there is need to develop new pharmacophores, for which identification of new drug targets is essential. This review focuses on targets arising from classical and unique metabolic pathways in the malaria parasite, highlighting the research being carried out in India in the context of the global scenario. A significant amount of research in India and elsewhere has provided new knowledge on parasite biology, that could pave the way for the development of new pharmacophores. However, it is a matter of regret to record that malaria being a poor man's disease does not enthuse pharmaceutical companies in general to invest and bring out new molecules. Developing countries like India should take a lead in developing new but affordable antimalarials.

Keywords. Chloroquine, Artemisinin, Heme, Hemozoin, Apicoplast

1 **Introduction** 124

2 **Classical Targets** 125

2.1 Folate Pathway 125
2.2 Purine Biosynthetic Pathway 125
2.3 Pyrimidine Biosynthetic Pathway 127
2.4 Carbohydrate Metabolic Pathway 129

3 **Unique Features of Malaria Parasite Biology** 130

3.1 Food Vacuole and Hemoglobin Degradation 130
3.2 Hemozoin Formation as Drug Target 130
3.3 De novo Biosynthesis of Heme in the Malaria Parasite 131
3.4 Other Heme-Based Drug Targets 132
3.5 Reversal of Drug Resistance 134
3.6 Apicoplast as a Drug Target 135

4 **Development of New Antimalarials in India** 138

5 **Parasite Invasion and Cytoadherence** 138

© Springer-Verlag Berlin Heidelberg 2003

6 Conclusion 139

7 References 140

Abbreviations

DHPS	Dihydropteroate synthase
DHFR	Dihydrofolate reductase
HGXPRT	Hypoxanthine-guanine (xanthinine)phosphoribosyl transferase
PRPP	Phosphoribosyl pyrophosphate
Pf	*Plasmodium falciparum*
ADSS	Adenylosuccinate synthase
PABA	*para*-Aminobenzoic acid
TIM	Triosephosphate isomerase
ALA	δ-Aminolevulinate
mdr	multidrug resistant gene
crt	cryptic gene
DOXP	1-Deoxy-D-xylulose-5-phosphate
ACP	Acyl carrier protein
HSPG	Hepatocyte heparan sulfate proteoglycans
GPI	Glycosylphosphatidylinositol
EMP-1	Erythrocyte membrane protein-1
ICGEB	International Centre for Genetic Engineering and Biotechnology
JNCASR	Jawaharlal Nehru Centre for Advanced Scientific Research

1 Introduction

The three major infectious diseases posing a serious threat to mankind are visualized to be tuberculosis, malaria and HIV for which effective vaccines are not available. Around 300–500 million people contract malaria every year and 1.7–2.5 million people succumb to the disease. The mortality includes 1 million children below five years of age, especially in Africa. The story of lost opportunities in the control of malaria in India needs to be kept in mind in all the efforts towards containing the disease in the country. The substantial dip in malaria incidence in the early nineteen seventies due to public health measures has given way to a substantial increase of greater than 2 million cases per year in recent years. The rebound is attributed to stoppage in insecticide spraying and improper implementation of control strategies that has led to vector and parasite resistance. Besides, global factors such as population migration across international and state borders, increased land use for irrigation, dam construction, industries and urbanization are responsible for the present scenario [1].

While, reliable estimates on mortality rates in India are not available (said to be less than a thousand), the loss of man hours and trauma due to morbidity are substantial. *P. vivax* used to be the dominant species (almost 85%) responsible for malaria in the country, but in recent years the figure for the deadly *P. falciparum* infections is reaching almost 40%. Almost all the mortality cases are due to *P. falciparum* malaria, with drug-resistance contributing to severe complications. Chloroquine-resistance has already been reported in *P. vivax* and this could spread and pose a problem, if adequate steps are not taken to contain it.

While hygiene and vector control are essential to contain malaria, the need of the hour is also to develop newer antimalarials to handle drug-resistant cases, since an effective vaccine for malaria is still eluding the research community. The development of new pharmacophores would depend on the identification of new drug targets in the parasite. Interestingly, newer features of malaria parasite biology have come to light in recent years and these provide exciting new opportunities to develop new antimalarials. This review briefly summarizes the research efforts in India in this direction, in the context of the global scenario.

2 Classical Targets

2.1 Folate Pathway

The de novo as well as salvage pathways for folate biosynthesis occur in the malaria parasite. Major drug targets in the folate pathway are dihydropteroate synthase (DHPS) and dihydrofolate reductase (DHFR). The sulfa drugs, sulfadoxine and dapsone, used as antimalarials are analogues of *p*-aminobenzoic acid, one of the two substrates for DHPS. Pyrimethamine and cycloguanil are potent inhibitors of DHFR. The malaria parasite has adapted well to the challenge of these inhibitors and although they provide a major second line of defence in chloroquine-resistant cases, there is need to develop newer antifolates. While several mutations in DHFR and DHPS have been recorded in experimentally induced resistance, the mutations in DHFR at (S108 N) and in DHPS (A437G) have been found to correlate with increased pyrimethamine/sulfadoxine resistance in field isolates. When multiple mutations are present in DHFR and DHPS enzymes simultaneously, the pyrimethamine/sulfadoxine combination would be rendered ineffective [2]. In the Indian context, mutations in DHFR gene were analyzed in *P. falciparum* patient blood samples of all age groups in areas surrounding Delhi and Assam at the Malaria Research Centre, Delhi. Mutations at positions 16, 51, 59, 108 and 164 were analyzed using appropriate PCR primers. In a sample study of 40 clinical isolates, 55% showed C59R and S108 N double mutations and 32.5% showed the S108 N single mutation. Parasites carrying double or single mutant forms showed a broad range of elevated MIC values for pyrimethanine (76–6754 nM) compared to the sensitive range (141–380 nM). Mutations at the 16, 51, and 164 positions were not detected [3].

2.2 Purine Biosynthetic Pathway

In the intraerythrocytic stage, both the red cell and the malaria parasite are incapable of de novo biosynthesis of purines. Uninfected and infected red cells take up preformed purines and the parasite is dependent on the salvage pathway for making purine nucleotides. Enzymes of the salvage pathway have all been demonstrated in the infected erythrocytes. These include, adenine phosphoribosyl transferase, AMP deaminase, 3′,5′-nucleotidase, adenosine deaminase, purine nucleoside phosphorylase, hypoxanthine-guanine phosphoribosyl transferase, adenylosuccinate synthase, adenylosuccinate lyase, IMP dehydrogenase and adenosine kinase (Fig. 1). Some of the enzymes have been characterized in detail and shown to possess unique properties. Adenosine deaminase from *P. falciparum* is sensitive to deoxycoformycin but not to *erythro*-9-(2-hydroxy-3-nonyl)adenine, unlike the enzyme from other sources [4]. Purine nucleoside phosphorylase from *P. falciparum* is similar to that from vertebrate sources. The parasite enzyme can only use inosine and hypoxanthine as substrates, whereas the enzyme from other sources can use guanine, guanosine, allopurinol and 6-thioinosine as substrates. Formycin B is a more potent inhibitor of the plasmodial enzyme than that from human erythrocyte [5]. The enzyme hypoxanthine-guanine-phosphoribosyl transferase (HGPRT) from *P. falciparum* (HGXPRT) is unique and can use xanthine in addition, to hypoxanthine and guanine as substrates, unlike the enzyme from *P. lophure*, *P. chabaudi* and humans [6]. Many substituted purine compounds have, therefore, been tested for antimalarial activity, but so far without success. Bredenin and mycophenolic acid, inhibitors of IMP dehydrogenase, and hadacidin, an inhibitor of adenylosuccinate lyase, have been found to inhibit growth of *P. falciparum* in vitro. The gene sequence of adenylosuccinate lyase from *P. falciparum* has been shown to

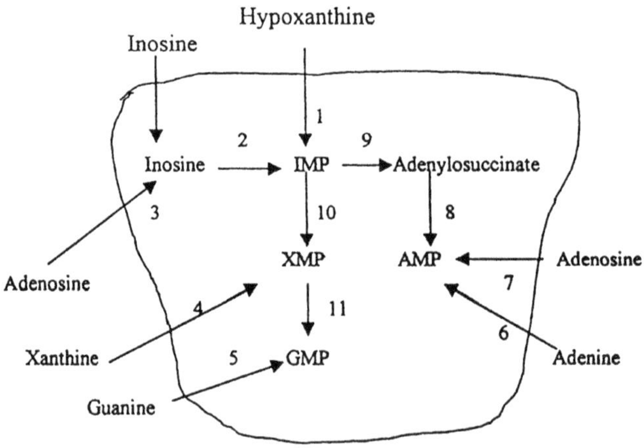

Fig. 1. Salvage purine biosynthetic pathway in the malaria parasite. The de novo pathway of purine biosynthesis does not exist in the malaria parasite. Preformed purines required from the host are utilized in the salvage pathway. 1, Hypoxanthine-guanine/xanthine phosphoribosyl transferase (HGXPRT); 2, Purine nucleoside phosphoribosylase; 3, Adenosine deaminase (ADA); 4, HGXPRT; 5, HGXPRT; 6, Adenine phosphoribosyl transferase (APRT) 7, Adenosine kinase (AK); 8, Adenylosuccinate lyase (AL); 9, Adenylosuccinate synthetase (AS); 10, IMP dehydrogenase; 11, GMP synthase

be different from that of the human enzyme and, therefore, is a potential drug target.

At the Jawaharlal Nehru Centre for Advanced Scientific Research (JNCASR), Bangalore, Hemalatha Balaram and her associates have been looking at some of the key steps in the parasite salvage pathway for purine biosynthesis, as potential drug targets. One of the problems with the malaria parasite HGXPRT has been the poor enzyme activity of the recombinant protein expressed in *E. coli*. Subbayya et al. [7] have worked out the conditions necessary for high levels of enzyme activity, which is critically dependent on the ratios of enzyme, phosphoribosyl pyrophosphate (PRPP), hypoxanthine and buffer conditions. Evidence has also been presented for the existence of different kinetic states of the enzyme during activation and storage. Although the enzyme also uses xanthine as substrate, hypoxanthine alone may be the biologically relevant purine base for this enzyme. The activity of HGXPRT is inhibited at high substrate concentrations and this could complicate the kinetic analysis in the presence of inhibitors. A novel bacterial screen has been developed to identify new inhibitors and substrates for HGXPRT. This is based on a combination of two *E. coli* strains, one (SΦ606) deficient in salvage pathway enzymes and another (S-609) deficient both in de novo and salvage pathway enzymes [8]. With the use of chimeric enzymes between humans and *P. falciparum*, it has been shown that the N-terminal 50 amino acids, although not proximal to the active site in the crystal structure, modulate the substrate specificity by making xanthine an additional substrate for the parasite enzyme [9].

Another enzyme of the purine salvage pathway being examined by the group at JNCASR is adenylosuccinate synthetase which, along with adenylosuccinate lyase, catalyzes the conversion of IMP to AMP. The cDNA for the parasite adenylosuccinate synthase (PfADSS) has been cloned and the deduced amino acid sequence of the *P. falciparum* enzyme exhibits 67% homology with that of the human enzyme. The PfADSS activity is inhibited by hadacidin, a known competitive inhibitor of the enzyme [10]. The cDNA has also been hyperexpressed in *E. coli* and the recombinant enzyme purified to homogeneity and its properties studied. The enzyme is a dimer in low ionic strength buffers and there is equilibration between the monomer and dimer in buffers of increased ionic strength [11]. The availability of purified recombinant enzymes of the purine salvage pathway with high activity would be of use in the design of novel inhibitors.

2.3 Pyrimidine Biosynthetic Pathway

Unlike the case of purine biosynthesis, the malarial parasite is capable of de novo biosynthesis of pyrimidines. The first three enzymes in the pathway (carbamyl phosphate synthetase, aspartate transcarbamylase and dihydroorotase) are separate proteins, unlike the eukaryotes where they exist as a 240 Kda multifunctional protein. Dihydroorotate dehydrogenase converts dihydrorotate to orotate and the electrons are passed on to the cytochromes and oxygen through ubiquinone. Orotate phosphoribosyltransferase and oroditidine 5′-decarboxylase are once again discrete proteins in *P. falciparum* unlike in mammalian cells where

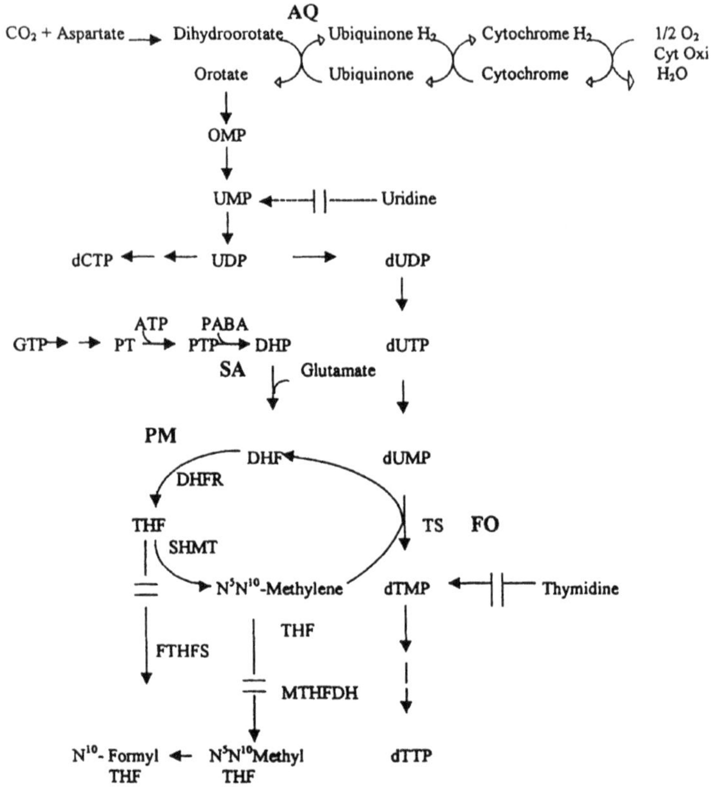

Fig. 2. The de novo pyrimidine and associated folate biosynthetic pathways in the malaria parasite. The specific drug targets are actually located in the associated electron transport and folate pathways. AQ, Atavaquone; SA Sulfanilic acid; FO, Fluorouracil; PM, Pyrimethamine. DHF, Dihydrofolate; THF, Tetrahydrofolate; SHMT, Serine/hydroxymethyltransferase; DHFR, Dihydrofolate reductase; MTHFDH, Methyl tetrahydrofolate dehydrogenase; FTHFS, Formyl tetrahydrofolate synthase

they exist mostly as bifunctional proteins. As already indicated, thymidylate synthetase from *Plasmodium* exists as a bifunctional protein in combination with dihydrofolate reductase. Figure 2 depicts the de novo pyrimidine biosynthetic pathway in *Plasmodium* and its interrelationship with the folate pathway and the ubiquinone-mediated electron transport chain.

As already indicated the sulfa drugs, analogues of PABA, and pyrimethamine act as antimalarial agents by interfering with the folate pathway. A promising development has been the introduction of atovaquone as an antimalarial. Early studies on hydroxynaphthoquinones as mitochondrial respiratory chain inhibitors and their structural resemblance to ubiquinone have eventually led to the development of atovaquone {2-[*trans*-4-(4′-chlorophenyl)cyclohexyl]-3-hydroxy-1,4-naphthoquinone} with potent antimalarial activity in vitro and in vivo. Evidence is available to indicate that atovaquone inhibits electron transport at the bc1 complex of parasite mitochondria (Fig. 2). It has also been suggested that atovaquone inhibits the parasite mitochondrial membrane potential [12]. Resistance to the drug arises readily when it is used as a monotherapy. The likely candidate to confer resistance appears to be the cytochrome b gene of the parasite with amino acid changes occurring in the QoII region. However, a combination of atovaquone and proguanil (2.5:1 ratio, registered as Malarone) gives 100% cure rates without evidence of recrudescence [13], although there is need for controlled use of this new drug. Studies on the mechanism of action of this combination reveal that the synergistic action of proguanil on atovaquone may be independent of its antifolate action. It may significantly enhance the ability of atovaquone to collapse the mitochondrial membrane potential. Further studies are needed in this regard, since site-specific, uncoupling activity of this combination can have far-reaching implications for antimicrobial drug development.

2.4 Carbohydrate Metabolic Pathway

Glucose is rapidly taken up by parasitized erythrocytes and it is estimated that a single infected red cell can consume 100 times more glucose than an uninfected cell [14]. Malaria parasites rely principally on anaerobic glycolysis and almost all the glucose is utilized through the anaerobic Embden-Meyerhoff-Parnas (EMP) pathway. Some glucose is also metabolized through the pentose phosphate pathway. Many of the enzymes of glycolysis occur as isozymes. The plasmodial LDH is 500 times more active with 3-acetylpyridine-adenine nucleotide than with NAD and is a good diagnostic marker and drug target [15].

In red blood cells infected with *P. falciparum*, there is a significant increase in phosphoglucose isomerase and enolase, both of which show plant-enzyme like properties and a pentapeptide motif and could constitute unique drug targets [16]. Although the malaria parasite has the mitochondrion, the citric acid cycle does not operate and energy is derived principally through anaerobic glycolysis. Some of the electron-transfer component enzymes are cytosolic, linked to the function of the pyridine biosynthetic pathway.

A significant number of studies has been carried with the *P. falciparum* triosephosphate isomerase (TIM) in Bangalore. The parasite gene has been cloned and expressed in *E. coli*. The 28 Kda protein was shown to be functionally active [17]. The crystal structure of the recombinant enzyme at 2.2 Å resolution has been reported. This was carried out by the molecular replacement method using trypanosomal TIM as the starting model. The Pf TIM has leucine at position 183 unlike most other TIMs which have glutamate at this position. Pf TIM has phenylalanine instead of Ser96 present in human TIM. Another interesting feature is the occurrence of a cysteine residue at the dimer interface of Pf TIM (Cys13) in contrast to human TIM, where this residue is a methionine [18]. Thus, the crystal structure of Pf TIM provides a framework for new therapeutic leads. The Pf TIM has been found to possess unusual stability. The eight strand 8/8 barrel seen is a robust structural motif and the TIM barrel can withstand several nicks in the polypeptide backbone with limited effects on structure and stability [19]. The protein has also served as a good model system to study folding of multimeric proteins. An intersubunit disulfide bonded mutant of Pf TIM, Y74C, has been studied. The replacement of tyrosine by cysteine leads to a cavity at the TIM dimer interface and results in dramatic destabilization. Stability can be restored by covalent tethering of the dimer interface by disulfide bonds [20]. This study has led to the designing of interface peptides to inhibit the dimeric enzyme, by disrupting the native multimeric state. This approach can provide lead structures for potential inhibitor design [21].

3 Unique Features of Malaria Parasite Biology

3.1 Food Vacuole and Hemoglobin Degradation

The mature trophozoites undergo invagination of the peripheral plasma and parasitophorous vacuolar membranes, occluding the erythrocyte cytoplasm efficiently. When cytostome seals into a vesicle, the parasitophorous vacuolar membrane becomes the inside partner, gets digested and the vesicle is now surrounded by a single membrane, the parasite plasma membrane, giving rise to the food vacuole [22].

The *P. falciparum* food vacuole contains at least three major proteases, falcipain (cysteine protease) and plasmepsins I and II. The precise role of each enzyme in hemoglobin degradation is not fully understood, although it is an ordered process. There is a debate regarding whether falcipain or plasmepsins initiate the degradation process and this could be determined by the redox environment of the food vacuole [23, 24]. Further degradation of the peptides could be carried out by additional exo- and endo-peptidases in the food vacuole and parasite cytosol. A metallopeptidase, designated as falcilysin, has been purified from *P. falciparum* [25]. Potent plasmepsin inhibitors have been discovered and these compounds inhibit plasmepsins at low nanomolar concentrations and are effective in culture at high nanomolar to low micromolar concentrations. However, the *P. falciparum* genome project has recently revealed that the parasite has at least ten genes that encode different aspartic proteases, which makes the unravelling of the in vivo target of aspartic protease inhibitors much more complicated than previously thought [26]. In addition, the challenge would be to design inhibitors that are specific for the parasite proteases without affecting the host enzymes. Thus, falcipain inhibitors that do not affect the host cathepsins L and B as well as plasmepsin inhibitors that do not act on the host cathepsin D would be ideal [27].

3.2 Hemozoin Formation as Drug Target

Large amounts of free heme would be released as a by-product of hemoglobin degradation in the food vacuole. Free heme is toxic and lyses malaria parasites [28]. The parasite aggregates the heme into an inert pigment referred to as hemozoin as a mechanism of detoxification. A recent report on the crystal structure of hemozoin indicates that the β-hematin molecules are linked into dimers through reciprocal iron-carboxylate bonds to one of the propionic acid side chains of each porphyrin ring and the dimers form chains linked by hydrogen bonds on the crystal [29].

Chloroquine as well as other quinoline antimalarials (mefloquine, halofantrine), alkaloids (quinine, quinidine) as well as artemisinin derivatives that all have antimalarial properties bind heme. It is visualized that chloroquine binds heme, and chemisorption of the drug into crystallized hemozoin can lead to inhibition of further heme aggregation [29, 30]. Quinine and mefloquine bind heme less avidly than chloroquine and inhibit hemozoin formation only at high micromolar concentrations that may not be reached with the standard dosings used. Therefore, the mechanisms of action of quinine and mefloquine may not be identical to those of chloroquine, although they may share some biological effects. The results also suggest that amodiaquine acts by a mechanism very similar to that of chloroquine.

The mechanism of hemozoin formation is not clear. Slater and Cerami [31] proposed that an enzyme, heme polymerase is involved. The enzyme has so far not been isolated. Spontaneous as well as autocatalytic heme "polymerization" has also been proposed [32,33]. The involvement of malaria histidine-rich protein (Pfhrp2) has been suggested [34]. A significant number of studies have been carried out in India on parasite hemozoin formation as a drug target. The contributions relate to the mechanisms of hemozoin formation and the interference of antimalarials in the process. Studies at the International Centre for Genetic Engineering and Biotechnology (ICGEB), New Delhi and elsewhere have described the necessity of a biological catalyst to mediate hemozoin formation [35, 36]. The large number of repeat sequences, Ala-His-His-ala-ala-Asp, in Pfhrp2 [37,38] have been shown to be the sites for binding heme. The heme binding studies indicate that up to 18 equivalents of heme could be bound by Pfhrp2 protein with an observed Kd of 0.94 μM. Chloroquine has a greater affinity for heme (Kd, 37 nM) than Pfhrp2. It is suggested that chloroquine may act by inhibiting heme-binding to Pfhrp2. It can extract bound heme from Pfhrp2 and form a complex that is toxic to the parasite [39]. Artemisinin and its analogues have also been found to inhibit hemozoin formation as well as hemoglobin degradation. Heme is also a potent inhibitor of cysteine protease and therefore, hemoglobin degradation is also interfered with [40]. However, there are results which indicate that, although artemisinin can form adducts with heme in vitro and in vivo, they may not be responsible for parasite killing [24]. There is a view that heme acts as a catalyst to activate artemisinin into an alkylating agent and a 25 Kda target protein with homology to the tumor protein family has been identified in *P. falciparum* [41].

3.3
De novo Biosynthesis of Heme in the Malaria Parasite

The malaria parasite has cytochromes and hemoproteins. The question of the source of supply of heme for the formation of these proteins, since the heme derived from the host hemoglobin is essentially stored as the inert hemozoin pigment in the food vacuole of the parasite, led to the demonstration in this laboratory of the biosynthesis of heme de novo in the parasite. The pathway was demonstrated using precursors such as [2-^{14}C]glycine and [4-^{14}C]δ-aminolevulinate (ALA) and assay of the enzymes ALA synthase, ALA dehydratase and ferrochelatase in the parasite. Interestingly, succinylacetone, a specific inhibitor of ALA dehydrase was found to inhibit the growth of *P. falciparum* in culture [42]. Attempts to clone the parasite ALA dehydrase based on PCR amplification using primers derived from the host gene were not successful. When the enzyme was isolated from the parasite and purified, it turned out to be the host enzyme. Biochemical and immunoelectron microscopic studies indicated that host ALA dehydrase is actually translocated into the parasite and that this enzyme is functional in the parasite, synthesizing heme [43]. More recent studies have indicated that, when an N- and C-terminal deleted mutant of host ALA dehydrase (~12 Kda) is added to a culture of *P. falciparum*, it is taken up by the parasite and there is a striking fall in parasite ALA dehydrase activity and heme synthesis followed by death of the parasite [44]. These results have highlighted certain new facets of parasite biology in terms of import of host proteins by the parasite to carry out biosynthesis of key molecules and the strategies that can be developed to interfere with this process. A larger question is the mechanism of import of proteins from the host red cell and circulation and the possible role and fate of such proteins in the parasite.

An integrated picture of the scope of heme metabolism as a drug target is given in Fig. 3.

3.4 Other Heme-Based Drug Targets

While hemozoin formation has been investigated as a major drug target, results are available to suggest that chloroquine and other heme-interacting antimalarials act simultaneously by a number of mechanisms. Therefore, these additional sites also form potential drug targets.

It has been calculated that only 30% of the heme in the food vacuole of *P. falciparum* is converted into hemozoin at the trophozoite stage. It has been suggested that non-polymerized heme exits the food vacuole and is subsequently degraded by glutathione. The parasite has been shown to have a large capacity for glutathione synthesis. Chloroquine and amodiaquine are visualized to accumulate in the food vacuole, inhibiting hemozoin formation and thereby increasing the efflux of heme out of the food vacuole. These drugs have been found to competitively inhibit the degradation of heme by glutathione, thus leading to accumulation of heme in parasite membranes and resulting in the death of the parasite [45, 46]. This would mean that the competition between glutathione and chloroquine has to take place in the parasite cytoplasm and it is not clear whether such a concentration of the drug would be available outside the food vacuole. Under conditions similar to those existing in the food vacuole, free heme has been shown to undergo peroxidative decomposition. Chloroquine and quinacrine inhibit this process leading to the build-up of a membrane-associated drug concentration [47].

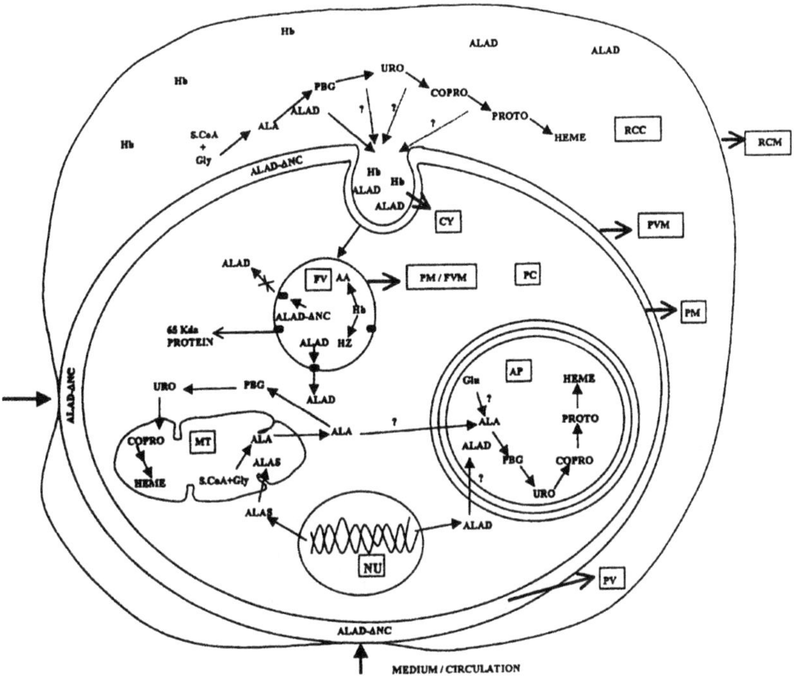

Fig. 3. An integrated picture of heme metabolism in the malaria parasite. Interference with the formation of hemozoin in the food vacuole is a site for major antimalarial action. The parasite also makes heme de novo, which is also a drug target. Yet another drug target is the import of host enzymes by the parasite to synthesize heme. Apicoplast is a possible site (not established) for de novo heme biosynthesis using parasite gene-coded plant-like enzymes. CY, cytostome; FV, Food vacuole; PM, Parasite membrane; FVM, Food vacuolar membrane; PVM, Parasitophorous vacuolar membrane; RCM, Red cell membrane; MT, Mitochondria; AP, Apicoplast; NU, Nucleus; ALAD, δ-Aminolevulinate deydrase; ALAD-ΔNC, An N- and C-terminal deleted mutant of ALAD

Studies in this laboratory showed that the parasite protein synthesis requires heme and that chloroquine inhibits heme-dependent protein synthesis [48]. Further studies have shown that inhibition of parasite protein synthesis by chloroquine precedes its inhibitory effect on hemozoin formation [49]. Inhibition of heme-dependent protein synthesis in the malaria parasite can offer a common mechanism of action for all the antimalarials interacting with heme, provided that an adequate concentration of the drug is available at the site of protein synthesis.

Iron chelators have been found to have antimalarial activity, although the exact source of chelator-bound iron in the parasite is not clear. Desferioxamine and a variety of other iron chelators such as aromatic chelators, reversed siderophores, hydroxypyridinones and others have been shown to have antimalarial activity [50, 51]. It is held that iron deposition may be partially responsible for severe malaria. Additionally, the iron chelator may protect the ischemic tissue from oxidative damage and also enhance parasite-directed immune responses [52].

3.5
Reversal of Drug Resistance

Chloroquine is the drug of choice to treat malaria. However, resistance to this drug has spread rapidly in the last decade. Resistance to other quinoline drugs as well as antifolates has become a major problem. Thus, identification of new drug targets and development of new drugs have become very necessary. Clinical resistance to artemisinin is yet to be reported, but this is a drug that needs to be sparingly used in cases of complicated/cerebral malaria, so that the parasite is not under pressure to develop resistance.

An important drug target would be to break the chloroquine resistance, so that the life of this unique drug is prolonged. The mechanism of chloroquine resistance has been under considerable debate. It was initially thought that the resistant strains of *P. falciparum* accumulate lesser amounts of chloroquine than the sensitive strains due to an active efflux mechanism mediated by the P-glycoprotein coded by the *Pfmdr1* gene [53]. Subsequently, it has become clear that chloroquine resistance is multigenic in character, although the role of *Pfmdr1* in mefloquine resistance is less controversial. The other mechanisms suggested are:

1. A chloroquine importer may actually be involved in the uptake of the drug and this may be manifesting a lower activity in resistance.
2. A CG2 allele coding for a 330 Kda protein in chromosome 7 with different patterns of polymophisms in Asia and Africa may be involved [54].
3. Cellular uptake of chloroquine is dependent on its binding to ferriprotoporphyrin IX and resistance is due to a decrease in the affinity of this interaction [55].

A recent study provides direct proof that mutations in *pfmdr1* gene can confer resistance to mefloquine, quinine and halofantrine. The same mutations influence parasite resistance towards chloroquine in a strain-specific manner and the level of sensitivity to the structurally unrelated compound, artemisinin [56]. More recent studies have shown that allelic modifications of Cg2 and Cg1 genes do not alter chloroquine response of drug-resistant *P. falciparum* and therefore are unlikely to be directly involved in resistance [57]. A new gene, *Pfcrt* with 13 exons has been identified near *Cg2* on chromosome 7 and this codes for a transmembrane protein in digestive vacuoles of malaria parasite. Sets of point mutations in *Pfcrt* were found to be associated with chloroquine resistance in vitro in laboratory lines of *P. falciparum* from Africa, South America and South East Asia. A lysine to threonine mutation (K76T) was detected in all the resistant isolates, but not in the chloroquine-sensitive isolates, tested in vitro. Furthermore, genetic transformation with the mutant *Pfcrt* conferred chloroquine resistance in three different sensitive isolates [58]. In a field study in Mali, the *Pfcrt* K76T mutation was detected in all 60 chloroquine-resistant cases against a baseline 41% in 116 random samples. The *Pfmdr1* N86Y mutation was detected in 48 out of 56 chloroquine-resistant cases. The *Pfcrt* K76T mutation was more strongly associated with chloroquine-resistance than the *Pfmdr1* N86Y mutation. It is suggested that the latter mutation may not have a primary role in chloroquine resistance, but may confer some advantage to the parasite in the presence of chloroquine because of a loss of fitness due to the *Pfcrt* mutation or augment the level of resistance. The function of *Pfcrt* protein is not clear as yet. It is felt that the K76T mutation in *Pfcrt* and the N86Y mutation in Pfmdrl can serve as molecular markers for chloroquine resistance in vivo and in vitro [59].

Calcium-channel blockers such as verapamil were found to counteract chloroquine resistance and this led to the original suggestion that resistance may be mediated by a verapamil-sensitive multidrug efflux pump as is the case in multidrug-resistant mammalian tumor cells [60]. However, the mechanism of verapamil's action itself has become a matter of debate and it has been suggested that it may inhibit the activity of a membrane channel indirectly regulating chloroquine transport into the parasite cytoplasm [61]. The concept of a chloroquine importer mediated by an Na^+/H^+ exchanger involving cytosolic pH regulation has been contested and the drug uptake in *P. falciparum* is held to be determined by its binding to ferriprotoporphyrin IX [55] as originally suggested by Fitch and coworkers [62]. It is suggested that chloroquine resistance is characterized by a reduced affinity of chloroquine-ferriprotoporphyrin IX binding and that this is reversible by verapamil. Diverse lysosomotropic compounds which could increase the pH of hemoglobin delivery vesicles to the food vacuole but not that of the cytosol, would also mimic the effect of verapamil, the net result being a decreased affinity of chloroquine-ferriprotoporphyrin IX binding [55]. The CG2 protein earlier and now the Pfcrt protein on the food vacuolar membrane could modulate this interaction and this could be a molecular marker for chloroquine sensitivity/resistance.

Irrespective of the mechanisms involved, many compounds have been shown to reverse chloroquine resistance and the search for a new class of resistance "modulator drugs" which could be administered with chloroquine to effectively potentiate its action and reverse resistance is underway (Table 1) [63]. In this context, it has been shown that aminoquinolines with 2–12 carbon side chains (ethyl, propyl to dodecyl) are as active as chloroquine against chloroquine-sensitive *P. falciparum*. Aminoquinolines with side chains shorter or longer than chloroquine were also found to be active against chloroquine-, mefloquine- and multiply-resistant *P. falciparum*. Verapamil had no effect on the activity of aminoquinolines that were active against the resistant parasites [64].

3.6 Apicoplast as a Drug Target

The identification of plastids in parasites of the phylum *Apicomplexa* has posed several interesting questions on their origin and possible new sites for antimalarial intervention. The plastid in apicomplexan parasites is of secondary endosymbiotic origin. It is visualized that the primary event is the endosymbiosis of a cyanobacterium-like prokaryotic cell by green or red algae. In secondary endosymbiosis the plastid-containing primary endosymbionts are themselves engulfed and retained by the recipient eukaryote. This leaves *P. falciparum*, for

Table 1. Reversal of drug resistance in vivo

Drug class	Drug name
Chloroquine-resistance reversal	
Antihypertensive	Verapamil, Nifedipine
Anti-anxiety	Desipramine
Antipsychotic	Chloropromazine
Anticancer	Vinblastine, Daunomycin
Anti-allergic	Chlorophenaramine Ketotifen Terfenadine Cypoheptadine
Antidepressant	Fluoxetine
Antifungal	Miconzole Econazole
Quinine-resistance reversal	
Antipsychotic	Chloropromazine
Mefloquine-resistance reversal	
Antipsychotic	Penfluoridol

example, with a three- or four-membrane enclosed apicoplast [66]. The 35 Kb plastid genome of *P. falciparum* is completely mapped and the genome is of wide spread occurrence over a range of malaria species [67].

The possible functions of the apicoplasts have come under intense investigation. The primary candidates are biosynthetic pathways leading to the production of amino acids, fatty acids and heme, besides the processes of transcription, translation and replication of the plastid genome [66]. All these processes have become ideal drug targets (Fig. 4).

The *P. falciparum* and *T. gondii* gene data bases indicate entries for chorismate synthase gene suggesting that the shikimate pathway for aromatic amino acid biosynthesis is present in apicomplexans, as in plants and algae. This pathway is not present in animals. Four enzymes of the shikimate pathway have been demonstrated in apicomplexan parasites. Glyphosate, a herbicide and inhibitor of 5-enoyl-pyruvyl shikimate-3-phosphate synthase has been shown to function as an antimalarial [68, 69]. A recent study has shown the *P. falciparum* chorismate synthase to be monofunctional like the plant enzyme, but it is localized in the parasite cytosol, like the fungal enzyme [70]. Genes for two enzymes involved in a non-mevalonate pathway of isoprenoid biosynthesis, reported previously only for eubacteria and plastids of algae and plants, have been identified in *P. falciparum*. These are 1-deoxy-D-xylulose-5-phosphate (DOXP) synthase and DOXP reductoisomerase. The recombinant DOXP reductoisomerase of *P. falciparum* has been shown to be inhibited by fosidomycin and a derivative called FR 90098, known inhibitors of the plant and algal enzymes. The compounds have been found to inhibit the growth of multidrug-resistant strains in vitro and to cure mice infected with the malarial parasite [71].

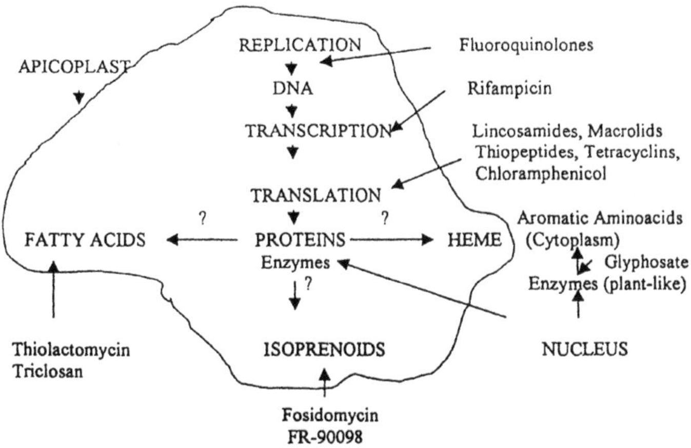

Fig. 4. Drug targets in the Apicoplast. The various metabolic transaction steps are good drug targets. The apicoplast is also visualized to be a site for fatty acid and isoprenoid synthesis. In addition, the parasite gene codes for plant-like enzymes synthesizing aromatic amino acids, although these enzymes have been shown to be localized in the cytoplasm

The identification of the apicoplast in the malaria parasite has added newer dimensions to the function of the heme biosynthetic pathway demonstrated in this laboratory. Although these studies have shown that the parasite ALA dehydrase and possibly some of the downstream enzymes are of host red cell origin [43, 44], the data base indicates that the parasite has putative genes for these enzymes. The parasite gene for ALA synthase has been cloned [72] and studies in this laboratory have shown that this enzyme is localized in the mitochondrion (unpublished data). Phylogenetic studies by Sato et al. [73] have indicated that the putative parasite gene for ALA dehydrase has plant-like features. An interesting question is whether the parasite carries out heme biosynthesis in the apicoplast. If so, the scenario would be that the parasite may be carrying out heme biosynthesis using ALA dehydrase from the host in the mitochondrial (ALA synthase)/cytoplasmic pathway, as is the case in liver, and using its own ALAD for the pathway in the apicoplast. Yet another question would be the source of ALA, the first committed product of heme biosynthesis. While studies in this laboratory have shown that the parasite ALA synthase is located in the mitochondrion, as is the case in liver, chloroplasts use the 5-carbon pathway involving glutamyl-tRNA to make ALA. The question is whether the apicoplast has the 5-carbon pathway machinery to make its own ALA or if the metabolite is obtained from the mitochondrion. Answers to these interesting questions would also provide new opportunities to develop antimalarials [74].

Concrete proof is now available to conclude that the parasite apicoplast is involved in fatty acid biosynthesis. The similarity of nuclear genes in *Plasmodium* and *Toxoplasma* coding for acyl carrier protein (ACP) β-ketoacyl-ACP synthase III (FabH) and β-hydroxyacyl ACP dehydratase (FabZ) to the plastid homologues has been established. The potential plastid-targeting sequences have been identified. The actual targeting of these enzymes to the apicoplast has also been demonstrated. Thiolactomycin, an inhibitor of FabH, has been shown to inhibit growth of *P. falciparum* in culture [75, 76].

Studies by Namita Surolia and coworkers at the JNCASR, Bangalore have actually led to the demonstration of fatty acid biosynthesis in *P. falciparum*. These workers have purified the enzyme, enoyl-ACP-reductase (FabI) from *P. falciparum*. Triclosan, an antimicrobial biocide, 5-chloro-2-(2,4-dichloro-phenoxy)phenol, inhibits this enzyme and is shown to be a powerful antimalarial [77].

4 Development of New Antimalarials in India

In the last two decades, several 8-aminoquinoline analogues have been synthesized at CDRI, Lucknow and the compound CDRI 80/53 (an analogue of primaquine) was selected for preclinical development because of its anti-relapse activity against the relapsing simian malaria parasite *P. cynomolgi* B. This compound is characterized by a high level of safety in LD_{50} assays, low level of methemoglobin toxicity in beagle dogs, as well as a high therapeutic index and safety in general pharmacological screens. Its major advantage over primaquine is its low propensity for increasing methemoglobin levels in treated cases. The drug is marketed as Bulaquine. Another promising antimalarial developed at CDRI is α/β-arteether. The drug has been shown to be a fast-acting schizonticide and is safe to use in animals and humans. The drug is marketed under the trade name E-mal [65].

5 Parasite Invasion and Cytoadherence

It would be ideal if therapeutic strategies were available to prevent the entry of the parasite into the liver and/or erythrocytes and adherence of the infected red cells to the endothelial cells.

Malaria sporozoites are transported by the blood to the liver microcirculation, where they come into contact with endothelial and Kupffer cells, before they have access to the hepatocytes. Hepatocyte heparan sulfate proteoglycans (HSPG) and other host receptors are visualized to interact with the parasite circumsporozoite protein, thrombospondin-related adhesive protein, etc. to facilitate selective targeting to hepatocytes.

Similarly, parasite entry into blood cells involves merozoite recognition and attachment to erythrocytes, erythrocyte entry and formation of parasitophorous vacuole as well as membrane closure and transformation from merozoites to trophozoites. Parasite proteins (MSP-1, MSP-2, MSP-3 and MSP-4) anchored in the plasma membrane by the post-translational addition of GPI (glycosyl phosphotidylinositol) interact with the Duffy blood group antigen on the erythrocyte membrane in the case of *P. vivax*. In the case of *P. falciparum*, the merozoites use the terminal sialic acid residues on erythrocyte membrane glycophorins as a major determinant of invasion.

The binding domains of *P. vivax* Duffy binding protein and *P. falciparum* erythrocyte binding proteins (region II) have been cloned and expressed at ICGEB, New Delhi. The recombinant proteins have been refolded to acquire their specific binding properties. This interaction is a molecular site for intervention.

The process by which infected erythrocytes adhere to microvascular endothelium is referred to as cytoadherence. This sequestration process is central to the pathophysiology of *P. falciparum* malaria and occurs predominantly in the venules of vital organs, especially the brain. The infected erythrocytes not only adhere to vascular epithelium, but also bind two or more uninfected erythrocytes giving rise to rosettes. The primary rosetting ligand is Pf EMP-1 (erythrocyte membrane protein). The host receptors include CD36 and ICAM-1. The Pf EMP-1s contain an N-terminal Duffy binding-like domain harboring motifs to bind heparin. Heparin sulfate and chondroitin sulfate inhibit this interaction. The feature of antigenic variation and switching of the infected erythrocyte surface is attributed to the polymorphism in PfEMP1 and var genes [78].

At ICGEB, New Delhi, a var gene expressed in an ICAM-1 binding *P. falciparum* field isolate has been cloned and expressed. The ICAM-1 binding DBL domain has also been identified. It may be possible to target functional DBL domains to inhibit cytoadherence and protect against syndromes such as cerebral malaria and placental malaria. In the latter case, it is known that hyaluronic acid, a high molecular weight GAG present on the syncytiotrophoblasts linking the placenta and chondroitin sulfate, are the receptors [79].

These receptor-mediated interactions are more amenable to vaccine intervention, than perhaps as drug-targets since these receptors may be mediating other normal physiological functions.

Parasite membrane biogenesis and the involvement of membranes such as the parasite tubovesicular membrane in nutrient supply are all candidates for drug intervention.

6 Conclusion

A significant amount of new knowledge has been gained in terms of parasite biology. New drug targets have thus emanated from this knowledge and a significant amount of research in the area is also being carried out in India. But, it is disappointing to record that there is very little interest to bring out a new antimalarial, since malaria is a poor man's disease and does not make economic sense for investments in major companies of the world. Perhaps developing countries should take this as a challenge and find a solution for this vexed scourge.

Acknowledgement. Studies carried out in the author's laboratory were supported by grants from the Department of Biotechnology, New Delhi.

References

1. Pattanayak S (1998) In: Gupta S, Sood OP (eds), The changing face of malaria. Ranbaxy Science Foundation, Gurgaon, India, p 17
2. Cowman AF (1999) In: Sherman IW (ed), Malaria-Parasite Biology, Pathogenesis and Protection. ASM Press, Washington, DC, p 317
3. Biswas S, Escalante A, Chaiyaroj S, Angkasekwinai P, Lal AA (2000) Trop Med Int Hlth 5:737
4. Cowman AF, Morry MJ, Biggs BA, Cross GAM, Foote SJ (1988) Proc Natl Acad Sci USA 85: 9109
5. Daddona PE, Wisemann WP, Milhouse W, Chern J-W, Townsend LB, Hershfield MS, Webster HK (1986) J Biol Chem 261:11667
6. Queen SA, Vander Jagt D, Reyes P (1988) Mol Biochem Parasitol 30:123
7. Subbayya INS, Balaram H (2000) Biochem Biophys Res Commun 279:433
8. Shivashankar K, Subbayya INS, Balaram H (2001) J Mol Microbiol Biotechnol 3:557
9. Subbayya INS, Sukumaran S, Shivashankar K, Balaram H (2000) Biochem Biophys Res Commun 272:596
10. Sumathy K, Jayalakshmi R, Shivayogi MS, Balaram H (2000) Curr Sci 78:610
11. Jayalakshmi R, Sumathy K, Balaram H (2002) Protein Expression and Purification, in press
12. Srivastava IK, Rottenberg H, Vaidya AB (1997) J Biol Chem 272:3961
13. Looareesuwan S, Viravan C, Webster HK, Kyle DE, Hutchinson DB, Canfield CJ (1996) Am J Trop Med Hyg 54:62
14. Rooth E Jr (1990) Blood Cells 16:453
15. Makler MT, Hinrichs DJ (1993) Am J Trop Med Hyg 48:205
16. Dzierszinski F, Popescu O, Toursel C, Slomianny C, Yahiaoui B, Tomavo S (1999) J Biol Chem 274:24888.
17. Ranie J, Kumar VP, Balaram H (1993) Mol Biochem Parasitol 61:159
18. Velankar S, Ray SS, Gokhale RS, Suma S, Balaram H, Balaram P, Murthy MRN (1999) Structure 5:751
19. Ray SS, Balaram H, Balaram P (1999) Chem Biol 6:625
20. Gopal B, Ray SS, Gokhale RS, Balaram H, Murthy MRN, Balaram P (1999) Biochem 38:478
21. Kumar Singh S, Maithal K, Balaram H, Balaram P (2001) FEBS Lett 501:19
22. Slomianny C (1990) Blood Cells 16:369
23. Goldberg DE, Slater AFG, Cerami A, Henderson GB (1990) Proc Natl Acad Sci USA 87:2931
24. Rosenthal PJ, Meshnick R (1998) In: Sherman IW (ed), Malaria – Parasite Biology, Pathogenesis and Protection. ASM Press, Washington, DC, p 355
25. Eggleson KK, Duttin KL, Goldberg DE (1999) J Biol Chem 274:32411
26. Coombs GH, Goldberg DE, Klemba M (2001) Trends Parasitol 17:532
27. Rosenthal PJ (1996) Mol Biochem Parasitol 75:255
28. Orjih A, Banyal HS, Chevli R, Fitch CD (1981) Science 214:667
29. Pagola S, Stephens PW, Scott-Bohle D (2000) Nature 404:307
30. Egan TJ, Mavuso WM, Neokazi KK (2001) Biochemistry 40:204
31. Slater AFG, Cerami A (1992) Nature 355:167
32. Egan T Ross DC, Adams PA (1994) FEBS Lett 352:54
33. Dorn A, Stoffel R, Matile H, Bubendorf A, Ridley RG (1995) Nature 374:269
34. Sullivan DJ, Gluzman IY, Goldberg DE (1996) Science 271:219
35. Pandey AV, Tekwani BL (1996) FEBS Lett 393:189
36. Blaner G, Akkawi M (2000) Biochem J 346:249
37. Sullivan DJ, Gluzman IY, Goldberg DE (1996) Science 271:219
38. Wellems TE, Howard RJ (1986) Proc Natl Acad Sci USA 83:6065
39. Pandey AV, Bisht H, Babbarwal VK, Srivastava K, Pandey KC, Chauhan VS (2001) Biochem J 355:333
40. Pandey AV, Tekwani BL, Singh RL, Chauhan VS (1999) J Biol Chem 274:19383
41. Bhisuttibhan J, Pan, HQ, Hossler PA, Walker DJ, Yowell CA, Carlton J, Dame JB, Meshnick SR (1998) J Biol Chem 273:16192
42. Surolia N, Padmanaban G (1992) Biochem Biophys Res Commun 187:744
43. Bonday ZQ, Taketani S, Gupta PD, Padmanaban G (1997) J Biol Chem 272:21839

44. Bonday ZQ, Dhanasekharan S, Rangarajan PN, Padmanaban G (2000) Nature Med 6:898
45. Ginsburg H, Famin O, Zhang J, Krugliak, M (1998) Biochem Pharmacol 15:1305
46. Atamna H, Ginsburg H (1997) Eur J Biochem 250:670
47. Loria P, Miller S, Foley M, Tilley L (1999) Biochem J 339:363
48. Surolia N, Padmanaban G (1991) Proc Natl Acad Sci USA 88:4788
49. Karthikeyan G (1997) PhD thesis, Indian Institute of Science Bangalore
50. Shanzer A, Libman J, Lytton SD, Glickstein H, Cabantchik ZI (1991) Proc Natl Acad Sci USA 88:6585
51. Lytton SD, Mester B, Dayan I, Glickstein H, Libman J, Shanzer A, Cabantchik ZI (1993) Blood 81:214
52. Mabeza GF, Biemba G, Gordenk VR (1996) Acta Haematol 95:78
53. Foote SJ, Kyle DE, Martin, RK, Oduola AM, Forsyth K, Kemp DJ, Cowman AF (1990) Nature 345:255
54. Su, X, Kirkman LA, Fujioka H, Wellems TE (1997) Cell 91:593
55. Bray PG, Janneh O, Raynes KJ, Mungthin M, Ginsburg H, Ward SA (1999) J Cell Biol 145:363
56. Reed MB, Saliba KJ, Caruana SR, Kirk K, Cowman AF (2000) Nature 403:906
57. Fidock DA, Nomura T, Cooper RA, Su X, Talley A, Wellems TE (2000) Mol Biochem Parasitol 110:1.
58. Fidock DA, Nomura T, Talley AK, Cooper RA, Dzekuna SM, Ferdig MT, Ursos LMB, Sidhu AS, Naude B, Deitsch KW, Su X-Z, Wooton JC, Roepe PD, Wellems TE (2000) Mol Cell 6:861
59. Djimde A, Doumbo OK, Cortese JF, Kayenato K, Doumbo S, Diourte Y, Dicko A, Su X-Z, Nomura T, Fidock DA, Wellems TE, Plowe CV (2001) New Eng J Med 344:257
60. Martin SK, Odoula AM, Milhous WK (1987) Science 235:899
61. Martiney JA, Cerami A, Slater AFG (1995) J Biol Chem 270:22393
62. Fitch CD, Chevli R, Banyal HS, Phillips G, Pfaller MA, Krogstad DJ (1982) Antimicrobial Agents Chemother 21:819
63. Milhous WK, Kyle DE (1998) In: Sherman IW (ed), Malaria - Parasite Biology, Pathogenesis and Protection. ASM Press, Washington DC, p 303
64. De D, Krogstad FM, Cogswell FB, Krogstad DJ (1996) Am J Trop Med Hyg 55:579
65. Kamboj VP, Dutta GP (1998) In: Gupta S, Sood OP (eds), The Changing Face of Malaria. Ranbaxy Science Foundation, Gurgaon, p 79
66. Mcfadden GI, Roos DS (1999) Trends Microbiol 7:328
67. Wilson RJM, Denny DW, Dreiser PR, Rangachari K, Roberts K, Roy A, Whyte A, Strath M, Moore D (1996) J Mol Biol 261:155
68. Roberts F, Roberts CW, Johnson JJ, Kyle DE, Krell T, Coggins JR, Coombs GH, Milhous WK, Tzipori S, Ferguson DJP, Chakraborti D, McLeod R (1998) Nature 393:801
69. Keeling PJ, Palmer JD, Donald RGK, Roos DS, Waller RF, McFadden GI (1999) Nature 397: 219
70. Fitzpatrick T, Ricken S, Lanzer M, Amrhein N, Macheroux P, Kappes B (2001) Mol Microbial 40:´65
71. Jomaa H, Wiesner J, Sanderbrand S, Altincicek B, Weidemeyer C, Hintz M, Turbachova I, Eberl M, Zeidler J, Lichtenthaler HK, Soldati D, Beck E (1999) Science 285:1573
72. Wilson CM, Smith AB, Baylon RV (1996) Mol Biochem Parasitol 75:271
73. Sato S, Tews I, Wilson RJM (2000) Internatl J Parasitol 30:427
74. Padmanaban G, Rangarajan PN (2001) Expert Opin Ther Targets 5:423
75. Waller RF, Keeling PJ, Donald RGK, Striepen B, Handman E, Lang-Unnasch N, Cowman AF, Besra GS, Roos DS, McFadden GI (1998) Proc Natl Acad Sci USA 95:12352
76. Waller RF, Reed MB, Cowman AF, McFadden GI (2000) Embo J 19:1794
77. Surolia N, Surolia A (2000) Nature Med 7:167
78. Baruch DI, Pasloske BL, Singh, HB, Bi X, Ma XC, Feldman M, Tasraschi TF, Howard RJ (1995) Cell 82:77
79. Beeson JG, Rogerson SJ, Cooke BM, Reeder JC, Chai W, Lawson AM, Molyneux ME, Brown GV (2000). Nature Med 6:86

Received: May 2002

Adv Biochem Engin/Biotechnol (2003) 84: 143–182
DOI 10.1007/b11040

Current Status of Malaria Vaccine Development

Virander Singh Chauhan [1] · Devesh Bhardwaj [2]

[1] Malaria Research Group, International Center for Genetic Engineering and Biotechnology, Aruna Asaf Ali Marg, New Delhi-110067, India. *E-mail: Virander@icgeb.res.in*

[2] Malaria Research Group, International Center for Genetic Engineering and Biotechnology, Aruna Asaf Ali Marg, New Delhi-110067, India. *E-mail: devesh@icgeb.res.in*

There is an urgent need to develop an effective vaccine against malaria – a disease that has approximately 10% of the world population at risk of infection at any given time. The economic burden this disease puts on the medico-social set-up of countries in Sub-Saharan Africa and South East Asia is phenomenal. Increasing drug resistance and failure of vector control strategies have necessitated the search for a suitable vaccine that could be integrated into the extended program of immunization for countries in the endemic regions. Malaria vaccine development has seen a surge of activity in the last decade or so owing largely to the advances made in the fields of genetic engineering and biotechnology. This revolution has brought sweeping changes in the understanding of the biology of the parasite and has helped formulate newer more effective strategies to combat the disease. Latest developments in the field of malaria vaccine development will be discussed in this chapter.

Keywords. *Plasmodium*, Malaria, Vaccine Adjuvant, DNA Vaccine, Vaccinomics

1 **Introduction** 145

2 **Life Cycle of Malaria Parasite** 146

2.1 Parasite Development in the Vector 146
2.2 Parasite Development in the Host 147
2.3 The Disease 147

3 **Rationale for Development of Malaria Vaccines** 148

4 **Target Stages for Malaria Vaccine Development** 149

4.1 Pre-Erythrocytic Stage Vaccines 149
4.1.1 The Irradiated Sporozoite Model 149
4.1.2 Pre-Erythrocytic Stage Antigens 150
4.1.2.1 Sporozoite Surface Proteins 150
4.1.2.1.1 Circumsporozoite Protein (CS) 150
4.1.2.1.2 Thrombospondin Related Adhesive Protein (TRAP/SSP2) 152
4.1.2.2 Proteins Expressed During Liver Stages 153
4.1.3 Human and Non-Human Primate Trials of Vaccines Based on Pre-Erythrocytic Stage Antigens 154

© Springer-Verlag Berlin Heidelberg 2003

4.2 Erythrocytic Stage Vaccines 158
4.2.1 The Naturally Acquired Immunity 158
4.2.2 Erythrocytic Stage Antigens 158
4.2.3 Human and Non-Human Primate Trials of Vaccines Based on Erythrocytic Stage Antigens 161
4.3 Sexual Stage Vaccines 164
4.4 Vaccines against *P. vivax* Malaria 165

5 Alternate Technologies for Delivery of Vaccine Antigens 166

6 Search for Human Compatible Adjuvant for Malaria Vaccine . . 168

7 Malaria Vaccine Development in the Post-Genomics Era 169

8 Hurdles in the Way to Develop Effective Malaria Vaccines 170

9 Conclusions 173

10 References 173

Abbreviations

ABRA Acidic basic repeat antigen
AMA Apical membrane antigen
CD Complementarity determinant
CDRI Central Drug Research Institute
CFA Complete Freund's adjuvant
CpG Cytosine-phospho-guanine
CS Circumsporozoite protein
CTL Cytotoxic T lymphocyte
DBL Duffy antigen binding ligand
DBP/DABP Duffy antigen binding protein
DNA Deoxyribose nucleic acid
EBA Erythrocyte binding antigen
EGF Epidermal growth factor
ELISA Enzyme linked immunosorbant assay
EMP Erythrocyte membrane protein
GMP Good manufacturing practice
HbsAg Hepatitis B surface antigen
HEP Hepatocyte exported protein
HLA Human leukocyte antigen
HSPG Heparin sulfate proteoglycans
ICAM-1 Inter cellular adhesion molecule-1
IFA Immunoflourescence assay
IFN-γ Interferon gamma
IgG Immunoglobulin G

IL-1	Interleukin 1
ISI	Inhibition of sporozoite invasion
LSA	Liver stage antigen
MAb	Monoclonal antibody
MAP	Multiple antigen peptide
MF	Microfluidized
MHC	Major histocompatibility complex
MSP	Merozoite surface protein
MVA	Modified vaccinia Ankara
NO	Nitric oxide
NYVAC	New York strain of Vaccinia virus
ODN	Oligodeoxynucleotide
ORF	Open reading frame
PbCS	CS of *Plasmodium berghei*
PCR	Polymerase chain reaction
PfCS	CS of *Plasmodium falciparum*
PvCS	CS of *Plasmodium vivax*
PyCS	CS of *Plasmodium yoelii*
QS	*Quillaja saponaria*
RAP	Rhoptry associated protein
RESA	Ring infected erythrocyte surface antigen
SALSA	Sporozoite and liver stage antigen
SBAS	Smith Kline Beechem adjuvant system
SERA	Serine repeat antigen
STARP	Sporozoite threonine and asparagine rich protein
TNF-α	Tumor necrosis factor alpha
TRAP	Thrombospondin related adhesive protein
TT	Tetanus toxoid

Single letter amino acid codes

N	Asparagine
A	alanine
P	proline
V	valine
D	aspartate
M	methionine

1 Introduction

Malaria ranks with acute respiratory infections, measles and diarrheal diseases as one of the major causes of mortality and morbidity worldwide, affecting nearly 40% of the total population and accounting for about 3–5 million deaths and 300–500 million new cases, annually. Besides widespread drug resistance in the parasite and insecticide resistance in the anopheles mosquito vector, inadequate infrastructure measures for delivery of control measures, uncontrolled population growth and increased mobility of non-immune populations to endemic areas have also contributed to the alarming situation posed by malaria today. Under the present scenario, development of either new anti-malarial drugs or of an effective vaccine that can generate parasite specific protective responses, present attractive alternatives. Historically, vaccines have been one of the most cost effective and easily administered means of controlling infectious diseases, but no licensed vaccines exist yet for malaria. A multitude of difficulties has hampered the development of vaccines for malaria; complex multi-stage life cycle involving a vast array of receptor-ligand interactions during invasion, incomplete knowledge of the effector mechanisms involved in malaria immunity, and last but not the least, lack of suitable experimental models to test the efficacy of new vaccine strategies and extrapolatable in vitro correlates of protection have all contributed significantly towards the slower than required pace of malaria vaccine development.

Unlike many acute viral diseases, which produce lifelong resistance to re-infection, malaria causes immunity only after several years of recurring infection and illness. Immunity to malaria acquired in this way is only partially effective and results in mild, some times asymptomatic infections in spite of the persistence of parasites. This immunity is rather short-lived, unless reinforced by frequent reinfections. Another feature of the acquired immunity is that it is highly stage and strain specific, owing largely to the capability of the parasite to alter critical antigenic structures even within a single host over many generations. Therefore an individual would need to encounter the whole spectrum of the locally transmitted repertoire of strains. In this scenario, a child who dies of malaria despite previous parasitization would have encountered a new strain although eventually there must be some degree of cross protection against different stages of the parasite life cycle.

2 Life Cycle of Malaria Parasite

Understanding the life cycle of the parasite is fundamental to all efforts to develop vaccines, efforts that in the end focus on reducing the effects of the asexual erythrocytic stage, the only stage of the life cycle that causes disease in the mammalian host. Malaria parasites belong to the class *Sporozoa*, order *Coccidia*, family *Plasmodiidae*, genus *Plasmodium*. The species infecting humans are *P. falciparum, P. vivax, P. malariae, and P. ovale.* Of these, *P. vivax* has the widest geographical range and is prevalent in many temperate, subtropical and tropical zones. *P. falciparum* is prevalent in tropical and subtropical zones mostly. *P. malariae* and *P. ovale* have rather restricted distributions.

2.1 Parasite Development in the Vector

The *Plasmodium* species that cause malaria are transmitted exclusively by a few mosquito species of the genus *Anopheles*. Both the prevalence and the capacity of the disease to spread are closely related to the biology of the vector [1]. Even closely related species of anopheline mosquitoes differ markedly in their capacity to act as vectors [2]. For example, the high level of transmission in Africa is to a large extent the result of the preference for peridomestic feeding and superior longevity of the *A. gambiae* mosquito, over most other anopheline vectors. The mosquito is not simply a flying syringe that mechanically transfers parasites from one human to another. It is an interactive host within which the parasite undergoes a complex process of differentiation and growth. Thus, a female mosquito becomes infected with *Plasmodium* when she ingests male and female gametocyte-stage parasites in the course of taking a blood meal from an infected human. Fertilization subsequently occurs in the mosquito mid gut and a zygote is formed which differentiates further into an ookinete that penetrates the mid gut wall and develops into an oocyst. Thousands of sporozoites are produced within the oocyst and eventually make their way to the salivary glands where they mature. The transmission cycle is completed when an infected mosquito takes its next blood meal and introduces sporozoites into the human host where more developmental changes take place.

2.2 Parasite Development in the Host

The sporozoites inoculated by an infected mosquito during feeding enter into host's peripheral circulation where they travel until they reach the liver and penetrate hepatocytes within 30 min of inoculation. Inside the hepatocyte, the parasite undergoes cycles of asexual replication (schizogony) for 2–10 days depending on the species, producing exo-erythrocytic schizonts each containing as many as 30,000 merozoites. The exo-erythrocytic schizont ruptures and the merozoites are released into the bloodstream to invade red blood cells. The parasite then undergoes erythrocytic schizogony, producing as many as 36 merozoites per schizont. After 48–72 h, the infected red blood cell ruptures and releases merozoites that invade fresh red blood cells. All clinical manifestations of malaria occur only during erythrocytic stages. The sexual stage is initiated when some erythrocytic parasites differentiate into male and female gametocytes, which are then taken up by the mosquito during a blood meal, to begin another round of life cycle.

2.3 The Disease

The pathogenic process in malaria occurs only during the erythrocytic cycle of parasite development with fever and chills being the first set of alerting symptoms as a result of erythrocyte disrupture by merozoites. These malaria-associated symptoms are believed to be caused by a parasite-released toxin that induces macrophages to secrete tumor necrosis factor-α (TNF-α) and interleukin-1 (IL-1) [3]. The most severe forms of the disease, because of their fatal outcome, cerebral malaria and severe anemia, are specific for *P. falciparum* infection. Both types of complications are almost always accompanied by high parasitemia [4], but they differ with respect to certain epidemiological parameters. Thus, cerebral malaria is a more frequent complication in areas of low endemicity where severe anemia is seldom found [4, 5]. Conversely, the latter clinical picture is the predominant complication in areas of very high endemicity [4, 5]. With regard to the pathogenesis of cerebral malaria, much knowledge has been gained over past few years. The observation that cerebral anorexia was commonly associated with adhesion of infected erythrocyte to cerebral capillaries led to the identification of the involvement of adhesion molecules, such as CD36, ICAM-1, thrombospondin, and E-selectin in the host, to which the parasitized erythrocytes bind [6]. Interestingly, parasite induced TNF-α may play a crucial role in the upregulation of ICAM-1 expression in cerebral blood vessels, thus leading to cytoadherence of infected erythrocytes [7]. Severe anemia is one of the major causes of malaria-related death in children with a high endemicity in Africa [8] which could be aggravated by other socio-economic and political factors.

3 Rationale for Development of Malaria Vaccines

There are two major experimental findings that have provided much of the support for the concept of a malaria vaccine. One finding is that in rodent and primate models and human trials, immunization with radiation-attenuated sporozoites induces sterile protective immunity against malaria [9–11]. The second finding is that transfer of immunoglobulin G (IgG) from semi-immune individuals provides some degree of protection against malaria [12, 13]. Obviously neither strategy, immunization with irradiated sporozoites nor passive transfer of immune sera, is practical for application on a large scale, but both imply that development of immunity to malaria vaccination is feasible. As depicted in Fig. 1, a malaria vaccine could act by preventing the invasion of hepatocytes by sporozoites (antibodies) or by preventing development of the exo-erythrocytic stages within the hepatocytes (T cells, cytokines and perhaps antibodies) [14]. Such a vaccine targets the pre-erythrocytic stage and would therefore preclude both the development as well as the transmission of disease.

By interfering with the asexual erythrocytic stage of the life cycle (antibodies and cellular products such as cytokines) a vaccine would prevent or reduce morbidity or mortality by eliminating or reducing the parasite load [15]. Furthermore, parasite-derived material released from the infected erythrocyte at the time of rupture and release of merozoites is thought to induce the human host to produce cellular factors that contribute to or exacerbate the pathogenesis of malaria. By inducing neutralizing antibodies to these parasite products or by inhibiting cytoadherence of infected erythrocytes implicated in cerebral and other forms of severe malaria, an asexual erythrocytic vaccine would reduce morbidity and mortality. Another type of vaccine, a transmission-blocking vaccine, would not protect the individual, but would interfere with parasite development within the mosquito [16]. This would reduce morbidity and mortality, and perhaps eliminate malaria in a geographically isolated area. A reduced parasite load associated with decreased transmission could also allow enhanced efficacy of vaccines targeted at other stages [17].

Because distinct immune mechanisms operate against the different stages of the parasite's life cycle and because most antigens are not expressed at all stages, malaria vaccine development strategies have focused primarily on induction of

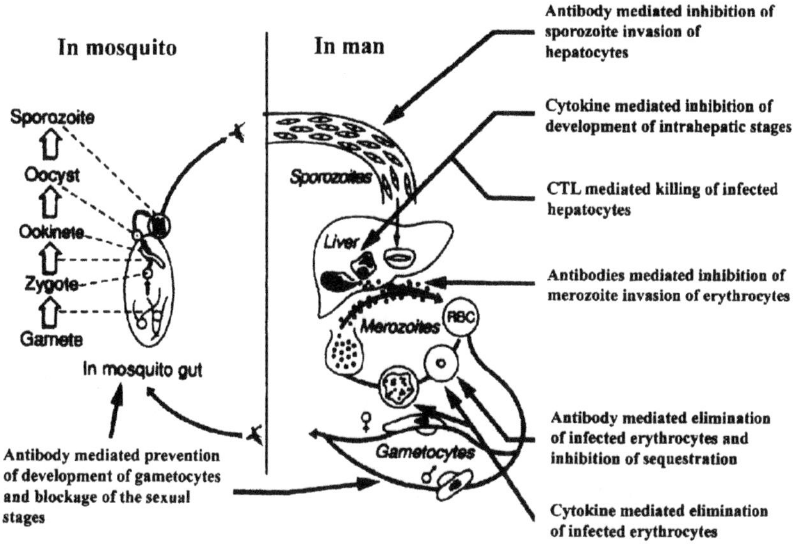

Fig. 1. Parasite life cycle and stages under immune attack. Intervention strategies aimed at different stages of the parasite development in reality can be anticipated to overlap broadly. Thus a pre-erythrocytic stage vaccine even if not 100% protective, could significantly reduce the number of merozoites being released into the bloodstream where immune responses elicited by erythrocytic stage vaccines could further reduce the parasite load and in turn help reduce the clinical severity of the disease. Responses against gametocyte stages could further reduce the transmission and rate of infection

immunity against single stages. However, given the improbability that a vaccine directed against a single antigen will be completely protective, a malaria vaccine will most likely need to produce different immune responses against multiple parasite antigens from several life cycle stages [18].

4 Target Stages for Malaria Vaccine Development

4.1 Pre-Erythrocytic Stage Vaccines

4.1.1 *The Irradiated Sporozoite Model*

Demonstration of development of stage and species specific protection in animals and humans immunized with radiation attenuated sporozoites (γ-sporozoites) led to attempts to identify the mechanisms of protection and the target antigens involved. There is evidence that both the neutralizing antibodies directed against the sporozoite surface proteins and the cellular effector mechanisms that destroy the liver stages are involved in induction of protective immunity. A major component of the humoral responses, elicited by immunization with γ-sporozoites was directed against the sporozoite surface proteins. These antibodies effectively inhibited parasite infectivity because viable sporozoites, when preincubated with these antibodies, were no longer infective in culture as well as when injected into a susceptible host [19, 20]. Moreover, passive transfer of anti-sporozoite antibodies protected naïve mice and macaques against viable sporozoite challenge [21, 22]. In addition to the evidence of a role for antibodies in the γ-sporozoite induced protection, there is a large body of data indicating that the protective immunity is primarily mediated by T cells which recognize malaria antigens presented on the surface, in the context of class I HLA proteins, by the infected hepatocytes. Adoptive transfer of spleen cells from immune mice protected naïve animals from sporozoite challenge [23]. Interestingly, in vivo depletion of $CD8^+$ T cells, but not $CD4^+$ cells, resulted in abrogation of the protection in the γ-sporozoite immunized mice, underlining the important role of $CD8^+$ T cells in protection from experimental malaria [24]. In addition to $CD8^+$ T cells, in some strains of mice, the roles of $CD4^+$ and γ/∂ T cells and the natural killer (NK) cells have been suggested [25]. More recent data, however, have suggested a critical role of cytokines as the leading effector mechanism operative in the γ-sporozoite induced protection. Thus, IFN-γ secreted from the antigen specific T lymphocytes is believed to stimulate infected hepatocytes to secrete nitric oxide (NO), which ultimately leads to arrest of parasite development [26, 27]. Although most of the data on the mechanistic aspects of protective immunity induced by γ-sporozoites have been raised using in-bred mice, one or more of these mechanisms is/are expected to work in the genetically diverse human population.

4.1.2 *Pre-Erythrocytic Stage Antigens*

4.1.2.1 *Sporozoite Surface Proteins*

4.1.2.1.1 Circumsporozoite Protein (CS)

Analysis of immune responses induced by γ-sporozoites identified two major sporozoite surface proteins, the circumsporozoite protein (CS) and the thrombospondin related adhesive protein (TRAP). Cloning of the CS genes revealed that the CS protein belonged to a family of 40–60 kDa proteins, contain hydrophobic signal and anchor sequences at the NH_2 and COOH termini, respectively, as well as two small regions-I and -II (R-I and R-II) that were conserved in all CS proteins [28]. The bulk of the protein, however, comprised a central region containing variable numbers of tandem amino acid repeat sequences, unique for each species. The observation that a conserved region of the CS protein, R-II, has a high degree of homology to a cell adhesion molecule, thrombospondin, and to other sulfatide binding proteins, suggested that the sporozoites might use R-II as a ligand for a sulfated cellular receptor on the hepatocyte surface [29]. Indeed,

R-II was shown to be the ligand that recognized heparan sulfate proteoglycans (HSPG) on the surface of hepatocytes [30]. Synthetic peptides as well as the antibodies against R-II specifically blocked this interaction. The peptides and the antibodies were also capable of inhibiting sporozoite invasion of cultured hepatocytes in vitro [31].

CS protein was the first antigen identified to be responsible for attenuated sporozoite-induced protective immunity. Antibodies to the central repeat regions of this protein were detected in human volunteers immunized with γ-sporozoites and their sera inhibited parasite infectivity [32]. Moreover, passive transfer of monoclonal antibodies (MAbs) specific to CS protein protected naïve mice from sporozoite challenge [22]. The vaccine potential of CS protein was experimentally demonstrated in studies where mice immunized with a repeat peptide of the *P. berghei* CS (PbCS) coupled to tetanus toxoid (TT) as carrier developed high titers of antibodies reactive with the native CS protein on *P. berghei* sporozoites and those with the highest levels of anti-repeat antibodies were protected [33]. The levels of anti-CS antibodies and protection were dependent on the method of conjugation and the type of carrier protein used [34–36]. Significantly however, we at ICGEB have shown that immunization of different strains of mice with synthetic peptides designed on the basis of *P. falciparum* CS (PfCS) protein elicited high titer antibodies even without a carrier protein. Serum obtained from mice immunized with $PfCS_{345-362}$ (P18), $PfCS_{331-362}$ (P32) and $PfCS_{331-390}$ (P60), inhibited sporozoite invasion of cultured hepatocytes in vitro and provided protection from a heterologous challenge of *P. berghei* sporozoites [31, 37]. This, along with the observation that sera from irradiated sporozoite immunized mice cross-reacted with P18 and P32, demonstrated that the structure of these peptides contained potent B-cell epitope(s). Moreover, approximately 30% of the serum samples collected from individuals living in *P. falciparum* endemic areas reacted with P18 and P32 in ELISA. Immunization with P32, which contains a single T cell and B-cell epitopes, resulted in an anamnestic antibody response without the use of a carrier protein.

As noted above, the vaccines designed to induce antibody responses in animals provided only partial protection from parasite challenge and it was realized that, along with humoral responses, cellular immunity also plays an important role in the protection provided by the irradiated sporozoites to animals and humans. The $CD8^+$ and $CD4^+$ CTLs specific for the epitopes within the CS protein were detected in the volunteers immunized with attenuated sporozoites [11]. Moreover, the presence of CS specific CTLs in the immune subjects living in the malaria endemic regions further supports the inclusion of CS in the malaria vaccine. A number of CTL epitopes were identified that sensitized target cells for lysis by $CD8^+$ T cells derived from the volunteers immunized with irradiated sporozoites [38]. Adoptive transfer of CTLs also protected naïve animals from viable sporozoite challenge [39]. However, the synthetic peptides were poor immunogens for induction of cellular immunity in humans and animals and alternate strategies for vaccine delivery were devised to facilitate induction of CTL responses. A high degree of vaccine-induced protection mediated by CS-specific $CD8^+$ CTLs was obtained in a mouse malaria model by immunization with recombinant influenza virus expressing a *P. yoelii* CS (PyCS) epitope, followed by a second immunizing dose consisting of a highly attenuated vaccinia virus expressing the same epitope [40, 41]. Oral immunization of mice with a recombinant *Salmonella typhimurium* expressing *P. berghei* CS (PbCS) induces protein-specific CTL and provides $CD8^+$ T cell-dependent protection of ~75% of mice against sporozoite challenge [42, 43]. Furthermore, immunization with a recombinant attenuated vaccinia virus expressing the PbCS-induced CS protein-specific CTL, anti-CS protein repeat antibodies and $CD8^+$ T cell-dependent protection against sporozoite challenge [44]. However, this protection was strain and species specific, since immunization of mice with recombinant *S. typhimurium*, vaccinia or pseudo-rabies virus [45] expressing PyCS protein failed to induce protective CTL responses against sporozoite challenge. Immunization with irradiated mastocytoma cells, P815, transfected with the PyCS gene induced protection in 50% of the challenged mice [46]. Synthetic peptides with lipid tails, lipopeptides or peptides without lipid tail, but when co-immunized with a universal T helper cell epitope, also elicited specific CTL responses and protected mice from sporozoite challenge in *P. yoelii* as well as *P. berghei* models.

4.1.2.1.2
Thrombospondin Related Adhesive Protein (TRAP/SSP2)

TRAP/SSP2, the second sporozoite surface protein, was identified by gene cloning using a blood stage genomic library [47]. An MAb derived from a *P. yoelii* sporozoite immunized mouse identified the homologous protein in the rodent malaria [48]. Later the *trap* gene was described from other murine, simian and human malaria species as well [49, 50]. TRAP has a multidomain organization, comprising the hydrophobic signal and anchor sequences at the NH_2 and COOH terminus, respectively, an adhesive A-domain that is homologous to similar domains found in the integrins. The protein shares a thrombospondin like sulfatide binding domain, R-II with the CS protein. The TRAP of human, simian and rodent malaria differ in the repeat regions in terms of the sequence as well as the number of repeats. Various studies have indicated the role of TRAP in sporozoite invasion of host hepatocyte. Knockout studies performed in *P. berghei* showed that TRAP was essential for gliding motility of the salivary gland sporozoite and invasion of hepatocytes [51]. Recombinant *P. falciparum* TRAP (PfTRAP) bound specifically to sulfated glycoconjugates and to hepatoma cells *in vitro* primarily in an R-II dependent manner, as those constructs lacking the R-II motif did not show this binding property. Binding of TRAP to hepatocyte microvilli was abolished by heparitinase treatment of the liver sections indicating that TRAP and CS used HSPGs as a common receptor on the hepatocyte surface [52, 53]. Although first described from the blood stages, the expression of TRAP during the erythrocytic stages of the parasite has been controversial. We have found that immunization of mice and rabbits with an 18-mer peptide of PfTRAP corresponding to aa 249–266, induced antibodies that reacted with blood stage parasites in IFA and identified a ~78 kDa protein in the parasite lysates [54]. Further experimental data generated by us have suggested the presence of a 1.6 kb TRAP-specific transcript in the trophozoite stage (Chauhan et al., unpublished data). This issue is still under intense investigation. If the presence of TRAP is finally confirmed during the blood stages it will be even more attractive target for malaria vaccine development.

Vaccine potential of TRAP was indicated by the observation that sporozoite-immunized volunteers developed antibodies reactive with recombinant PfTRAP [48]. T cells from the volunteers showed proliferative activity when stimulated with recombinant protein [48]. CTLs specific for epitopes derived from PfTRAP were also found in the circulation of sporozoite-immunized volunteers [55] and in patients naturally exposed to malaria in the endemic regions of Africa [56]. Although antibodies against *P. yoelii* TRAP (PyTRAP) or PfTRAP do not dramatically inhibit sporozoite invasion and liver stage development [48, 57], the finding that adoptive transfer of PyTRAP-specific CTL clones protected naïve recipient mice against sporozoite challenge [46] suggested that a CTL response to TRAP is an important component of irradiated sporozoite-induced immunity. Immunization with mastocytoma cells, expressing PyTRAP also protected mice against viable sporozoite challenge [46, 58]. Moreover, there was an additive effect on protection when mice were immunized with cells that expressed both PyCS and PyTRAP [58]. This protection was abrogated upon treatment of the mice with anti-$CD8^+$ antibodies suggesting that $CD8^+$ T cells were mediating the protection. Similarly, antigen-specific $CD8^+$ CTLs and IFN-γ mediated protection were obtained in mice following immunization with DNA vaccines for priming and recombinant MVA for boosting [59].

4.1.2.2
Proteins Expressed During Liver Stages

Several proteins are expressed during the liver stage development of the parasite. The *l*iver *s*tage *a*ntigen-1 (LSA-1), is a 200 kDa protein expressed within the parasitophorous vacuole throughout liver stage schizogony [60]. CTL and antibodies specific for epitopes in LSA-1 were identified in γ-sporozoite-immunized volunteers and naturally exposed individuals of Africa [61, 62]. In Gambia, the presence of the class I HLA allele, Bw53, is associated with resistance to severe malaria and a conserved CTL epitope restricted by Bw53 was detected in naturally exposed individuals [61, 63]. Antibody and T cell responses were also found in naturally exposed individuals in India directed against the relatively invariant region of PfLSA-1 [64].

The *ex*ported *p*rotein, Exp-1, another protein expressed during liver stage infection, is present on the parasitophorous vacuole membrane of infected hepatocytes and in the cytoplasm of host cells [65]. A *P. yoelii* homologue, PyHEP-17, was also identified and shown to have a similar expression pattern. γ-Sporozoite-immunized volunteers produced CTL specific for several epitopes derived from PfEXP-1 [62]. A monoclonal antibody specific for PyHEP-17 was shown to eliminate *P. yoelii* infected hepatocytes from culture [66].

A growing number of additional antigens has been described in the liver stages, for which there is as yet no relation with the role in γ-sporozoite-induced immunity. Screening of *P. falciparum* expression libraries with hyperimmune sera identified many proteins, for example, the *s*porozoite *t*hreonine and *a*sparagine *r*ich *p*rotein (STARP, 78 kDa), the *s*porozoite *a*nd *l*iver *s*tage *a*ntigen

(SALSA, 70 kDa), the *l*iver *s*tage *a*ntigen-3 (LSA-3, 205 kDa) are expressed on the sporozoite surface and the liver stage [67–69].

4.1.3
Human and Non-Human Primate Trials of Vaccines Based on Pre-Erythrocytic Stage Antigens (Table 1)

The successful immunization of non-human primates and humans with the radiation-attenuated sporozoites led to a search for effective sub-unit vaccines for malaria. A number of sub-unit vaccine prototypes was designed to induce humoral and cellular immune responses based on the antigens from sporozoite and liver stages of the parasite (Table 1). The first malaria vaccine was a chemically synthesized peptide based on the repeat region of the CS protein, $(NANP)_3$ (amino acid codes in single letter format), and was conjugated to the tetanus toxoid carrier protein [19, 70]. Phase I and II trials of $(NANP)_3$-TT demonstrated safety and immunogenecity when administered adsorbed to alum as adjuvant [71, 72], in a dose-dependent manner. Of the 35 volunteers immunized with this vaccine, three with highest antibody titers against sporozoites and (NANP) were challenged. One out of three volunteers was totally protected, while the remaining two vaccinees demonstrated partial protection as measured by the delay in the prepatent period [71]. The suboptimal response to the peptide-protein conjugate vaccine was considered to be due, in part, to low epitope density and possibly epitope suppression as a result of the use of TT as a carrier protein [73]. A more efficacious vaccine would be obtained by the inclusion of parasite-derived T cell epitopes to elicit anamnestic responses in malaria-primed individuals living in endemic areas. A multiple antigen peptide (MAP) construct containing a T cell epitope from the minor repeat region, T1, and the $(NANP)_3$ B epitope (T1B4) was shown to be highly immunogenic and elicited antibody responses in three species of *Aotus* monkeys [74] with Freund's adjuvant as well as other formulations acceptable for human use such as alum or alum plus QS21. Phase I/IIa trials have been initiated with T1B4 MAP vaccine in an effort to determine if the broad immunogenecity and high levels of antibodies observed in primates can also be elicited in humans. Another MAP containing universal T cell epitopes in combination with a *P. vivax* repeat epitope elicited protective antibody responses in primates [75, 76].

Several trials were done with various recombinant forms of the CS protein with and without extraneous T helper cell epitopes in monkeys as well as humans. Immunization of six volunteers with a recombinant protein consisting of 32 copies of PfCS repeats [MDP$(NANP)_{15}$NVDP$(NANP)_{15}$NVDP] formulated with alum elicited protective immune responses in only one of the six [77]. In an immunogenicity trial with a similar construct formulated in monophosphoryl A and squalane, eight out of 16 volunteers developed high titer antibodies [78]. In another study with this vaccine two out of 11 volunteers were protected and two others had delayed onset of parasitemia [79]. Surprisingly however, there was no correlation between protection and anti-sporozoite titer, and inhibition of sporozoite invasion (ISI) results. Similar results were obtained with another recombinant vaccine based on the repeat sequences of PfCS conjugated to the detoxified *Pseudomonas arugenosa*

Table 1. Human and non-human primate trials of vaccines based on the pre-erythrocytic stage candidate antigens

Antigen	Form of vaccine	Description	Host	Protection	Ref.
CSP	Synthetic peptide	MAC containing repeats and P2/P30 of tt	Monkey	Partial	[83]
		Repeat region peptide conjugated to diptheria toxoid	Monkey	None	[217]
		MAP containing T cell epitopes from CS along with P30 of tt	Monkey	ND	[76]
		Peptides panning the non-repeat regions from N- and C-terminus	Monkey	ND	[218]
		Padre 45 with CpG oligonucleotides	Monkey	ND	[214]
		MAP (T1B)4	Monkey	ND	[74]
		(NANP)3 conjugated to tt	Human	Partial	[71, 219]
		C-terminal corresponding to aa 282–383	Human	ND	[220]
		Polyoxime synthetic MAP containing T helper and B cell epitopes	Human	ND	[221, 222]
		Peptides spanning the non-repeat regions from the N- and C-terminus	Human	ND	[223]
	Recombinant	Repeat region and the flanking sequences produced in *E. coli* and yeast	Monkey	Partial	[81, 82]
		Multiple antigen construct	Monkey	Partial	[75,83]
		Repeat region fused to 32 residues from bacterial tetracycline resistance protein translated out of frame	Monkey	ND	[224]
		32 copies of repeat sequence fused to 32 residues of tetracycline resistance protein translated out of frame	Human	Partial	[77]
		Repeat region covalently coupled to detoxified *Pseudomonas* exotoxin A	Human	Partial	[80]
		Repeat region fused with 81 residues of the NS protein of Influenza A virus	Human	Partial	[79]
		RTS,S: aa 210–398 fused with the HbsAg	Human	Partial	[84, 87, 88, 211, 212]

Table 1 (continued)

Antigen	Form of vaccine	Description	Host	Protection	Ref.
		Repeat region covalently coupled to detoxified *Pseudomonas* exotoxin A	Human	None	[225, 226]
		Repeatless protein containing only N- and C-terminus	Human	None	[215]
		Almost complete protein produced in yeast Complete protein produced in baculovirus	Human	ND	[227, 228]
		Repeat sequence fused to HbsAg	Human	ND	[229]
		Repeat region fused with 81 residues of the NS protein of Influenza A virus	Human	ND	[78]
		32 copies of repeat sequence fused to 32 residues of tetracycline resistance protein translated out of frame	Human	ND	[230]
		RTS,S: aa 210–398 fused with HbsAg	Human	ND	[210]
	DNA	DNA vaccine plasmid	Monkey Monkey Human	Partial ND ND	[200] [194] [193]
TRAP	Recombinant	Ectodomain produced in *E. coli*	Monkey	Partial	[a]
	DNA	DNA vaccine plasmid	Monkey Monkey	Partial None	[200] [b]
LSAs	Synthetic peptide	Peptides with and without lipid tails	Monkey	ND	[90]
		Lipopeptides and recombinant protein	Chimpanzee	Partial ND	[91] [69]
	DNA	DNA vaccine plasmid	Monkey	ND	[194]

[a] Chauhan et al. unpublished data.
[b] Bhardwaj et al. unpublished data.

toxin A [80]. Several other trials in humans with recombinant constructs failed to induce any protective immune responses (Table 1).

Studies in the squirrel monkeys with recombinant vaccines based on the CS protein of *P. vivax* were carried out [81, 82]. These vaccines elicited protective immune responses against sporozoite challenge, however, once again there was no correlation between the observed protection and humoral or cellular immune responses in these trials. In other trials using *Saimiri* monkeys, protective immune responses were elicited by immunization with multiple antigen constructs based on the *P. vivax* CS (PvCS) protein repeat region and a universal T helper epitope of tetanus toxin formulated with different adjuvants [75, 83].

A promising pre-erythrocytic stage malaria vaccine, called RTS, S, described first in 1995 is currently scheduled to undergo large scale human trials in Africa. The vaccine consists of sequences from the CS protein of *P. falciparum* strain 3D7 (aa 210–398) fused to 226 amino acids corresponding to HBsAg (adw serotype) [84]. This region of CS protein contains the central repeat region as well as non-repeat B cell epitopes, two potential T helper epitopes, aa 326–343 (Th2R) [85] and aa 361–380 (Th3R) [85], and a CTL epitope, aa 368–390 [86]. The vaccine was produced in yeast and formulated in three different adjuvants; RTS, S with alum and monophosphoryl lipid A (SBAS4), RTS, S in an oil-in-water emulsion (SBAS3), and RTS, S in this emulsion plus monophosphoryl lipid A and QS21 (SBAS2). In a limited study, six out of seven subjects who received RTS, S with only one of the formulations, SBAS2 were protected from a challenge with *P. falciparum* sporozoites. Protection was dependent on and correlated well with the levels of the anti-CS repeat antibodies [87]. In a test of long-term efficacy 6 months after the first challenge, the protected subjects were re-challenged; 1 out of 5 subjects was completely protected and the remaining showed a significant delay in the pre-patent period. Surprisingly though, this time protection did not correlate with the levels of the anti-CS repeat antibodies [88]. Subsequently, it was reported that RTS, S/SBAS2 induced $CD4^+$ dependent IFN-γ secretion as well as the lymphoproliferative responses and antibodies in the immunized subjects, highlighting a possible immune mechanism of protection. Importantly though, CS-specific $CD8^+$ CTLs were not detected [89]. In a recently conducted dose-response phase I/IIa trial of this formulation, an overall efficacy of 41% was reported. Phase III trials with RTS, S are currently under progress, the results of which are expected to be available in 2–3 years time.

There have not been any trials in human with vaccines based on other pre-erythrocytic stage antigens. In a recently concluded trial conducted by the Malaria Research Group at ICGEB in collaboration with CDRI, Lucknow immunization with recombinant *P. cynomolgi* TRAP induced protective immunity in rhesus monkeys and one out of eight animals was completely protected whereas four others had significantly delayed onset of parasitemia. The vaccine, formulated with Freund's adjuvant, induced both humoral and CTL responses against the protein which correlated well with the protection (Chauhan et al. unpublished results). Similarly, immunization of monkeys and chimpanzees with lipopeptides and recombinant proteins corresponding to different regions of LSA-3 was shown to provide protection [90, 91]. Recently, DNA vaccines have provided an exciting new tool for the induction of effective CTLs against the encoded antigens and are discussed in detail later.

Despite enormous efforts no pre-erythrocytic stage vaccine is available yet that can effectively prevent either sporozoite invasion of hepatocytes or the development of intrahepatocytic stages of the parasite. The current research efforts are directed towards identification of more and/or better candidate antigens involved in eliciting protective immune responses in the γ-sporozoite model. The completion of the sequencing of the malaria genome is expected to give a major boost to the search for such candidate antigens.

4.2
Erythrocytic Stage Vaccines

4.2.1
The Naturally Acquired Immunity

Naturally acquired immunity to malaria has often been underestimated. It is highly effective in reducing the severity of disease, morbidity and mortality, even though it rarely, if ever, produces sterile immunity. In areas of intense, year-round malaria transmission, death from malaria is much reduced in older children after the age of five and very rare once the child is past the age of ten years. In these areas, during the first one to three years of life the incidence of fatal malaria is at its peak and starts to decline rapidly thereafter and by the age of adolescence the average density of parasitemia also starts to decline. This perhaps suggests that there are two very different mechanisms of protection at work, anti-disease immunity, which develops rapidly and anti-parasite immunity, which is acquired rather more gradually. The erythrocytic stages of the parasite are responsible for the pathology induced by malaria and present a particular challenge for vaccine development. Blood-stage parasites are surrounded by several membrane systems and develop within erythrocytes, which lack class I and class II MHC antigens and thus remain 'invisible' to the immune system of the host. However, blood-stage antigens have the potential to induce protection in primates against *P. falciparum* challenge [92], indicating the possibility to develop effective blood-stage vaccines. With regard to individuals living in endemic regions, vaccine-induced immunity does not necessarily have to be sterile. Controlled malaria infections could rather serve the purpose of repeatedly boosting the immunity.

4.2.2
Erythrocytic Stage Antigens

A number of different stage-specific antigens are considered as potential vaccine candidates, the ***m****erozoite* ***s****urface* ***p****rotein-1* (MSP-1) antigen being the most studied of these [93, 94]. MSP-1 is initially synthesized during schizogony as a high molecular weight 195 kDa precursor protein and is expressed on the surface of the merozoites [95]. The complete amino acid sequence of the polymorphic MSP-1 molecule is now documented for an impressively large number of geographical strains of *P. falciparum* and clearly demonstrates blocks of variable, constant and dimorphic regions [96, 97]. The precursor protein undergoes proteolytic processing at the maturation of schizonts to give 83 kDa, 28–30 kDa, 38 kDa and 42 kDa fragments [98]. The membrane bound 42 kDa fragment undergoes another phase of processing to produce 33 kDa and 19 kDa fragments. During the invasion of erythrocytes, the 19 kDa fragment is retained on the surface of the merozoite and the same is also found present in the newly invaded erythrocyte [99] leading to the suggestion that the 19 kDa protein may be playing critical role in merozoites invasion of erythrocytes [99].

The data suggesting a role for MSP-1 in inducing protective immunity are extensive. A number of sero-epidemiological studies has suggested an association between humoral, and in some cases cellular, responses to the C-terminal MSP-1_{19} or MSP-1_{42} fragments of MSP-1 [100–104]. Recent studies have also indicated that mainly the antibodies directed against the 19-kDa C-terminal fragment are responsible for protective immunity [105]. Affinity purified native MSP-1, induced protection against *P. falciparum* and *P. yoelii* parasite challenge in primate and rodent model systems, respectively [92, 98]. Similarly, MSP-1 isolated from cultured parasites protected *Aotus* and *Saimiri* monkeys against challenge with *P. falciparum* [106, 107]. The work done by us and others with constructs based on B- and T-cell epitopes from conserved MSP-1 sequences, has shown that synthetic hybrid multiple antigen peptides based on *P. falciparum* sequences are highly immunogenic and partially protect mice against a heterologous challenge [108, 109]. Studies in the murine model have also provided evidence that both cellular and humoral responses generated against MSP-1 may have crucial role to play vis a vis protection [110–112].

Other *P. falciparum* antigens identified on the surface of merozoites and shown to be targets of parasite-neutralizing immune responses are MSP-2, MSP-3, AMA-1, EBA-175 and SERA. The merozoite surface protein-2 (MSP-2; 45–55 kDa) is anchored to the merozoite surface membrane by glycosyl phosphatidyl inositol (GPI) [113] and its involvement in protective immune responses was indicated by the merozoite invasion inhibitory effect of MAbs to the antigen [113, 114]. While merozoite surface protein-3 (MSP-3; 50 kDa) is found associated with the merozoite surface only after schizont rupture, it exists in the parasitophorous vacuole when in the infected erythrocyte [115]. The antigen was identified as a major target for antibody-dependent cell immunity (ADCI) in vitro, as demonstrated with IgG from malaria immune adults, acting in cooperation with monocytes [115].

The *a*pical *m*embrane *a*ntigen-1 (AMA-1) is a 66–83 kDa integral membrane protein located in the rhoptries of the apical complex and on the surface of merozoites [96, 116]. There is relatively indirect evidence that AMA-1 is a target of protective immunity. Although there is high prevalence of anti-AMA-1 antibodies in West Africans exposed to malaria, no correlation was found with parasitemia or clinical malaria [117].The *e*rythrocyte *b*inding *a*ntigen-175 (EBA-175) of *P. falciparum* (PfEBA-175) and the *d*uffy *a*ntigen *b*inding *p*rotein (DBP) of *P. vivax* and *P. knowlesi* (Pv/PkDBP) also are typical integral membrane proteins and are present in the micronemes of merozoites. The extracellular domains of these proteins can be divided into six regions based on sequence homology, out of which region II was identified as the critical binding domain in the in vitro erythrocyte binding assays [118]. Antibodies against recombinant forms of PfEBA-175 R-II were shown to inhibit merozoite invasion of erythrocytes in vitro [119]. Antibodies raised in rabbits against *E. coli* expressed and refolded PvDBPRII were shown to inhibit erythrocyte binding to COS cells bearing PvDBPRII on the surface [120]. Furthermore, sera from New Guinean adults or *P. vivax* infected North Americans contained antibodies reactive with recombinant PvDBP [121]. It has been reported, at least in the case of PfEBA-175, that some parasites can invade through a sialic acid-independent pathway [122]. Thus immunization with EBA-175 may be circumvented if this molecule is not absolutely required for invasion, suggesting that probably EBA-175 alone may not be suitable as an anti-*P. falci-*

parum vaccine but, nonetheless, there is clear rationale for receptor blocking vaccines designed to inhibit receptor-ligand interactions of this protein.

The *s*erine *r*epeat *a*ntigen (SERA) is a 120 kDa soluble protein localized in the parasitophorous vacuole of late erythrocytic stage parasites. There is only indirect evidence about the role of SERA in naturally acquired immunity. Antibodies against SERA were shown to agglutinate merozoites and prevent invasion and growth in vitro [123, 124]. No seroepidemiological evidence is available to indicate a role of SERA in the acquired immunity. Nonetheless a number of vaccine studies has been conducted in animals and the data support the inclusion of SERA in a multicomponent malaria vaccine.

The *r*hoptry-*a*ssociated *p*roteins-1 and -2 (RAP-1 and-2) are localized to the rhoptries and are tightly associated in the form of a complex [125]. Antibodies against this protein are reported to inhibit the merozoite invasion of erythrocytes. More evidence of involvement of RAP-1/2 in naturally acquired immunity was provided by an immunological survey in Papua New Guinea where over 80% adults were found to have antibodies against this complex [126].

The *a*cidic *b*asic *r*epeat *a*ntigen (ABRA) is another important merozoite surface protein and is found in the immune clusters of merozoites upon release from erythrocytes. The protein, as its name suggests, is rich in acidic and basic amino acid residues. Moreover, the sequence of this protein is highly conserved among field isolates. We have been interested in characterization of this unusually highly conserved protein and to develop it as a potential vaccine candidate. We showed that recombinant fragments of ABRA possess chymotrypsin-like serine protease activity [126a] and that polyclonal antibodies raised against synthetic peptides derived from the most hydrophilic regions of ABRA and the recombinant fragments based on different parts of the protein inhibited merozoite invasion of erythrocytes in vitro in 50–90% [126b, c]. Our further studies have shown that the recombinant N-terminal fragment of ABRA, which contains all five of the cysteine residues, interacts with the erythrocyte surface protein, band 3 (Chauhan et al, unpublished results). Similarly, the native ABRA protein isolated from *P. falciparum* cultures also shows binding with band 3 protein. Recently, ABRA orthologues have been reported from *P. vivax* and the simian malaria parasites, *P. cynomolgi* and *P. knowlesi* and were named as MSP-9, placing this protein in a growing family of merozoite surface proteins [126d]. Most interestingly, proteins from all these species have positionally conserved cysteines underlining an important but as yet unidentified role for this antigen. Further unraveling of the biochemical function of ABRA will warrant its induction into the multicomponent malaria vaccine.

Parasitization of the erythrocyte leads to changes of the cell surface, involving both modified host structures and the expression of *P. falciparum* *e*rythrocyte *m*embrane *p*rotein-1 (PfEMP-1) (200–400 kDa) [127]. PfEMP-1 is the molecule primarily involved in the binding to vascular endothelium, and would thus seem an ideal target for the immune attack. Unfortunately, the parasite has developed an extraordinary capacity for phenotypic variation of PfEMP-1, presumably as a means of immune evasion [128]. The *r*ing infected *e*rythrocyte *s*urface *a*ntigen (RESA) is a 155 kDa protein localized in the dense granules of the apical organeller complex in merozoites and is translocated to the cytoplasmic side of the erythrocyte membrane after invasion [129]. Antibodies against RESA inhibited invasion in in vitro assays [130] and protected monkeys in passive transfer experiments [131]. Moreover, a number of longitudinal seroepidemiological studies suggested a correlation between anti-RESA antibodies and reduced parasitemia or clinical episodes [132–135].

4.2.3
Human and Non-Human Primate Trials of Vaccines Based on Erythrocytic Stage Antigens (Table 2)

A number of animal immunization studies have been carried out with native and recombinant MSP-$1_{19/42}$ antigen with mixed results. Using an affinity-purified preparation of MSP-1 polypeptides, *Aotus* monkeys were protected from malaria with a formulation containing Freund's adjuvant, whereas the same polypeptides did not protect animals in an alternate adjuvant formulation [92, 136]. Similar results were obtained with recombinant constructs in monkeys [106, 137, 138]. Recombinant C-terminal fragments of MSP-1 expressed in baculovirus (PfMSP-1_{42}) or yeast (PfMSP-1_{19}) were shown to partially protect *Aotus* monkeys [139–143]. Similarly, rhesus monkeys were protected against *P. cynomolgi* infection by immunization with *E. coli* produced recombinant PcMSP-1_{19} [144]. Partial protection was also obtained in monkeys immunized with recombinant fragments of MSP-1, fused with universal T helper epitopes [145–147], but a similar construct failed to protect monkeys in another study [148]. A similar construct was tested in a phase I trial in humans to study immunogenicity and safety [149]. In various animal and human studies, protection obtained with recombinant MSP-1 fragments, however, was found to be adjuvant-dependent. Vaccines formulated in complete Freund's adjuvant only induced protective immunity and animals immunized with same antigen but formulated with six different adjuvants suitable for human use were not protected [142]. These studies thus suggest that C-terminal fragments of MSP-1 are very promising targets for human vaccination, provided that an adequately effective and tolerable adjuvant is available for human use. The C-terminal 19 kDa fragment is rich in cysteine residues that are highly conserved, and contains epidermal growth factor (EGF)-like domains and antibodies directed against this fragment in both *P. yoelii* and *P. falciparum* can inhibit the invasion of merozoites [150]. The inhibitory antibodies are directed to conformational epitopes present in one or both the EGF-like domains [104, 151–153]. The antigenicity of the EGF-like domains, however, is conformation dependent since reduction of disulfide bonds abolishes immunogenicity [154]. It has also been shown that the integrity of both the domains is vital to protection [155, 156].

Several studies have explored protective efficacy of other erythrocytic stage antigens in primates and humans. Partial protection was observed in rhesus monkeys immunized with native *P. knowlesi* AMA-1 formulated in saponin adjuvant [157]. Similarly, 4 of 5 *Saimiri* monkeys were partially protected against the *P. fragile* challenge after immunization with baculovirus expressed PfAMA-1 formulated in Montanide ISA 720 [158], and the degree of protection appeared to correlate with the antibody response following immunization. A recombinant

Table 2. Human and non-human primate trials of vaccines based on the erythrocytic stage candidate antigens

Antigen	Form of vaccine	Description	Host	Protection	Ref.
SPf66	Synthetic peptide		Monkey	Partial	[231–233]
			Human	Partial	[173–176, 234–238]
			Human	None	[177–180]
			Human	ND	[239–241]
MSP-1	Native	Parasite derived affinity purified antigen	Monkey	Partial	[92, 136]
	Synthetic	Corresponding to the N-terminus	Monkey	Partial	[244]
	Recombinant	From N- and C-terminus	Monkey	Partial	[137]
		Corresponding to aa 147–321, 147–321 and 1060–1195	Monkey	Partial	[138]
		C terminal fragment of MSP-1 fused with universal T epitope CS.T3	Monkey	Partial	[145, 146]
		42- and 19-kDa C terminal fragments produced in *E. coli* and yeast, respectively	Monkey	Partial	[140]
		42-kDa C-terminal fragment produced in baculovirus	Monkey	Partial	[139]
		42- and 19-kDa C-terminal fragments produced in baculovirus	Monkey	Partial	[144]
		19-kDa C terminal fragment fused with P2 and P30 produced in yeast	Monkey	Partial	[147]
		19-kDa C terminal fragment fused with P2 and P30	Monkey	Partial	[143, 142]
		42- and 19-kDa C-terminal fragments fused with P2 and P30 produced in baculovirus and yeast, respectively	Monkey	Partial	[141]
		Recombinant vaccinia virus expressing MSP-1	Monkey	None	[167]
		19-kDa fragment produced in *E. coli*	Monkey	None	[243]
		19-kDa fragment fused with P2 and P30 produced in yeast	Monkey	None	[148]
		C-terminal fragment fused with universal T helper epitope CS.T3	Human	None	[170]
			Human	ND	[149, 169]
		Fusion protein comprising of MSP-1, SERA and HRP-II	Monkey	Partial	[242]

Table 2 (continued)

Antigen	Form of vaccine	Description	Host	Protection	Ref.
	DNA	DNA vaccine plasmid	Monkey Monkey	Partial None	[200] [a]
RESA	Recombinant	Recombinant protein produced in *E. coli*	Monkey	Partial	[165, 245]
		Recombinant vaccinia virus	Monkey	None	[167]
		Recombinant protein	Human Human	None ND	[170] [169, 246]
AMA-1	Native	Affinity purified	Monkey	Partial	[157]
	Recombinant	Recombinant protein produced in baculovirus	Monkey	Partial	[158]
		Recombinant protein produced in yeast	Monkey	None	[159]
		Recombinant vaccinia virus	Monkey	None	[167]
	DNA	DNA vaccine plasmid	Monkey Monkey	Parial None	[200] [a]
MSP-2	Recombinant	Fusion proteins comprising of fragments of MSP-2, SERA and HRP-II	Monkey	Partial	[242]
		Recombinant vaccinia virus	Monkey	None	[167]
		Recombinant protein	Human Human	None ND	[170] [169]
SERA	Recombinant	Recombinant fragments corresponding to aa 24–285 and aa 24–506	Monkey	Partial	[162–164]
RAP1/2	Native	Gel purified parasite protein	Monkey	Partial	[247]
	Recombinant	Recombinant protein	Monkey	Partial	[166]

[a] Bhardwaj et al. unpublished data.

P. vivax AMA-1 protein produced in *E. coli*, however, failed to protect rhesus monkeys from a heterologous challenge with *P. cynomolgi* parasites [159]. The protection conferred by AMA-1 was shown to be dependent on the formation of correct disulfide bonds in the ectodomain of *E. coli*-expressed recombinant *P. chabaudi* AMA-1 [160] since reduction and alkylation of the refolded recombinant protein abrogated the protection in mice, underlining once again the need to produce recombinant material in a properly refolded conformation for the vaccine to be effective.

Several studies have assessed the vaccine potential of native and recombinant SERA in animals. Immunization of *Saimiri* monkeys with native SERA formulated with either CFA or alum and of *Aotus* monkeys with recombinant proteins based on SERA using either CFA or an oil-in-water emulsion provided partial protection from parasite challenge [161–164]. Recombinant vaccines based on

RESA, and RAP-1/2 antigens also provided partial protection in monkeys [165, 166]. However, in later studies RESA did not produce protective immunity in monkeys [167, 168]. Human trials in Papua New Guinea with three recombinant antigens, MSP-1, MSP-2 and RESA, formulated in Montanide ISA 720 were recently concluded. The volunteers developed low antibody responses and, upon challenge with rather low dose inoculum, developed patent parasitemia. In this study there was no significant difference in the immunized and control subjects in the initial parasite growth rates [169, 170] indicating the lack of efficacy of the vaccines and/or their formulations with Montanide adjuvant.

Several animal and human trials were carried out with a synthetic peptide construct named SPf66 beginning in late 1980s (Table 2). This vaccine was based on sequences derived from three erythrocytic stage antigens that were selected from a battery of peptides tested in *Aotus* monkeys for their ability to confer partial or significant protection from the parasite challenge. Three peptides from 35, 55 and 83 kDa proteins were linked together by the repeat region peptide PNANP of the PfCS protein [171]. The vaccine has proved to be safe and was well tolerated without any severe side effects in many animal and human trials including young children and infants. Several field trials in Latin America were done with this vaccine, with an efficacy of 38.8–60.2% [172–174]. Surprisingly however, there was no correlation between the antibody titers against SPf66 and the incidence of malaria episodes. Moreover, SPf66 was found to be a poor immunogen that provided little or no protection against *P. falciparum* infection in further trials conducted at other sites [175–180]. Although safe and variably immunogenic, the mode of action, if any, of the SPf66 vaccine is still largely unknown and must be established in order to facilitate any improvements in the design of the vaccine. Despite its shortcomings, SPf66 was the first chemically synthesized molecule based on multiple antigenic determinants that was found to be safe and immunogenic in humans. This must be considered a landmark achievement in the field of synthetic subunit vaccine development in general and particularly in malaria vaccine against the blood stage parasite.

4.3 Sexual Stage Vaccines

The sexually reproductive stages of the malaria parasite, which are responsible for disease transmission by the mosquito vector, are targets for transmission blocking vaccines. Transmission may be blocked by non-specific factors such as cytokines or by antibodies that specifically recognize unique antigens expressed by gametocytes, thereby preventing parasite differentiation in the mosquito mid gut [17]. Transmission blocking capacity was shown in experimental in vivo models by vaccination of turkeys and chickens with the extracellular male or female gametes that induced immunity in the birds and that could completely prevent the ability of a subsequent malaria infection of the same species of parasite from the infecting mosquitoes. The target antigens for this immunity were identified on the surface of the gametes, zygote and ookinete stages of the parasites. These include Pfs230 [181] and Pfs48/45 or Pfs25 and Pfs28 [182], which appear on the surface of gametes or zygotes, respectively. All these antigens are rich in cysteines and are expected to have a highly ordered structure. Of these, Pfs25 and Pfs28 have four EGF-like domains each. One of these antigens, Pfs25 was expressed in yeast and formulated with alum as an adjuvant for testing in phase I clinical trials. Serum from these subjects when tested, had moderate, but suboptimal, transmission blocking effects – a 50% reduction in infectivity of *P. falciparum* to mosquitoes compared to controls. Further testing of Pfs25 and its *P. vivax* homologue is being planned with alternative adjuvant systems.

4.4 Vaccines against *P. vivax* Malaria

Efforts to develop vaccines for *P. vivax* have largely been overshadowed due to the enormous toll of mortality caused by *P. falciparum* in Africa and relatively less investment has been made in this equally important area. *P. vivax* imposes a serious burden of morbidity and is a serious threat to non-immune persons traveling to endemic areas from malaria-free regions. There may be 50–75 million cases of *P. vivax* malaria annually according to the World Health Organization. *P. vivax* causes an infrequently fatal but debilitating disease that seriously impairs quality of life and economic productivity. A number of *P. vivax* strains resistant to chloroquine have been identified in India, Indonesia, Myanmar etc., eventually these strains will spread widely, leading to an increase in the worldwide prevalence of *P. vivax*.

Development of a *P. vivax* vaccine has been slowed down mainly due to the non-availability of a convenient method of in vitro cultivation of *P. vivax*, which has made it difficult to obtain the biological raw material for vaccine development and testing. Lack of a reliable challenge system for humans also contributes to the slower than required pace of vaccine development. In the absence of in vitro production of gametocytes that could be used to infect laboratory reared mosquitoes, it is difficult to conduct volunteer challenge studies and the effectiveness of a vaccine can be measured only by conducting clinical trials in an endemic area.

As with *P. falciparum*, vaccines against *P. vivax* may be desgined for different stages of the life cycle. The rationale for a *P. vivax* vaccine is largely based on the analogy with the successful immunization of human volunteers with γ-sporozoites of *P. falciparum* or with the naturally acquired immunity to *P. falciparum* infection seen in the residents of areas of endemic infection. Most of the studies done with *P. vivax* candidate antigens have been carried out in monkeys and only a few candidate vaccines have reached the stage of human trials. These are summarized in Tables 1 and 2. Thus, vaccines based on *P. vivax* CS and TRAP of pre-erythrocytic stages, MSP-1, AMA-1 and the DBP of the erythrocytic stage and the those based on the sexual stage antigens are being developed and tested at various stages of development. In addition, an increasing number of *P. vivax* orthologues of already identified and characterized *P. falciparum* genes is being discovered. It is likely that, as with *P. falciparum*, an effective, broadly useful vaccine against *P. vivax* will require the induction of several types of immune responses against multiple parasite antigens and many different technologies are being employed for this purpose.

The Malaria research group at ICGEB with support from the World Health Organization, the Malaria Vaccine Initiative, Government of India and in collaboration with industrial partners has developed optimized protocols for production of recombinant 19- and 42-kDa fragments of PvMSP-1 and the RII of PvDBP. The infrastructure and procedures to produce gram amounts of these proteins under GMP conditions in biochemically active forms are in place. The immunogenicity studies in animals and toxicity studies are currently underway. Human clinical trial sites are being developed and trials should start in due course. In parallel efforts are being made to develop a non-human primate model for *P. vivax* malaria. *P. cynomolgi*, the simian agent of malaria naturally infects rhesus monkeys and bears close resemblance with *P. vivax* biologically, immunologically and patho-physiologically. The *P. cynomolgi*-rhesus model may allow a reliable and reproducible sporozoite challenge to test the efficacy of vaccine candidate antigens orthologous to known and well characterized *P. vivax* antigens, vaccine formulations and immunization schedules. In our studies with recombinant *P. cynomolgi*TRAP (PcTRAP), partial protection was obtained in the rhesus monkeys. Similar studies with other antigens such as PcMSP-$1_{19/42}$, RII of PcEBA (orthologue of PvDBP) and PcCS with different adjuvant formulations and as DNA vaccines are planned for the immediate future.

5 Alternate Technologies for Delivery of Vaccine Antigens

Although a number of malaria vaccine candidate antigens have been identified, extensive research is still being done to find the most suitable ways to deliver the same in order to elicit appropriate immune responses. In fact malaria vaccine development efforts have been the torchbearer in the field of vaccine development per se. For example, the first synthetic peptide-based vaccine that was administered in humans was a malaria vaccine, which opened the way for such efforts in other disease situations. It is now well known that cytotoxic cellular immune responses play a key role in providing protection against the liver stages. Also it is being increasingly realized that an effective malaria vaccine would need to induce long-lasting immune responses against the different stages of the parasite life-cycle. Hence a multi-antigen/multi-stage vaccine would perhaps be the most effective way to counter malaria. Conventional approaches of making recombinant vaccines impose a restraint on the number of proteins or fragments thereof which can be included in a multi-antigen vaccine owing largely to the logistic difficulties in production of more than a few recombinant proteins in the correctly folded form. From the viewpoint of a developing country such as India such a proposition is not economically feasible.

DNA vaccination is the latest and promising development in the field of vaccinology and has revolutionized the science of vaccine development and offers many logistic advantages over the traditional forms of vaccines. The technology of DNA vaccination also provides a platform for construction of multi-component/multi-antigen vaccines in a cost effective manner compared to a multi-component recombinant protein vaccine. The observation that immunization with naked DNA could lead to strong and persistent cellular and humoral immune responses initially surprised many scientists working with conventional protein vaccines [183–186]. Moreover, DNA vaccines provided protection from pathogens in various disease models [187–190]. The ability of DNA vaccines to induce cellular immune responses against the encoded antigens was particularly useful for malaria researchers and was put to test in the murine models. Immunization with plasmid DNA bearing PyCSP gene provided protection to 68% of the immunized mice from sporozoite challenge [191] in a genetically restricted manner. Antigen specific CTLs were induced in the DNA immunized animals and protection was dependent on the presence of $CD8^+$ T cells and IFN-γ [191]. The genetic restriction of this DNA vaccine in mice was circumvented by immunization with a mixture of two plasmids encoding PyCSP and PyHEP17 antigens. The protection was completely dependent on the generation of IFN-γ secreting $CD8^+$ T-cells, and/or nitric oxide [192]. In a recent human trial, immunization with naked DNA was shown to be safe and resulted in antigen-specific low to moderate CTL responses [193]. Rhesus monkeys were immunized simultaneously with four DNA plasmids encoding different *P. falciparum* proteins from the sporozoite and liver stages to study immune responses against multi-antigen DNA immunization. The results obtained in rhesus monkeys [194], provided the foundation for ongoing clinical trials in humans with a multigene *P. falciparum* DNA vaccine [195]. Promising results were also obtained with DNA vaccines encoding blood-stage antigens (PyMSP-1) in murine models [196–198]. Animal studies are being done with other important erythrocytic stage antigens such as SERA, RAP-1/2 also.

Although a great tool, DNA vaccination at best has provided incomplete protection only. Ways and means to improve the immunogenecity and efficacy of DNA vaccines are being explored. These include heterologous prime-boost, co-immunization with cytokine plasmids and better and alternate methods and routes of DNA delivery or a combination of these. In this context boosting of DNA primed mice with recombinant poxvirus expressing same antigen has yielded the best results so far [59, 199]. Better protection was observed in mice immunized with either the PbCS or PyCS DNA vaccines when heterologous boosting with the recombinant vaccinia viruses expressing the same antigens followed the priming with DNA plasmids. The heterologous prime-boost approach resulted in higher CTL response and IFN-γ secreting $CD8^+$ T cells which correlated well with protection. However, in a recent trial with a multivalent DNA vaccine encoding four antigens from the pre- and erythrocytic stages of *P. knowlesi* provided no protection to monkeys from sporozoite challenge in a DNA-vaccinia heterologous prime-boost approach [200]. Poor results were also obtained in *Aotus* monkeys where DNA primed animals were boosted with recombinant PfEBA-175 protein [201]. Some studies have reported higher immunogenicity when the DNA plasmids were coated on gold particles and delivered by a gene gun; even so, little data is present on the effect of the enhanced immune responses on protection in primates or humans. It appears that the success of DNA vaccines will largely depend on the ability to efficiently deliver the vaccine DNA in vivo [202]. Use of virus-like particles is another alternate vaccine delivery system and has been demonstrated to be effective in boosting protective immune responses in mice [203–205]. Once again, there are no data so far in primates or humans to indicate the efficacy of such vaccines.

Vaccination with recombinant vaccinia viruses presents an attractive alternative for the delivery of antigens and recombinant viruses expressing antigens from pathogenic agents have been shown to be capable of inducing both humoral and cellular protective immune responses. In an early study with recombinant vaccinia virus expressing four proteins from the erythrocytic stage of *P. falciparum* failed to protect *Saimiri* monkeys from parasite challenge [167]. Later, another malaria vaccine was engineered in NYVAC, a highly attenuated strain of vaccinia, to express seven proteins from all stages of the *Plasmodium* life cycle; three exo-erythrocytic stage antigens (CS, TRAP and LSA-1), three erythrocytic stage antigens (MSP-1, AMA-1 and SERA), and one sexual stage antigen (Pfs25) were cloned. Each of these antigens was expressed in cultured cells infected with recombinant vaccinia, NYVAC-Pf7 [206]. The vaccine was found to be safe and immunogenic in rhesus monkeys. A phase I/IIa immunogenecity and efficacy trial was done with three immunizations with two different dosages. The vaccine was safe and variably immunogenic, with generally poor antibody responses. However, >90% of the volunteers elicited cellular immune responses and 1 out of 35 challenged subjects was completely protected, and there was a significant delay in time to parasite patency in the groups of volunteers who received either the low or high dose of vaccine compared with control volunteers [207].

In an attempt to create a multiantigen/multistage vaccine against *P. falciparum*, a string-of-beads approach was employed. A synthetic gene was constructed to encode 12 B cell, 6 T cell proliferative and 3 CTL epitopes derived from nine *P. falciparum* stage specific antigens corresponding to the sporozoite, erythrocytic and sexual stages and a universal T cell epitope from tetanus toxoid [208, 209]. The protein expressed in the baculovirus expression system induced high levels and long-lasting antibody responses against the vaccine in rabbits immunized with different adjuvant formulations. All animals had strong and comparable IFA titers against sporozoites, moderate against merozoites, and low against gametocytes. The antibodies to this vaccine almost completely inhibited invasion (98% inhibition) at a concentration of 50 μg/mL. This vaccine offers multiple layers of protective immune responses, capable of preventing infection at multiple stages in the parasite's life cycle. Evaluation of the protective efficacy of this vaccine in non-human primates is currently under progress.

6
Search for Human Compatible Adjuvant for Malaria Vaccine

Adjuvants, "substances used in combination with a specific antigen that produce more immunity than the antigen alone", are common components of the vaccines. Presently aluminium salts are the only adjuvants approved for human use. Although safe and non-toxic, alum is a weak adjuvant for antibody induction to protein subunits and poor for cell-mediated immunity. The problem is compounded by the fact that most of the malaria vaccine candidates based on the conserved features of the antigens are rather poor immunogens. This problem was recognized early in the field of malaria vaccine development and is being addressed by developing new and better adjuvants suitable for human use. In fact in a large way malaria vaccine field is responsible for the search for better formulations and delivery technologies for human use. Over the past few years many experimental adjuvants have advanced to the stage of clinical trials and some have shown high potency. In this regard, an oil-in-water emulsion containing monophosphoryl lipid A and QS21 has undergone a number of preclinical and clinical studies with the RTS, S vaccine and has been found suitable for induction of humoral and cellular immune responses [87, 210]. The formulation, although safe and non-reactogenic, was found to be capable of inducing only partial protection in humans [88, 211, 212]. In a phase I clinical trial in Papua New Guinea another adjuvant, Montanide ISA 720 was used to study immunogenecity and safety of three recombinant malaria antigens namely MSA-1, MSA-2 and RESA. Antibody and T cell responses were detected against all three antigens and the vaccine formulation was found to be safe and only mildly reactogenic [169, 213], but did not raise protective immune responses as assessed by the differences in the initial growth rates of parasite in the control versus immunized volunteers [170]. In *Aotus* monkeys, an MSP-1 based vaccine produced in baculovirus was tested with MF59 lipid particulate adjuvant in comparison with Freund's adjuvant. However, protection was seen only in animals immunized with the formulation containing Freund's adjuvant [141]. Similarly, in another study with MSP-1_{19} based vaccine formulated in six different adjuvants, protection was seen only in the animals immunized with vaccine formulated in Freund's adjuvant [142]. Other adjuvants being tried for malaria vaccines include CpG oligonucleotides. CpG dinucleotides are abundantly present as unmethylated motifs in the bacterial DNA and are capable of stimulating innate immune response in vertebrates acting as "danger signals." Synthetic oligonucleotides containing these motifs in the "context" can mimic the danger signal and have demonstrated great utility as vaccine adjuvants and immunopotentiators. *Aotus* monkeys were immunized with synthetic peptides based on sequences of PfCS protein formulated in Montanide ISA 720 with and without CpG ODN. The animals receiving vaccine along with CpG exhibited significantly greater immune responses as compared to the ones that were immunized with a vaccine without the CpG [214]. In a separate study humans were immunized with a vaccine based on the PfCS antigen encapsulated in liposomes with monophosphoryl lipid A. Although safe and immunogenic, the vaccine did not protect immunized volunteers from sporozoite challenge [215]. So far, results with formulations based on different adjuvants have been rather disappointing and emphasize the need to develop more effective adjuvants.

7
Malaria Vaccine Development in the Post-Genomics Era

Although malaria vaccine development has focused on nearly 20 putative vaccine candidates for as many years, as yet only two-RTS, S and SPf66 have undergone extensive clinical trials. With the sequencing of the *Plasmodium* genome nearing completion, the number of candidate antigens to be screened is expected to increase manifold. The effort to screen all, approximately 5000, genes will require some extraordinary techniques to be employed. The DNA vaccines, or "vaccinomics" may provide us with the right tools to complete this daunting task [216].

The most feasible approach could be identification of the proteins that are expressed by the radiation-attenuated sporozoites in the infected hepatocytes by DNA microarray techniques using the cDNA. However, because in vitro infection of hepatocytes with *P. falciparum* sporozoites is extremely inefficient, obtaining adequate amounts of mRNA will be difficult. In another strategy, PCR amplification of all open reading frames (ORFs) and cloning into DNA vaccine plasmids or linear expression elements can be performed. Screening of antisera obtained from mice immunized with individual constructs against radiation-attenuated hepatocytes could then identify the corresponding DNA sequences (ORFs). The corresponding epitopes capable of binding the HLA class I molecules from these genes will be subsequently predicted from the existing algorithms and validated for binding to class I HLA proteins in vitro. Assembling these peptides, typically tens of thousands of them, in a "string-of-beads" fashion on DNA vaccine plasmid(s) would then provide us with the reagents to test for their ability to induce protective immunity in humans. A different approach would be needed to develop an antibody-based erythrocytic stage vaccine. Antisera generated by immunization with the ORFs can be used to localize surface expressed and secreted proteins in the erythrocytic stages of *Plasmodium*. The DNA plasmids that induce such antibodies can be used in a mixture of all of them to immunize volunteers in the malaria endemic areas to prime the immune system in subjects living in these areas followed by experimental challenge studies. Conventional protein-by-protein or gene-by-gene approaches cannot be envisioned to handle the vast amount of data generated by genome sequencing efforts.

8 Hurdles in the Way to Develop Effective Malaria Vaccines

Malaria vaccine development is a complex and formidable challenge; the basis for malaria immunity, in spite of a tremendous amount of research remains ill understood. The malaria parasite has a fascinating but hugely complex life cycle. It goes through different stages of development with little if anything, in common between them. Different proteins are expressed at different stages, some on the surface of the infected cell whereas some may be secreted out at different times. Scientifically, two issues stand out with reference to malaria vaccine development. The first one is of extensive antigenic polymorphism observed in the case of *Plasmodium* proteins. In particular, blood stage antigens including leading vaccine candidates, are polymorphic and the antigenic diversity generated as a result of these polymorphism remains a problem in the way of malaria vaccine development. It is well known in the case of major vaccine candidates that immunization with recombinant proteins of one strain of parasite while protective against the homologous strain fail to provide protection from a different strain of the parasite. On the other hand, however, most malaria proteins have within their structures regions that are relatively more conserved. Interestingly, some of these conserved regions constitute functional domains of these proteins, for example, the RII^+ of CS and TRAP and the DBL domains of EBA-175/DABP. Vaccines that are being developed based on these antigens, are indeed focused on the conserved antigens. Rational use of these conserved regions in vaccine constructs is also likely to address the problem of cryptic epitopes as has been shown in the case of the CS and MSP-1.

Another major hurdle in the way of malaria vaccine development is the lack of appropriate in vitro correlates of protection. Results of in vitro experiments have not always correlated with protection in vivo or vice-versa. The sterile immunity achieved in animals and humans by immunization of γ-sporozoites is the gold standard so far, however, in the absence of a complete knowledge of the mechanisms of protection and due to the fact that many antigens are expressed in the hepatocytes infected by the attenuated sporozoites, it has been difficult to devise suitable in vitro tests to predict the correlates of protection. Anti-sporozoite antibodies have provided protection against viable sporozoite challenge in passive transfer experiments but the mere presence of these antibodies induced by vaccination does not correlate with protection. Similarly, although passive transfer of CS and TRAP specific CTL clones protects animals from viable sporozoite challenge, once again the presence of antigen-specific CTL has not correlated well with protection. Numerous antigen delivery systems induce high levels of CTL response but do not protect the immunized animals/humans. Assays which directly measure a relevant biological activity of humoral and cellular immune responses can be good candidate in vitro correlates of protection. Three such tests, the inhibition of sporozoite invasion inhibition assay, the inhibition of liver stage development assay, and the enumeration of antigen specific T cells secreting IFN-γ are being developed as the assays of choice to predict levels of protection in response to the vaccination pending extensive validation. At the same time, a better understanding of the mechanisms of protection in the γ-sporozoite vaccination model might give useful leads and better in vitro assays could then be developed to predict protective immunity.

Naturally acquired immunity to malaria does not provide sterile protection, but is remarkably efficient at moderating the clinical effects of infection. Although there is overwhelming body of evidence that protection is primarily mediated by antibodies to various surface and/or secreted merozoite antigens, data also indicate towards an involvement of T cells in providing protection. In the case of erythrocytic stage vaccines too, no valid in vitro tests have been possible to develop for prediction of protection induced by vaccines. Assays that measure the ability of antibodies to inhibit the invasion of erythrocytes by merozoites or the development of intra-erythrocytic parasites have been widely used to assess humoral immunity induced by vaccination. Another test, the antibody dependent cellular inhibition assay could also provide information on the potential mechanism of protection mediated by antibodies. However, no studies have yet evaluated the ability of well-studied candidate vaccine antigens to elicit antibodies active in these assays or attempted to correlate vaccine-induced protection in non-human primates with activity in these assays. In a related and exciting development, a robust neutralization assay has been described recently for MSP-1_{19}- a leading vaccine candidate [105]. The authors described a pair of matching parasite lines differing only in the MSP-1_{19} region from the parent lines. Use of these hybrid parasites allowed an accurate measurement of the effect of anti-MSP-1_{19} antibodies in in vitro invasion and growth inhibition tests. The technique of allelic recombination used to replace orthologous gene segments in *P. falciparum* with those of murine parasite *P. chabaudi* can potentially be used, where possible to create similar parasite lines for other leading erythrocytic stage vaccine candidates. Development of such new in vitro correlates that can be validated for vaccine potential of candidate vaccines will be most welcome step in the direction of malaria vaccine development.

Similarly lack of any suitable animal models for testing the efficacy of vaccine candidate antigens is a major problem. Models of murine malaria although very useful to study and understand parasite biology have not been relevant to human malaria vaccine development. At best these can act as useful initial screens or indicator for homologous antigens to be developed as vaccine for human malaria. Availability of a suitable primate model can greatly facilitate development of drugs and vaccines for human use. Once again however, there is no satisfactory monkey model for human malaria *P. falciparum*. Simian malarias, *P. knowlesi* and *P. cynomolgi* bear some similarities to human malaria *P. vivax* and can lead to useful comparative information but cannot be utilized with conviction for testing the vaccine efficacy. Although most protein homologues exist in all *Plasmodia* species, differences exist with respect to size, structure and kind of repeats and the sequence itself making it difficult to extrapolate results from one species to the other. The New World monkeys, for example, *Aotus* and *Saimiri* spp have been adapted for infection with some clones/isolates of *P. falciparum* but they are not natural hosts for this parasite and infection in these animals may result in inconsistent infection rates. At the same time trials of vaccines based on *P. falciparum* antigens in New World monkeys (*Aotus* and/or *Saimiri* spp.) may offer a way to test clinical grade antigen formulations to prove efficacy against homologous and heterologous strains. However, results of *P. falciparum* based vaccine trials in the New World monkeys have not clarified the debate whether or not experiments in these monkeys should be in the critical path of the development of a vaccine candidate prior to testing in large field trials.

Production of malaria vaccine proteins or their conserved functional fragments is also not straightforward. Most vaccine candidate antigens, selected on the basis of their localization and/or possible involvement in functions crucial for parasite survival and immunogenicity, more often than not contain a large number of cysteine residues. These residues are linked in a unique fashion to give a highly ordered structure which in many cases has already been shown to be crucial for appropriate protective immune responses. Thus, producing these vaccines by recombinant means after in vitro refolding of the overexpressed protein, particularly when *E. coli* is used as the expression system of choice, presents a formidable task. Production of recombinant material for clinical trials under GMP conditions is expensive and, given the status of malaria vaccine development, has been a stumbling block until recently. Even after sufficient amounts of the desired antigen can be produced, planning and carrying out clinical trials is a mammoth task. Natural malaria immunity is complex and to evaluate the effect of any intervention including a vaccine is complicated. There are no clear-cut markers and one may have to evaluate parameters such as reduced number of episodes, lowered levels of parasitemia post vaccination. It is well known that in field situations a large number of individuals harbor mature parasites without showing any signs of clinical disease. The nature of acquired immunity thus is not very well understood. It becomes necessary, therefore, to target vaccine trials in infants after establishing the safety and immunogenicity – not an easy task.

9 Conclusions

An effective malaria vaccine when available will hopefully be able to control the scourge of malaria in the very poor societies of the world, mostly in Africa and Asia. To realize this goal, additional resources are being mobilized on a global scale and increased government and philanthropic interests in this field is helping the research community in a great way. Efforts such as Malaria Vaccine Initiative, with funds from the Bill and Melinda Gates Foundation, and active interest of large pharmaceutical companies such as Glaxo SmithKline with the World Health Organization in the forefront have started bearing fruits and a number of phase 1/2a studies are either planned or already underway in different parts of the world. At the same time, more and more efforts are directed towards understanding the natural mechanisms of protection in the adult population of endemic areas and the one provided by the γ-sporozoites in animal models and humans. These models of protection provide motivation in the search for suitable and effective vaccine candidates, delivery systems and adjuvant formulations suitable for human use. However, even under the most optimistic scenario with all hurdles overcome and sufficient resources at disposal, it will be many years before we have effective formulations that are at least as good as or better than the natural immune mechanisms of protection against malaria.

10 References

1. Richman AM, Fotis KC (1998) Nat Med 4:552
2. Coluzzi M (1992) Parasitol Today 8:113
3. Playfair JH, Taverne J, Bate CA, de Souza JB (1990) Immunol Today 11:25
4. Marsh K (1992) Parasitology 104:S53
5. Greenwood J, Willis AE, Perham RN (1991) J Mol Biol 220:821
6. Roberts DJ, Craig AG, Berendt AR, Pinches R, Nash G, Marsh K, Newbold CI (1992) Nature 357:689
7. Bate CA, Taverne J, Playfair JH (1988) Immunology 64:227
8. Lackritz EM, Campbell CC, Ruebush TK, Hightower AW, Wakube W, Steketee RW, Were JBO (1992) Lancet 340:524
9. Clyde DF (1975) Am J Trop Med Hyg 24:397
10. Gwadz RW, Cochrane AH, Nussenzweig V, Nussenzweig RS (1979) Bull World Health Organ 57:165
11. Herrington D, Davis J, Nardin E, Beier M, Cortese J, Eddy H, Losonsky G, Hollingdale M, Sztein M, Levine M, Nussenzweig RS, Clyde D, Edelman R (1991) Am J Trop Med Hyg 45:539
12. McGregor IA, Carrington SC, Cohen S (1963) Trans Roy Soc Trop Med Hyg 170
13. McGregor IA (1974) Bull WHO 50:259
14. Sinnis P, Nussenzweig V (1996) In: Hoffman SL (ed), Malaria Vaccine Development: a Multi-Immune Response Approach. Washington DC, p 15
15. Hoffman SL, Miller LH (1998) Nat Med 4:520
16. Kaslow DC (1993) Curr Opin Immunol 5:557

17. Kaslow DC (1997) Int J Parasitol 27:183
18. Doolan DL, Hedstrom RC, Wang R, Sedegah M, Scheller LF, Hobart P, Norman JA, Hoffman SL (1997) Indian J Med Res 106:109
19. Zavala F, Tam JP, Hollingdale MR, Cochrane AH, Quakyi I, Nussenzweig RS, Nussenzweig V (1985) Science 228:1436
20. Nardin EH, Nussenzweig V, Nussenzweig RS, Collins WE, Harinasuta KT, Tapchaisri P, Chomcharn Y (1982) J Exp Med 156:20
21. Potocnjak P, Yoshida N, Nussenzweig RS, Nussenzweig V (1980) J Exp Med 151:1504
22. Charoenvit Y, Mellouk S, Cole C, Bechara R, Leef MF, Sedegah M, Yuan LF, Robey FA, Beaudoin RL, Hoffman SL (1991) J Immunol 146:1020
23. Verhave JP, Strickland GT, Jaffe HA, Ahmed A (1978) J Immunol 121:1031
24. Weiss WR, Sedegah M, Beaudoin RL, Miller LH, Good MF (1988) Proc Natl Acad Sci USA 85:573
25. Tsuji M, Romero P, Nussenzweig RS, Zavala F (1990) J Exp Med 172:1353
26. Doolan DL, Hoffman SL (2000) J Immunol 165:1453
27. Doolan DL, Hoffman SL (1997) Philos Trans R Soc Lond B Biol Sci 352:1361
28. Nussenzweig V, Nussenzweig RS (1989) Bull Mem Acad R Med Belg 144:493
29. Cerami C, Frevert U, Sinnis P, Takacs B, Clavijo P, Santos MJ, Nussenzweig V (1992) Cell 70:1021
30. Cerami C, Kwakye-Berko F, Nussenzweig V (1992) Mol Biochem Parasitol 54:1
31. Chatterjee S, Wery M, Sharma P, Chauhan VS (1995) Infect Immun 63:4375
32. Egan JE, Hoffman SL, Haynes JD, Sadoff JC, Schneider I, Grau GE, Hollingdale MR, Ballou WR, Gordon DM (1993) Am J Trop Med Hyg 49:166
33. Zavala F, Tam JP, Barr PJ, Romero PJ, Ley V, Nussenzweig RS, Nussenzweig V (1987) J Exp Med 166:1591
34. Lal AA, de la Cruz VF, Welsh JA, Charoenvit Y, Maloy WL, McCutchan TF (1987) J Biol Chem 262:2937
35. Egan JE, Weber JL, Ballou WR, Hollingdale MR, Majarian WR, Gordon DM, Maloy WL, Hoffman SL, Wirtz RA, Schneider I, Woollett GR, Young JF, Hockmeyer WT (1987) Science 236:453
36. Reed RC, Louis-Wileman V, Wells RL, Verheul AF, Hunter RL, Lal AA (1996) Vaccine 14:828
37. Bharadwaj A, Sharma P, Joshi SK, Singh B, Chauhan VS (1998) Infect Immun 66:3232
38. Aidoo M, Udhayakumar V (1999) Parasitol Today 16:50
39. Weiss WR, Berzofsky JA, Houghten RA, Sedegah M, Hollindale M, Hoffman SL (1992) J Immunol 149:2103
40. Li S, Rodrigues M, Rodriguez D, Rodriguez JR, Esteban M, Palese P, Nussenzweig RS, Zavala F (1993) Proc Natl Acad Sci USA 90:5214
41. Rodrigues M, Li S, Murata K, Rodriguez D, Rodriguez JR, Bacik I, Bennink JR, Yewdell JW, Garcia-Sastre A, Nussenzweig RS (1994) J Immunol 153:4636
42. Sadoff JC, Ballou WR, Baron LS, Majarian WR, Brey RN, Hockmeyer WT, Young JF, Cryz SJ, Ou J, Lowell GH, Chulay J (1988) Science 240:336
43. Aggarwal A, Kumar S, Jaffe R, Hone D, Gross M, Sadoff J (1990) J Exp Med 172:1083
44. Lanar DE, Tine JA, de Taisne C, Seguin MC, Cox WI, Winslow JP, Ware LA, Kauffman EB, Gordon D, Ballou WR, Paoletti E, Sadoff JC (1996) Infect Immun 64:1666
45. Sedegah M, Chiang CH, Weiss WR, Mellouk S, Cochran MD, Houghten RA, Beaudoin RL, Smith D, Hoffman SL (1992) Vaccine 10:578
46. Khusmith S, Charoenvit Y, Kumar S, Sedegah M, Beaudoin RL, Hoffman SL (1991) Science 252:715
47. Robson KJ, Hall JR, Jennings MW, Harris TJ, Marsh K, Newbold CI, Tate VE, Weatherall DJ (1988) Nature 335:79
48. Rogers WO, Malik A, Mellouk S, Nakamura K, Rogers MD, Szarfman A, Gordon DM, Nussler AK, Aikawa M, Hoffman SL (1992) Proc Natl Acad Sci USA 89:9176
49. Sijwali PS, Malhotra P, Puri SK, Chauhan VS (1997) Mol Biochem Parasitol 90:371
50. Templeton TJ, Kaslow DC (1997) Mol Biochem Parasitol 84:13

51. Sultan AA, Thathy V, Frevert U, Robson KJ, Crisanti A, Nussenzweig V, Nussenzweig RS, Menard R (1997) Cell 90:511
52. Robson KJ, Frevert U, Reckmann I, Cowan G, Beier J, Scragg IG, Takehara K, Bishop DH, Pradel G, Sinden R, Saccheo S, Müller H-M, Crisanti A (1995) Embo J 14:3883
53. Muller HM, Scarselli E, Crisanti A (1993) Parassitologia 35 Suppl 69
54. Sharma P, Bharadwaj A, Bhasin VK, Sailaja VN, Chauhan VS (1996) Infect Immun 64:2172
55. Wizel B, Houghten RA, Parker KC, Coligan JE, Church P, Gordon DM, Ballou WR, Hoffman SL (1995) J Exp Med 182:1435
56. Aidoo M, Lalvani A, Allsopp CE, Plebanski M, Meisner SJ, Krausa P, Browning M, Morris-Jones S, Gotch F, Fidock DA, Takiguchi M, Robson KJH, Greenwood BM, Druihle P, Whittle HC, Hill AVS (1995) Lancet 345:1003
57. Charoenvit Y, Fallarme V, Rogers WO, Sacci JB Jr, Kaur M, Aguiar JC, Yuan LF, Corradin G, Andersen E, Wizel B, Houghten RA, Oloo A, De la Vega P, Hoffman SL (1997) Infect Immun 65:3430
58. Khusmith S, Sedegah M, Hoffman SL (1994) Infect Immun 62:2979
59. Schneider J, Gilbert SC, Blanchard TJ, Hanke T, Robson KJ, Hannan CM, Becker M, Sinden R, Smith GL, Hill AV (1998) Nat Med 4:397
60. Fidock DA, Gras-Masse H, Lepers JP, Brahimi K, Benmohamed L, Mellouk S, Guerin-Marchand C, Londono A, Raharimalala L, Meis JF (1994) J Immunol 153:190
61. Hill AV, Elvin J, Willis AC, Aidoo M, Allsopp CE, Gotch FM, Gao XM, Takiguchi M, Greenwood BM, Townsend AR, McMichael AJ, Whittel HC (1992) Nature 360:434
62. Doolan DL, Hoffman SL, Southwood S, Wenworth PA, Sidney J, Chesnut RW, Keogh E, Appella E, Nutman TB, Lal AA, Gordon DM, Oloo A, Sette A (1997) Immunity 7:97
63. Hill AV, Allsopp CE, Kwiatkowski D, Anstey NM, Twumasi P, Rowe PA, Bennett S, Brewster D, McMichael AJ, Greenwood BM (1991) Nature 352:595
64. Joshi SK, Bharadwaj A, Chatterjee S, Chauhan VS (2000) Infect Immun 68:141
65. Sanchez GI, Rogers WO, Mellouk S, Hoffman SL (1994) Exp Parasitol 79:59
66. Charoenvit Y, Mellouk S, Sedegah M, Toyoshima T, Leef MF, De la Vega P, Beaudoin RL, Aikawa M, Fallarme V, Hoffman SL (1995) Exp Parasitol 80:419
67. Fidock DA, Bottius E, Brahimi K, Moelans II, Aikawa M, Konings RN, Certa U, Olafsson P, Kaidoh T, Asavanich A, Guerin-Marchand C, Druillhe P (1994) Mol Biochem Parasitol 64:219
68. Bottius E, BenMohamed L, Brahimi K, Gras H, Lepers JP, Raharimalala L, Aikawa M, Meis J, Slierendregt B, Tartar A, Thomas A, Druilhe P (1996) J Immunol 156:2874
69. BenMohamed L, Gras-Masse H, Tartar A, Daubersies P, Brahimi K, Bossus M, Thomas A, Druilhe P (1997) Eur J Immunol 27:1242
70. Nussenzweig V, Nussenzweig RS (1989) Adv Immunol 45:283
71. Herrington DA, Clyde DF, Losonsky G, Cortesia M, Murphy JR, Davis J, Baqar S, Felix AM, Heimer EP, Gillessen D, Nardin E, Nussenzweig RS, Hollingdale M, Levine M (1987) Nature 328:257
72. Etlinger HM, Felix AM, Gillessen D, Heimer EP, Just M, Pink JR, Sinigaglia F, Sturchler D, Takacs B, Trzeciak A, Matile H (1988) J Immunol 140:626
73. DiJohn D, Torres J, Murillo J, Losonsky GA, Herrington DA, Sturcher D, Levine MM (1989) Lancet 1:1415
74. Moreno CA, Rodriguez R, Oliveira GA, Ferreira V, Nussenzweig RS, Moya Castro ZR, Calvo-Calle JM, Nardin E (1999) Vaccine 18:89
75. Collins WE, Sullivan JS, Morris CL, Galland GG, Jue DL, Fang S, Wohlhueter R, Reed RC, Yang C, Hunter RL, Lal AA (1997) Am J Trop Med Hyg 56:200
76. Herrera S, De Plata C, Gonzalez M, Perlaza BL, Bettens F, Corradin G, Arevalo-Herrera M (1997) Parasite Immunol 19:161
77. Ballou WR, Hoffman SL, Sherwood JA, Hollingdale MR, Neva FA, Hockmeyer WT, Gordon DM, Schneider I, Wirtz RA, Young JF, Wasserman GF, Reeve P, Diggs CL, Chulay JD (1987) Lancet 1:1277
78. Rickman LS, Gordon DM, Wistar R, Jr., Krzych U, Gross M, Hollingdale MR, Egan JE, Chulay JD, Hoffman SL (1991) Lancet 337:998

79. Hoffman SL, Edelman R, Bryan JP, Schneider I, Davis J, Sedegah M, Gordon D, Church P, Gross M, Silverman C (1994) Am J Trop Med Hyg 51:603
80. Fries LF, Gordon DM, Schneider I, Beier JC, Long GW, Gross M, Que JU, Cryz SJ, Sadoff JC (1992) Infect Immun 60:1834
81. Collins WE, Nussenzweig RS, Ballou WR, Ruebush TK 2nd, Nardin EH, Chulay JD, Majarian WR, Young JF, Wasserman GF, Bathurst I, Gibson HL, Barr PJ, Hoffman SL, Wasserman SS, Broderson JR, Skinner JC, Procell PM, Filipski VK, Wilson CL (1989) Am J Trop Med Hyg 40:455
82. Collins WE, Nussenzweig RS, Ruebush TK 2nd, Bathurst IC, Nardin EH, Gibson HL, Campbell GH, Barr PJ, Broderson JR, Skinner JC, Wilson C, Stanfill PS, Procell PM (1990) Am J Trop Med Hyg 43:576
83. Yang C, Collins WE, Xiao L, Saekhou AM, Reed RC, Nelson CO, Hunter RL, Jue DL, Fang S, Wohlhueter RM, Udhayakumar V, Lal AA (1997) Vaccine 15:377
84. Gordon DM, McGovern TW, Krzych U, Cohen JC, Schneider I, LaChance R, Heppner DG, Yuan G, Hollingdale M, Slaoui M, Hauser P, Voet P, Sadoff JC, Ballou WR (1995) J Infect Dis 171:1576
85. Good MF, Maloy WL, Lunde MN, Margalit H, Cornette JL, Smith GL, Moss B, Miller LH, Berzofsky JA (1987) Science 235:1059
86. Malik A, Egan JE, Houghten RA, Sadoff JC, Hoffman SL (1991) Proc Natl Acad Sci USA 88:3300
87. Stoute JA, Slaoui M, Heppner DG, Momin P, Kester KE, Desmons P, Wellde BT, Garcon N, Krzych U, Marchand M (1997) N Engl J Med 336:86
88. Stoute JA, Kester KE, Krzych U, Wellde BT, Hall T, White K, Glenn G, Ockenhouse CF, Garcon N, Schwenk R, Lanar DE, Sun P, Momin P, Wirtz RA, Golenda C, Slaoui M, Wortmann G, Holland C, Dowler M, Cohen J, Ballou WR (1998) J Infect Dis 178:1139
89. Lalvani A, Moris P, Voss G, Pathan AA, Kester KE, Brookes R, Lee E, Koutsoukos M, Plebanski M, Delchambre M, Flanagan KL, Carton C, Slaoui M, Van Hoecke C, Ballou WR, Hill AV, Cohen J (1999) J Infect Dis 180:1656
90. Perlaza BL, Arevalo-Herrera M, Brahimi K, Quintero G, Palomino JC, Gras-Masse H, Tartar A, Druilhe P, Herrera S (1998) Infect Immun 66:3423
91. Daubersies P, Thomas AW, Millet P, Brahimi K, Langermans JA, Ollomo B, BenMohamed L, Slierendregt B, Eling W, Van Belkum A, Dubreuil G, Meis JF, Guerin-Marchand C, Cayphas S, Cohen J, Gras-Masse H, Druilhe P, Mohamed LB (2000) Nat Med 6:1258
92. Siddiqui WA, Tam LQ, Kramer KJ, Hui GS, Case SE, Yamaga KM, Chang SP, Chan EB, Kan SC (1987) Proc Natl Acad Sci USA 84:3014
93. Holder AA (1988) Prog Allergy 41:72
94. Cooper JA (1993) Parasitol Today 9:50
95. McBride JS, Heidrich HG (1987) Mol Biochem Parasitol 23:71
96. Peterson MG, Coppel RL, McIntyre P, Langford CJ, Woodrow G, Brown GV, Anders RF, Kemp DJ (1988) Mol Biochem Parasitol 27:291
97. Snewin VA, Herrera M, Sanchez G, Scherf A, Langsley G, Herrera S (1991) Mol Biochem Parasitol 49:265
98. Holder AA, Blackman MJ, Burghaus PA, Chappel JA, Ling IT, McCallum-Deighton N, Shai S (1992) Mem Inst Oswaldo Cruz 87:37
99. Blackman MJ, Heidrich HG, Donachie S, McBride JS, Holder AA (1990) J Exp Med 172:379
100. Riley EM, Allen SJ, Wheeler JG, Blackman MJ, Bennett S, Takacs B, Schonfeld HJ, Holder AA, Greenwood BM (1992) Parasite Immunol 14:321
101. Riley EM, Morris-Jones S, Blackman MJ, Greenwood BM, Holder AA (1993) Parasite Immunol 15:513
102. Egan AF, Morris J, Barnish G, Allen S, Greenwood BM, Kaslow DC, Holder AA, Riley EM (1996) J Infect Dis 173:765
103. al-Yaman F, Genton B, Kramer KJ, Chang SP, Hui GS, Baisor M, Alpers MP (1996) Am J Trop Med Hyg 54:443
104. Shi YP, Sayed U, Qari SH, Roberts JM, Udhayakumar V, Oloo AJ, Hawley WA, Kaslow DC, Nahlen BL, Lal AA (1996) Infect Immun 64:2716

105. O'Donnell RA, de Koning-Ward TF, Burt RA, Bockarie M, Reeder JC, Cowman AF, Crabb BS (2001) J Exp Med 193:1403
106. Herrera S, Herrera MA, Perlaza BL, Burki Y, Caspers P, Dobeli H, Rotmann D, Certa U (1990) Proc Natl Acad Sci USA 87:4017
107. Perrin LH, Merkli B, Loche M, Chizzolini C, Smart J, Richle R (1984) J Exp Med 160:441
108. Chatterjee S, Sharma P, Kumar S, Chauhan VS (1995) Vaccine 13:1474
109. Kabilan L, Sharma VP, Kaur P, Ghosh SK, Yadav RS, Chauhan VS (1994) Infect Immun 62:685
110. Holder AA, Freeman RR (1981) Nature 294:361
111. Freeman RR, Holder AA (1983) J Exp Med 158:1647
112. Majarian WR, Daly TM, Weidanz WP, Long CA (1984) J Immunol 132:3131
113. Clark JT, Donachie S, Anand R, Wilson CF, Heidrich H-G, McBride JS (1989) Mol Biochem Parasitol 32:15
114. Ramasamy R, Jones G, Lord R (1990) Immunol Lett 23:305
115. Oeuvray C, Bouharoun-Tayoun H, Gras-Masse H, Bottius E, Kaidoh T, Aikawa M, Filgueira MC, Tartar A, Druilhe P (1994) Blood 84:1594
116. Waters AP, Thomas AW, Deans JA, Mitchell GH, Hudson DE, Miller LH, McCutchan TF, Cohen S (1990) J Biol Chem 265:17974
117. Thomas AW, Trape JF, Rogier C, Goncalves A, Rosario VE, Narum DL (1994) Am J Trop Med Hyg 51:730
118. Chitnis CE, Miller LH (1994) J Exp Med 180:497
119. Narum DL, Haynes JD, Fuhrmann S, Moch K, Liang H, Hoffman SL, Sim BK (2000) Infect Immun 68:1964
120. Singh S, Pandey K, Chattopadhayay R, Yazdani SS, Lynn A, Bharadwaj A, Ranjan A, Chitnis C (2001) J Biol Chem 276:17111
121. Fraser T, Michon P, Barnwell JW, Noe AR, Al-Yaman F, Kaslow DC, Adams JH (1997) Infect Immun 65:2772
122. Okoyeh JN, Pillai CR, Chitnis CE (1999) Infect Immun 67:5784
123. Chulay JD, Lyon JA, Haynes JD, Meierovics AI, Atkinson CT, Aikawa M (1987) J Immunol 139:2768
124. Barr PJ, Inselburg J, Green KM, Kansopon J, Hahm BK, Gibson HL, Lee-Ng CT, Bzik DJ, Li WB, Bathurst IC (1991) Mol Biochem Parasitol 45:159
125. Ridley RG, Takacs B, Lahm HW, Delves CJ, Goman M, Certa U, Matile H, Woollett GR, Scaife JG (1990) Mol Biochem Parasitol 41:125
126. Stowers A, Taylor D, Prescott N, Cheng Q, Cooper J, Saul A (1997) Infect Immun 65: 2329
126. a) Kushwaha A, Rao PP, Duttu VS, Malhotra P, Chauhan VS (2000) Mol Biochem Parasitol 106:213; b) Sharma P, Kumar A, Singh B, Bharadwaj A, Sailaja VN, Adak T, Kushwaha A, Malhotra P, Chauhan VS (1998) Infect Immun 66:2895; c) Kushwaha A, Rao PP, Suresh P, Chauhan VS (2001) Parasite Immunol 23:435; d) Vargas-Serrato E, Barnwell JW, Ingravallo P, Perler FB, Galinski MR (2002) Mol Biochem Parasitol 120:41
127. Baruch DI, Pasloske BL, Singh HB, Bi X, Ma XC, Feldman M, Taraschi TF, Howard RJ (1995) Cell 82:77
128. Su XZ, Heatwole VM, Wertheimer SP, Guinet F, Herrfeldt JA, Peterson DS, Ravetch JA, Wellems TE (1995) Cell 82:89
129. Berzins K, Perlmann H, Udomsangpetch R, Wahlin B, Wahlgren M, Troye-Blomberg M, Carlsson J, Bjorkman A, Perlmann P (1985) Dev Biol Stand 62:99
130. Berzins K, Perlmann H, Wahlin B, Carlsson J, Wahlgren M, Udomsangpetch R, Bjorkman A, Patarroyo ME, Perlmann P (1986) Proc Natl Acad Sci USA 83:1065
131. Berzins K, Perlmann H, Wahlin B, Ekre HP, Hogh B, Petersen E, Wellde B, Schoenbechler M, Williams J, Chulay J, Perlmann P (1991) Infect Immun 59:1500
132. Petersen E, Hogh B, Marbiah NT, Perlmann H, Willcox M, Dolopaie E, Hanson AP, Bjorkman A, Perlmann P (1990) Trans R Soc Trop Med Hyg 84:339
133. Riley EM, Allen SJ, Troye-Blomberg M, Bennett S, Perlmann H, Andersson G, Smedman L, Perlmann P, Greenwood BM (1991) Trans R Soc Trop Med Hyg 85:436

134. al-Yaman F, Genton B, Anders R, Taraika J, Ginny M, Mellor S, Alpers MP (1995) Parasite Immunol 17:493
135. Beck HP, Felger I, Genton B, Alexander N, al-Yaman F, Anders RF, Alpers M (1995) Infect Immun 63:596
136. Siddiqui WA, Tam LQ, Kan SC, Kramer KJ, Case SE, Palmer KL, Yamaga KM, Hui GS (1986) Infect Immun 52:314
137. Holder AA, Freeman RR, Nicholls SC (1988) Parasite Immunol 10:607
138. Etlinger HM, Caspers P, Matile H, Schoenfeld HJ, Stueber D, Takacs B (1991) Infect Immun 59:3498
139. Chang SP, Case SE, Gosnell WL, Hashimoto A, Kramer KJ, Tam LQ, Hashiro CQ, Nikaido CM, Gibson HL, Lee-Ng CT, Barr PJ, Yokota BT, Hut GS (1996) Infect Immun 64:253
140. Kumar S, Yadava A, Keister DB, Tian JH, Ohl M, Perdue-Greenfield KA, Miller LH, Kaslow DC (1995) Mol Med 1:325
141. Stowers AW, Cioce V, Shimp RL, Lawson M, Hui G, Muratova O, Kaslow DC, Robinson R, Long CA, Miller LH (2001) Infect Immun 69:1536
142. Kumar S, Collins W, Egan A, Yadava A, Garraud O, Blackman MJ, Guevara Patino JA, Diggs C, Kaslow DC (2000) Infect Immun 68:2215
143. Egan AF, Blackman MJ, Kaslow DC (2000) Infect Immun 68:1418
144. Perera KL, Handunnetti SM, Holm I, Longacre S, Mendis K (1998) Infect Immun 66:1500
145. Herrera S, Herrera MA, Certa U, Corredor A, Guerrero R (1992) Mem Inst Oswaldo Cruz 87:423
146. Herrera MA, Rosero F, Herrera S, Caspers P, Rotmann D, Sinigaglia F, Certa U (1992) Infect Immun 60:154
147. Yang C, Collins WE, Sullivan JS, Kaslow DC, Xiao L, Lal AA (1999) Infect Immun 67:342
148. Collins WE, Kaslow DC, Sullivan JS, Morris CL, Galland GG, Yang C, Saekhou AM, Xiao L, Lal AA (1999) Am J Trop Med Hyg 60:350
149. Keitel WA, Kester KE, Atmar RL, White AC, Bond NH, Holland CA, Krzych U, Palmer DR, Egan A, Diggs C, Ballou WR, Hall BF, Kaslow D (1999) Vaccine 18:531
150. Chappel JA, Holder AA (1993) Mol Biochem Parasitol 60:303
151. Burns JM, Jr., Majarian WR, Young JF, Daly TM, Long CA (1989) J Immunol 143:2670
152. Egan AF, Chappel JA, Burghaus PA, Morris JS, McBride JS, Holder AA, Kaslow DC, Riley EM (1995) Infect Immun 63:456
153. Udhayakumar V, Anyona D, Kariuki S, Shi YP, Bloland PB, Branch OH, Weiss W, Nahlen BL, Kaslow DC, Lal AA (1995) J Immunol 154:6022
154. Spetzler JC, Rao C, Tam JP (1994) Int J Pept Protein Res 43:351
155. Daly TM, Long CA (1993) Infect Immun 61:2462
156. Tian JH, Miller LH, Kaslow DC, Ahlers J, Good MF, Alling DW, Berzofsky JA, Kumar S (1996) J Immunol 157:1176
157. Deans JA, Knight AM, Jean WC, Waters AP, Cohen S, Mitchell GH (1988) Parasite Immunol 10:535
158. Collins WE, Pye D, Crewther PE, Vandenberg KL, Galland GG, Sulzer AJ, Kemp DJ, Edwards SJ, Coppel RL, Sullivan JS, Morris CL, Anders RE (1994) Am J Trop Med Hyg 51:711
159. Kocken CH, Dubbeld MA, Van Der Wel A, Pronk JT, Waters AP, Langermans JA, Thomas AW (1999) Infect Immun 67:43
160. Crewther PE, Matthew ML, Flegg RH, Anders RF (1996) Infect Immun 64:3310
161. Delplace P, Bhatia A, Cagnard M, Camus D, Colombet G, Debrabant A, Dubremetz JF, Dubreuil N, Prensier G, Fortier B, Haq A, Weber J, Vernes A (1988) Biol Cell 64:215
162. Inselburg J, Bzik DJ, Li WB, Green KM, Kansopon J, Hahm BK, Bathurst IC, Barr PJ, Rossan RN (1991) Infect Immun 59:1247
163. Inselburg J, Bathurst IC, Kansopon J, Barr PJ, Rossan R (1993) Infect Immun 61:2048
164. Inselburg J, Bathurst IC, Kansopon J, Barchfeld GL, Barr PJ, Rossan RN (1993) Infect Immun 61:2041
165. Collins WE, Anders RF, Pappaioanou M, Campbell GH, Brown GV, Kemp DJ, Coppel RL, Skinner JC, Andrysiak PM, Favaloro JM, Corcoran LM, Broderson JR, Mitchell GF, Campbell CC (1986) Nature 323:259

166. Collins WE, Walduck A, Sullivan JS, Andrews K, Stowers A, Morris CL, Jennings V, Yang C, Kendall J, Lin Q, Martin LB, Diggs C, Saul A (2000) Am J Trop Med Hyg 62: 466
167. Pye D, Edwards SJ, Anders RF, O'Brien CM, Franchina P, Corcoran LN, Monger C, Peterson MG, Vandenberg KL, Smythe JA, Westley SR, Coppel RL, Webster TL, Kemp DJ, Hampson AW, Langford CJ (1991) Infect Immun 59:2403
168. Collins WE, Anders RF, Ruebush TK 2nd, Kemp DJ, Woodrow GC, Campbell GH, Brown GV, Irving DO, Goss N, Filipski VK, Coppel RL, Broderson JR, Thomas LM, Pye DO, Skinner JC, Wilson C, Stanfill PS, Procell PM (1991) Am J Trop Med Hyg 44:34
169. Saul A, Lawrence G, Smillie A, Rzepczyk CM, Reed C, Taylor D, Anderson K, Stowers A, Kemp R, Allworth A, Anders RF, Brown GV, Pye D, Schoofs P, Irving DO, Dyer SL, Woodrow GC, Briggs WR, Reber R, Sturchler D (1999) Vaccine 17:3145
170. Lawrence G, Cheng QQ, Reed C, Taylor D, Stowers A, Cloonan N, Rzepczyk C, Smillie A, Anderson K, Pombo D, Allworth A, Eisen D, Anders R, Saul A (2000) Vaccine 18:1925
171. Moreno A, Paarroyo ME (1989) Blood 74:537
172. Pattarroyo ME, Amador R, Calvijo P, Moreno A, Guzman F, Romero P, Tascon FA, Murillo LA, G. P (1988) Nature 332:158
173. Amador R, Moreno A, Valero V, Murillo L, Mora AL, Rojas M, Rocha C, Salcedo M, Guzman F, Espejo F, Nunez F, Patarroyo ME (1992) Vaccine 10:179
174. Valero MV, Amador LR, Galindo C, Figueroa J, Bello MS, Murillo LA, Mora AL, Patarroyo G, Rocha CL, Rojas M, Aponte JJ, Sarmiento LE, Losada DM, Coronell CG, Ortega NM, Rosas JE, Alonso PL, Patarroyo ME (1993) Lancet 341:705
175. Alonso PL, Smith T, Schellenberg JR, Masanja H, Mwankusye S, Urassa H, Bastos de Azevedo I, Chongela J, Kobero S, Menendez C, Hurt N, Thomas MC, Lyimo E, Weiss NA, Hayes R, Kitua AY, Lopez MC, Kilama WL, Teuscher T, Tanner M (1994) Lancet 344: 1175
176. Alonso PL, Tanner M, Smith T, Hayes RJ, Schellenberg JA, Lopez MC, Bastos de Azevedo I, Menendez C, Lyimo E, Weiss N, Kilama WL, Teuscher T (1994) Vaccine 12:181
177. D'Alessandro U, Leach A, Drakeley CJ, Bennett S, Olaleye BO, Fegan GW, Jawara M, Langerock P, George MO, Targett GA, Greenwood BM (1995) Lancet 346:462
178. Nosten F, Luxemburger C, Kyle DE, Ballou WR, Wittes J, Wah E, Chongsuphajaisiddhi T, Gordon DM, White NJ, Sadoff JC, Heppner DG (1996) Lancet 348:701
179. Acosta CJ, Galindo CM, Schellenberg D, Aponte JJ, Kahigwa E, Urassa H, Schellenberg JR, Masanja H, Hayes R, Kitua AY, Lwilla F, Mshinda H, Menendez C, Tanner M, Alonso PL (1999) Trop Med Int Health 4:368
180. Urdaneta M, Prata A, Struchiner CJ, Tosta CE, Tauil P, Boulos M (1998) Am J Trop Med Hyg 58:378
181. Quakyi IA, Carter R, Rener J, Kumar N, Good MF, Miller LH (1987) J Immunol 139: 4213
182. Duffy PE, Kaslow DC (1997) Infect Immun 65:1109
183. Fynan EF, Robinson HL, Webster RG (1993) DNA Cell Biol 12:785
184. Davis HL, Michel ML, Whalen RG (1993) Hum Mol Genet 2:1847
185. Raz E, Watanabe A, Baird SM, Eisenberg RA, Parr TB, Lotz M, Kipps TJ, Carson DA (1993) Proc Natl Acad Sci USA 90:4523
186. Ulmer JB, Donnelly JJ, Parker SE, Rhodes GH, Felgner PL, Dwarki VJ, Gromkowski SH, Deck RR, De Witt CM, Friedman A, Hawe LA, Leander KR, Martinez D, Perry HC, Shiver JW, Montgomery DC, Liu MA (1993) Science 259:1745
187. Ulmer JB, Deck RR, De Witt CM, Friedman A, Donnelly JJ, Liu MA (1994) Vaccine 12: 1541
188. Donnelly JJ, Ulmer JB, Liu MA (1995) Ann N Y Acad Sci 772:40
189. Donnelly JJ, Martinez D, Jansen KU, Ellis RW, Montgomery DL, Liu MA (1996) J Infect Dis 173:314
190. Boyer JD, Ugen KE, Wang B, Agadjanyan M, Gilbert L, Bagarazzi ML, Chattergoon M, Frost P, Javadian A, Williams WV, Refaeli Y, Ciccarelli RB, McCallus D, Coney L, Weiner DB (1997) Nat Med 3:526

191. Sedegah M, Hedstrom R, Hobart P, Hoffman SL (1994) Proc Natl Acad Sci USA 91:9866
192. Doolan DL, Sedegah M, Hedstrom RC, Hobart P, Charoenvit Y, Hoffman SL (1996) J Exp Med 183:1739
193. Wang R, Doolan DL, Le TP, Hedstrom RC, Coonan KM, Charoenvit Y, Jones TR, Hobart P, Margalith M, Ng J, Weiss WR, Sedegah M, de Taisne C, Norman JA, Hoffman SL (1998) Science 282:476
194. Wang R, Doolan DL, Charoenvit Y, Hedstrom RC, Gardner MJ, Hobart P, Tine J, Sedegah M, Fallarme V, Sacci JB, Jr., Kaur M, Klinman DM, Hoffman SL, Weiss WR (1998) Infect Immun 66:4193
195. Doolan DL, Hoffman SL (2001) Int J Parasitol 31:753
196. Wunderlich G, Moura IC, del Portillo HA (2000) Infect Immun 68:5839
197. Kang Y, Calvo PA, Daly TM, Long CA (1998) J Immunol 161:4211
198. Becker SI, Wang R, Hedstrom RC, Aguiar JC, Jones TR, Hoffman SL, Gardner MJ (1998) Infect Immun 66:3457
199. Sedegah M, Weiss W, Sacci JB, Jr., Charoenvit Y, Hedstrom R, Gowda K, Majam VF, Tine J, Kumar S, Hobart P, Hoffman SL (2000) J Immunol 164:5905
200. Rogers WO, Baird JK, Kumar A, Tine JA, Weiss W, Aguiar JC, Gowda K, Gwadz R, Kumar S, Gold M, Hoffman SL (2001) Infect Immun 69:5565
201. Jones TR, Narum DL, Gozalo AS, Aguiar J, Fuhrmann SR, Liang H, Haynes JD, Moch JK, Lucas C, Luu T, Magill AJ, Hoffman SL, Sim BK (2001) J Infect Dis 183:303
202. Dupuis M, Denis-Mize K, Woo C, Goldbeck C, Selby MJ, Chen M, Otten GR, Ulmer JB, Donnelly JJ, Ott G, McDonald DM (2000) J Immunol 165:2850
203. Gilbert SC, Plebanski M, Harris SJ, Allsopp CE, Thomas R, Layton GT, Hill AV (1997) Nat Biotechnol 15:1280
204. Gilbert SC, Schneider J, Plebanski M, Hannan CM, Blanchard TJ, Smith GL, Hill AV (1999) Biol Chem 380:299
205. Allsopp CE, Plebanski M, Gilbert S, Sinden RE, Harris S, Frankel G, Dougan G, Hioe C, Nixon D, Paoletti E, Layton G, Hill AV (1996) Eur J Immunol 26:1951
206. Tine JA, Lanar DE, Smith DM, Wellde BT, Schultheiss P, Ware LA, Kauffman EB, Wirtz RA, De Taisne C, Hui GS, Chang SP, Church P, Hollingdale MR, Kaslow DC, Hoffman S, Guito KP, Ballou WR, Sadoff JC, Paoletti E (1996) Infect Immun 64:3833
207. Ockenhouse CF, Sun PF, Lanar DE, Wellde BT, Hall BT, Kester K, Stoute JA, Magill A, Krzych U, Farley L, Wirtz RA, Sadoff JC, Kaslow DC, Kumar S, Church LW, Crutcher JM, Wizel B, Hoffman S, Lalvani A, Hill AV, Tine JA, Guito KP, de Taisne C, Anders R, Ballou WR (1998) J Infect Dis 177:1664
208. Shi YP, Hasnain SE, Sacci JB, Holloway BP, Fujioka H, Kumar N, Wohlhueter R, Hoffman SL, Collins WE, Lal AA (1999) Proc Natl Acad Sci USA 96:1615
209. Shi YP, Das P, Holloway B, Udhayakumar V, Tongren JE, Candal F, Biswas S, Ahmad R, Hasnain SE, Lal AA (2000) Vaccine 18:2902
210. Doherty JF, Pinder M, Tornieporth N, Carton C, Vigneron L, Milligan P, Ballou WR, Holland CA, Kester KE, Voss G, Momin P, Greenwood BM, McAdam KP, Cohen J (1999) Am J Trop Med Hyg 61:865
211. Bojang KA, Milligan PJ, Pinder M, Vigneron L, Alloueche A, Kester KE, Ballou WR, Conway DJ, Reece WH, Gothard P, Yamuah L, Delchambre M, Voss G, Greenwood BM, Hill A, McAdam KP, Tornieporth N, Cohen JD, Doherty T (2001) Lancet 358:1927
212. Kester KE, McKinney DA, Tornieporth N, Ockenhouse CF, Heppner DG, Hall T, Krzych U, Delchambre M, Voss G, Dowler MG, Palensky J, Wittes J, Cohen J, Ballou WR (2001) J Infect Dis 183:640
213. Pye D, Vandenberg KL, Dyer SL, Irving DO, Goss NH, Woodrow GC, Saul A, Alving CR, Richards RL, Ballou WR, Wu MJ, Skoff K, Anders RF (1997) Vaccine 15:1017
214. Jones TR, Obaldia N, 3rd, Gramzinski RA, Charoenvit Y, Kolodny N, Kitov S, Davis HL, Krieg AM, Hoffman SL (1999) Vaccine 17:3065
215. Heppner DG, Gordon DM, Gross M, Wellde B, Leitner W, Krzych U, Schneider I, Wirtz RA, Richards RL, Trofa A, Hall T, Sadoff JC, Boerger P, Alving CR, Sylvester DR, Porter TG, Ballou WR (1996) J Infect Dis 174:361

216. Gardner MJ, Tettelin H, Carucci DJ, Cummings LM, Aravind L, Koonin EV, Shallom S, Mason T, Yu K, Fujii C, Pederson J, Shen K, Jing J, Aston C, Lai Z, Schwartz DC, Pertea M, Salzberg S, Zhou L, Sutton GG, Clayton R, White O, Smith HO, Fraser CM, Adamas MD, Venter JC, Hoffman SL (1998) Science 282:1126
217. Millet P, Grady KK, Olsen M, Galland GG, Sullivan JS, Morris CL, Richardson BB, Collins WE, Hunter RL (1995) Am J Trop Med Hyg 52:328
218. Lopez JA, Gonzalez JM, Kettner A, Arevalo-Herrera M, Herrera S, Corradin G, Roggero MA (1997) Ann Trop Med Parasitol 91:253
219. Herrington DA, Clyde DF, Davis JR, Baqar S, Murphy JR, Cortese JF, Bank RS, Nardin E, DiJohn D, Nussenzweig RS, Nussenzweig V, Torres JR, Murillo J, Cortesia M, Sturchler D, Hollingdale MR, Levine MM (1990) Bull World Health Organ 68:33
220. Roggero MA, Weilenmann C, Bonelo A, Audran R, Renggli J, Spertini F, Corradin G, Lopez JA (1999) Parassitologia 41:421
221. Nardin EH, Oliveira GA, Calvo-Calle JM, Castro ZR, Nussenzweig RS, Schmeckpeper B, Hall BF, Diggs C, Bodison S, Edelman R (2000) J Infect Dis 182:1486
222. Nardin EH, Calvo-Calle JM, Oliveira GA, Nussenzweig RS, Schneider M, Tiercy JM, Loutan L, Hochstrasser D, Rose K (2001) J Immunol 166:481
223. Lopez JA, Weilenman C, Audran R, Roggero MA, Bonelo A, Tiercy JM, Spertini F, Corradin G (2001) Eur J Immunol 31:1989
224. Richards RL, Swartz GM, Jr., Schultz C, Hayre MD, Ward GS, Ballou WR, Chulay JD, Hockmeyer WT, Berman SL, Alving CR (1989) Vaccine 7:506
225. Brown AE, Singharaj P, Webster HK, Pipithkul J, Gordon DM, Boslego JW, Krinchai K, Su-archawaratana P, Wongsrichanalai C, Ballou WR, Permpanich B, Kain KC, Hollingdale MR, Wittes J, Que JU, Gross M, Cryz SJ, Sadoff JC (1994) Vaccine 12:102
226. Sturchler D, Berger R, Rudin C, Just M, Saul A, Rzepczyk C, Brown G, Anders R, Coppel R, Woodrow G, Pye D, Sorenson F, Gillessen G, Matile E, Reber-Liske R (1995) Am J Trop Med Hyg 53:423
227. Herrington DA, Nardin EH, Losonsky G, Bathurst IC, Barr PJ, Hollingdale MR, Edelman R, Levine MM (1991) Am J Trop Med Hyg 45:695
228. Herrington DA, Losonsky GA, Smith G, Volvovitz F, Cochran M, Jackson K, Hoffman SL, Gordon DM, Levine MM, Edelman R (1992) Vaccine 10:841
229. Vreden SG, Verhave JP, Oettinger T, Sauerwein RW, Meuwissen JH (1991) Am J Trop Med Hyg 45:533
230. Sherwood JA, Oster CN, Adoyo-Adoyo M, Beier JC, Gachihi GS, Nyakundi PM, Ballou WR, Brandling-Bennett AD, Schwartz IK, Were JB (1991) Trans R Soc Trop Med Hyg 85: 336
231. Patarroyo ME, Romero P, Torres ML, Clavijo P, Moreno A, Martinez A, Rodriguez R, Guzman F, Cabezas E (1987) Nature 328:629
232. Rodriguez R, Moreno A, Guzman F, Calvo M, Patarroyo ME (1990) Am J Trop Med Hyg 43:339
233. Ruebush TKn, Campbell GH, Moreno A, Patarroyo ME, Collins WE (1990) Am J Trop Med Hyg 43:355
234. Patarroyo ME, Amador R, Clavijo P, Moreno A, Guzman F, Romero P, Tascon R, Franco A, Murillo LA, Ponton G, Trujillo, G (1988) Nature 332:158
235. Noya O, Gabaldon Berti Y, Alarcon de Noya B, Borges R, Zerpa N, Urbaez JD, Madonna A, Garrido E, Jimenez MA, Borges RE, Garcia P, Reyes I, Prieto W, Colmenares C, Pabon R, Barraez T, de Caceres LG, Godoy N, Sifontes R (1994) J Infect Dis 170:396
236. Sempertegui F, Estrella B, Moscoso J, Piedrahita L, Hernandez D, Gaybor J, Naranjo P, Mancero O, Arias S, Bernal R, Cordova ME, Suarez J, Zicker F (1994) Vaccine 12:337
237. Alonzo PL, Smith T, Schellenberg JR, Masanja H, Mwankusye S, Urassa H, Bastos De Azevedo I, Chongela J, Kobero S, Menendez C, Hurt N, Thomas MC, Lyimo E, Weiss NA, Hayes R, Kitua AY, Lopez MC, Kilama WL, Teuscheer T, Tanner M (1995) Med Trop (Mars) 55:41
238. Valero MV, Amador R, Aponte JJ, Narvaez A, Galindo C, Silva Y, Rosas J, Guzman F, Patarroyo ME (1996) Vaccine 14:1466

239. Patarroyo G, Franco L, Amador R, Murillo LA, Rocha CL, Rojas M, Patarroyo ME (1992) Vaccine 10:175
240. Amador R, Moreno A, Murillo LA, Sierra O, Saavedra D, Rojas M, Mora AL, Rocha CL, Alvarado F, Falla JC, Orozco M, Coronell C, Ortega N, Molano A, Velasquez JF, Valero MV, Franco L, Guzman F, Salazar LM, Espezo F, Mora E, Farfan R, Zapata N, Rosas J, Calvo JC, Castro J, Quinones T, Nunez F, Patarroyo ME (1992) J Infect Dis 166:139
241. Migasena S, Heppner DG, Kyle DE, Chongsuphajaisiddhi T, Gordon DM, Suntharasamai P, Permpanich B, Brockman A, Pitiuttutham P, Wongsrichanalai C, Srisuriya P, Phonrat B, Pavanand K, Viravan C, Ballou WR (1997) Acta Trop 67:215
242. Enders B, Hundt E, Knapp B (1992) Mem Inst Oswaldo Cruz 87:413
243. Burghaus PA, Wellde BT, Hall T, Richards RL, Egan AF, Riley EM, Ballou WR, Holder AA (1996) Infect Immun 64:3614
244. Cheung A, Leban J, Shaw AR, Merkli B, Stocker J, Chizzolini C, Sander C, Perrin LH (1986) Proc Natl Acad Sci USA 83:8328
245. Collins WE, Pappaioanou M, Anders RF, Campbell GH, Brown GV, Kemp DJ, Broderson JR, Coppel RL, Skinner JC, Procell PM, Favaloro JM, Corcoran LM, Ma N-F, Mitchell GF, Campbell CC (1988) Am J Trop Med Hyg 38:268
246. Genton B, Al-Yaman F, Anders R, Saul A, Brown G, Pye D, Irving DO, Briggs WR, Mai A, Ginny M, Adiguma T, Rare L, Giddy A, Reber-Liske R, Stuerchler D, Alpers MP (2000) Vaccine 18:2504
247. Ridley RG, Takacs B, Etlinger H, Scaife JG (1990) Parasitology 101 Pt 2:187

Received: May 2002

Adv Biochem Engin/Biotechnol (2003) 84: 183–209
DOI 10.1007/b11041

Intracellular Delivery of Drugs to Macrophages

Amitabha Mukhopadhyay[1] · Sandip K. Basu[2]

[1] National Institute of Immunology, Aruna Asaf Ali Marg, New Delhi 110067, India
E-mail: amitabha@nii.res.in
[2] National Institute of Immunology, Aruna Asaf Ali Marg, New Delhi 110067, India
E-mail: sandip@nii.ernet.in

Toxic side effects which often complicate successful therapy in a number of diseases possibly arise due to the fact that at therapeutically effective concentrations the non-target cells in the body are also exposed to the cytotoxic effects of the drugs. Minimization of such adverse reactions might be feasible through drug delivery modalities that would reduce the uptake of the drugs by non-target cells and selectively deliver the drug only to the target cells (and/or intracellular sites) at relatively low extracellular concentrations. The current generic approach to site-specific drug delivery consists of attaching the therapeutic agent to a carrier recognized only by the cells where the pharmacological action is desired. Two types of recognition elements on the surface of target cells are being exploited for this purpose, viz., (i) antigens capable of generating specific, non-cross reactive antibodies, and (ii) receptors on the cell surface capable of efficient transport of the ligands. In general, incomplete specificity for the target cells and poor internalization of antibody-drug conjugates still limit the usefulness of antibodies for site-specific drug delivery applications necessitating exploration of alternatives. The alternate possibility is to exploit the exquisite cell type specificity and high efficiency of endocytosis of macromolecules mediated by specific receptors present on the surface of target cells for delivering drugs. A large number of infectious, metabolic, and neoplastic diseases are associated with macrophages leading to morbidities and mortalities to millions of people worldwide, thus an appropriate design of a drug delivery system to macrophages will be of tremendous help.

Keywords. Endocytosis, Receptor, Ligand, Site specific drug delivery

1 **Introduction** 185

2 **Genesis of the Work** 185

3 **Design of a Drug Carrier** 186

4 **Receptor-Mediated Endocytosis: an Overview** 187

5 **Intracellular Trafficking of Endocytosed Ligands** 189

6 **Receptor-Mediated Endocytosis: Characteristics Important for Drug Delivery** 192

© Springer-Verlag Berlin Heidelberg 2003

7 Macrophages as Target Cells 192

8 Types of Receptors Present on Macrophages 192

8.1 Mannose Receptor 193
8.2 Scavenger Receptor 193

9 Receptor-Mediated Intracellular Drug Delivery to Macrophages . . 194

9.1 To Eliminate Intracellular Pathogens 195
9.2 To Eliminate Neoplastic Cells 201
9.3 To Modulate Immune Responses 203

10 Conclusions 206

11 References 207

Abbreviations

LDL	Low-density lipoprotein
GTP	Guanine-nucleotide triphosphate
ATP	Adenosine triphosphate
GDP	Guanine-nucleotide diphosphate
ER	Endoplasmic reticulum
TGN	trans Golgi network
NSF	*N*-ethylmaleimide sensitive fusion protein
GDI	Guanine-nucleotide dissociation inhibitor
GDF	GDI displacement factor
GEF	Guanine-nucleotide exchange factor
SNAP	Soluble NSF attachment protein
V-SNARE	SNAP receptor on vesicles
T-SNARE	SNAP receptor on target
GAP	GTPase activating protein
HIV	Human immunodeficiency virus
AIDS	Acquired immunodeficiency syndrome
Ig	Immunoglobulin
IFN-γ	Interferon-γ
CRD	Carbohydrate-recognition domains
SR	Scavenger receptor
MBSA	Maleylated bovine serum albumin
MTX	Methotrexate
PAS	*para*-Aminosalicyclic acid
PolyG	Polyguanylic acid
VSV	Vesicular stomatitis virus
ANS	Antisense oligonucleotide
DNM	Daunomycin

DXR	Doxorubicin
MDR	Multidrug resistance
MDP	Muramyl dipeptide
IL	Interleukin
TNF-α	Tumor necrosis factor-α
EGF	Epidermal growth factor
EGFR	EGF receptor
CFU	Colony forming unit
mg	Milligram
Kg	Kilogram
μg	Microgram
mL	Milliliter
kDa	Kilodalton
BSA	Bovine serum albumin

1 Introduction

The enormous strides in improvement of human health and healthcare over the 20th century owe much to Paul Ehrlich's concept of chemotherapy, the use of a specific drug to eliminate the causal agent of a disease without undue side effects. However, the euphoria that chemotherapy generated was rather short-lived as the unacceptable side effects of many drugs began manifesting and as organisms or cancers, initially sensitive, started displaying resistance against the erstwhile highly effective drugs [1]. Toxic side effects which often complicate successful therapy in a number of diseases possibly arise due to the fact that at therapeutically effective concentrations the non-target cells in the body are also exposed to the cytotoxic effects of the drugs. This is an inherent limitation of current pharmacological practice, the central dogma of which was first enunciated by Brody in the 1930s, viz., the effect of a drug is proportional to its concentration in the blood or other body fluids. However, in any disease condition, only a relatively small fraction of the 10^{13} cells that comprise the human body needs intervention with drugs to cure or curb the disease. Unfortunately, many of the drugs in current therapeutic use have no special affinity for the diseased target cells and are available to the normal cells as well. Therefore, a relatively high dose of the drug needs to be present in the body fluids in order to attain therapeutically effective concentration of the drug inside the cells, which aggravates the possibility of toxic side reactions. The problem of delivering therapeutic agents to the target cells intended for intervention, excluding the vast majority of the 10^{13} cells constituting a human body, thus continues to be a prime challenge in therapeutics.

2 Genesis of the Work

Macrophages generate multiple defensive responses that serve to protect mammals against a variety of invading microorganisms and developing cancers. However, in many instances these defensive responses are overwhelmed, circumvented or even misdirected so that these protective cells become the focal points in a large number of diseases that include the entire spectrum of infectious, metabolic and neoplastic diseases that afflict millions of people worldwide. A wide variety of pathogens has evolved to use macrophages as the host causing various intracellular infections that take a major toll in terms of morbidity and mortality especially in the third world. Management of such diseases is difficult because the intracellular location of the causative agents provides them a measure of protection both from the host immune system and from chemotherapeutic agents. Therefore, a generalized delivery system for targeting a variety of therapeutic agents to the cells of macrophage lineage is likely to be useful. As early as 1906 Paul Ehrlich conceived the idea of using antibodies, which had been discovered only a few years previously, as carriers to target drugs selectively to the cells where pharmacological action is desired [2]. The current generic approach to site-specific drug delivery thus consists of attaching the therapeutic agent to a carrier recognized only by the target cells. Two types of recognition elements on the surface of target cells are being exploited for this purpose, viz., (i) antigens capable of generating specific, non-cross reactive antibodies, and (ii) receptors on the cell surface capable of efficient transport of the ligands. In general, incomplete specificity for the target cells and poor internalization of antibody-drug conjugates still limit the usefulness of antibodies for site-specific drug delivery applications necessitating exploration of alternatives. We have been exploring the alternate approach of exploiting the exquisite cell type specificity and high efficiency of endocytosis mediated by cell surface receptors present exclusively on the target cells for achieving such a selective drug delivery. Our conceptual approach to this problem consists of linking an appropriate drug to a macromolecular carrier recognized by receptors present exclusively on the surface of the macrophages so that the drug conjugate can bind to the receptor with high affinity and undergo efficient internalization. Subsequent degradation of the drug conjugate in the lysosomes should lead to intracellular release of the drug only in the receptor-bearing cells resulting in a high intracellular concentration of drug in the target cells at relatively low extracellular drug levels thereby minimizing toxic side reactions. In the present review, we have focused our discussion on the current perspective of selective and efficient intracellular drug delivery to macrophages by receptor-mediated processes, mainly highlighting work done in this area in India.

3 Design of a Drug Carrier

The key feature of a targeted drug is a carrier by which the drug would be recognized and taken up only by the target cells thereby diminishing the toxic side effects due to the interaction of the drug with normal cells. An ideal drug carrier should: (a) bypass the anatomic barriers, (b) be recognized only by the target cells, (c) discharge the drug at or inside the target cells and not elsewhere, (d) not cause toxic or immune reactions, and (e) the drug-carrier complex should be manufacturable under sterile and apyrogenic conditions.

To date, no single carrier meets all the above criteria. In general, two different types of carriers are being developed, viz., (i) particulate and (ii) soluble macromolecules. Particulate carriers are largely limited to topical applications because they cannot cross various anatomic barriers while soluble macromolecular carriers can be used for systemic applications [3,4]. Regardless of the particulate or soluble nature of the carrier, the carrier for targeting it to specific cells can either be an antibody specific for the target cell, or a ligand for the receptor molecules present only on the target cells [5]. In theory, monoclonal antibody-mediated drug delivery appears to be a versatile and universally applicable approach for site-specific drug delivery. However, the initial promise of this approach is still somewhat clouded because: (a) most so-called specific antibodies also react with non-target cells, (b) often the access of the antibodies to the target cells is limited, (c) most antibody-drug combinations are inefficiently internalized by the cells, and (d) antibodies and antibody-drug conjugates often produce immunological reactions.

The use of monoclonal antibodies for targeting purposes, however, is an extremely active area of research and many of these problems are being overcome. For instance, monoclonal antibodies specific for some cancer cells have been attached to highly potent bacterial or plant toxins to generate immunotoxins that selectively destroy cancer cells. Genetic engineering approaches are also being explored to increase the specificity or reduce the immunogenicity of such molecules [6].

4
Receptor-Mediated Endocytosis: an Overview

Biochemical, ultrastructural and genetic studies in a number of systems during the last two decades have elucidated the mechanism by which cells import macromolecules for their nutrition, defense, transport and other cellular processing. The principles of this process, called 'receptor-mediated endocytosis' (Fig. 1), was first enunciated by Goldstein, Brown and coworkers during 1975–1983 while trying to understand the molecular basis of atherosclerosis. These studies revealed how mammalian cells transport large molecules such as low-density lipoprotein (LDL, the main cholesterol-carrying protein in man) from the exterior to the interior of the cells [7,8]. They showed that specific proteins, the LDL receptors, on the surface of human cells bind LDL molecules with high affinity and specificity. LDL molecules bound to these receptors are internalized efficiently by the cells and are ultimately delivered through several vesicular compartments to the lysosomes, where the protein part of the LDL molecules are degraded. LDL receptor molecules, however, are not degraded along with LDL but are returned to the cell surface for use in multiple rounds of LDL delivery. Although first established with LDL, it later became apparent that cells use this process for transporting large molecules for a variety of purposes such as nutrition, hormone action, antigen processing and so on.

In general, ligands first bind to the cell surface receptor with high affinity and receptor-ligand complexes get clustered into coated pits. The major component of the coat is a 180 kDa protein, known as clathrin [9] which interacts with the

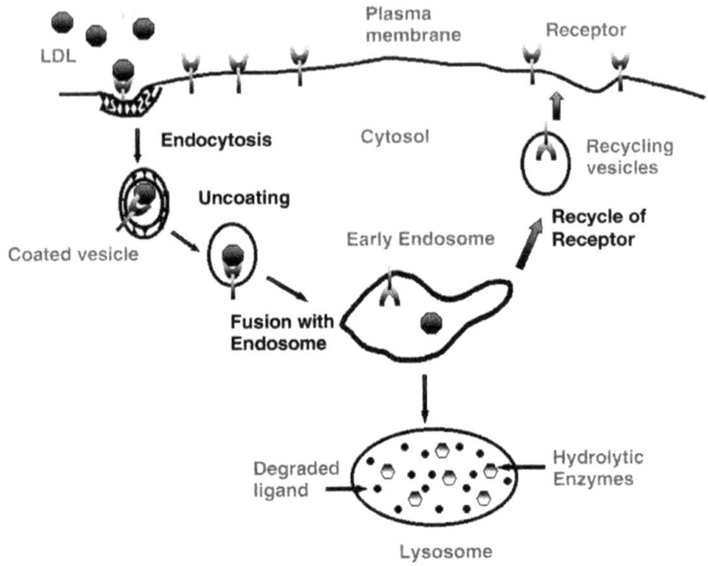

Fig. 1. Receptor-mediated endocytosis of LDL. Redrawn from [78]

cytoplasmic tail of the receptor via a 100/50/16 kDa protein complex, termed adaptors [10]. Movement of the receptors into the coated pits could be spontaneous or ligand induced. The next important step in receptor-mediated endocytosis is the internalization of the receptor-ligand complex. Studies with the LDL receptor first suggested that information for the internalization signal lies within the cytoplasmic tail of the receptor which is recognized by the adaptor on the cytoplasmic side of the plasma membrane [11]. Analysis of the genomic DNA encoding the cytoplasmic domain of the LDL receptor isolated from skin fibroblasts of a patient suffering from familial hypercholesterolemia revealed that a single base substitution converting a tyrosine to a cysteine codon at residue 807 abrogated the internalization ability of the LDL receptor [12, 13]. Similarly, tyrosine residues in the cytoplasmic domain of the other plasma membrane receptors were found to be critical for their internalization [14, 15]. However, once the coated pit is fully invaginated, a membrane scissors is required to release the coated vesicle. Studies with the temperature sensitive *shibire* mutant of *Drosophila melanogaster* showed the accumulation of coated pits [16] which led to the discovery of mammalian homologue of the *shibire* gene product known as dynamin. Dynamin is a 100 kDa protein with GTPase activity which acts as a membrane scissors to cleave the coated pits from the plasma membrane [17, 18]. Subsequently, disassembly of the clathrin coat is mediated by a 70 kDa uncoating ATPase [19] present in the cytosol leading to the formation of uncoated vesicles which are rapidly delivered together with their contents to the endosomal sorting compartment [20]. The endosomal compartment is a complex structure of tubules and vesicles [21] which mediate the sorting of the internalized cargo to the subsequent compartment through vesicle fusion depending on the nature of the receptor. The early endosomal compartments are acidified by a proton pump that resides on the membrane of the respective [22] vesicles and this acidic environment is responsible for the dissociation of most of the ligands from their receptors [23]. Receptor-ligand complexes after internalization appear to follow multiple pathways, which vary depending on the signal carried by the receptor. Therefore, depending on the nature of the receptor, the receptor-mediated endocytosis are classified into one of the following four types:

1. Receptors recycled and ligand degraded in the lysosomes; e.g., low density lipoprotein, mannose or galactose terminated oligosaccharides/glycoproteins [24, 25].
2. Both receptors and ligands recycled; e.g., transferrin, major histocompatibility antigens [26, 27].
3. Both receptors and ligands degraded; e.g., Fc receptors, epidermal growth factor [15, 28].
4. Receptors degraded and ligands transported across the cell; e.g., immunoglobulin A and immunoglobulin M [29].

5 Intracellular Trafficking of Endocytosed Ligands

After endocytosis of the receptor-ligand complex, transport of materials from one intracellular compartment to another occurs through vesicle budding and fusion. Vesicles containing cargo bud from the donor compartment and deliver their contents to the acceptor compartment by fusion with the latter. During the past few years, it has been shown that the processes of vesicle budding, docking and fusion with the acceptor compartment are regulated by a series of small GTP binding proteins [30]. All these proteins regulate membrane trafficking events by altering between two conformations in a nucleotide-dependent fashion where GTP bound form turns the protein "on" and hydrolysis of the GTP to GDP turns the protein "off". These activities are regulated by a series of generic and compartment-specific proteins. These proteins are specifically localized on distinct intracellular compartments and presumably control a specific transport step [31, 32]. Among the monomeric small GTPases, the Rab family of Ras related proteins [33] are well characterized as regulators of intracellular trafficking during endocytosis and secretion. Rab proteins are specifically localized to the cytoplasmic surface of the intracellular compartments that they subserve (Fig. 2). Rab 1 and Rab 2 are found to be localized in ER and cis-Golgi. Similarly, Rab 6 and Rab 9 are localized in medial Golgi and trans Golgi, respectively [34, 35]. Rab 4, Rab 5, Rab 11, Rab 18 and Rab 24 have all been shown to be associated with early endocytic compartments, suggesting that the early endocytic compartment is highly complex with multiple functions [36–38]. Internalized receptor and ligand first enter the peripheral sorting endosome where the membrane proteins destined for degradation or trans Golgi network (TGN) are sorted away from membrane proteins targeted for recycling back to the plasma membrane. A substan-

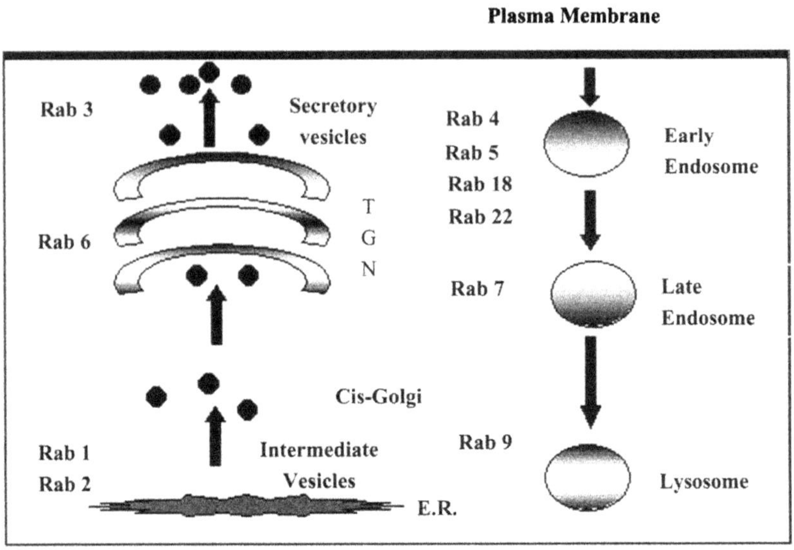

Fig. 2. Schematic diagram showing the localization of different Rab proteins and a particular transport event mediated by them

tial body of evidence indicates that Rab 5 regulates endocytosis from the plasma membrane to the early endosomes and is also involved in homotypic fusion between the early endosomes [39]. Rab 4 appears to control the recycling from the early endosomes to the plasma membrane. Recent studies indicate that Rab 11 regulates the recycling through the perinuclear endosomes. Rab 7 and Rab 9, and perhaps others still to be identified, are associated with the late endosomal compartments. Rab 9 has been shown to regulate the traffic from the late endosome to TGN and Rab 7 has been shown to regulate the traffic from the early to late endosomes/lysosome [40, 41]. It has also been shown that Rab 7 regulates the transport between the late endosome to a lysosome-like compartment [42]. Among the endocytic Rabs, Rab 5 has been most extensively studied. Rab 5 is active in GTP bound form and is regulated by multiple factors like *N*-ethylmaleimide sensitive fusion proteins (NSF), PI3 kinase, phospholipase A2 etc. [43–45]. Rab 5 and Rab 7 both regulate endocytosis but Rab 7 functions downstream of Rab 5. However, the exact relationship among all these factors in the overall mechanism of intracellular trafficking during endocytosis is largely unknown.

A current model (Fig. 3) suggests that Rab proteins are predominantly found on cellular membranes but a significant fraction is also present in the cytosol as a complex with a protein called guanine nucleotide dissociation inhibitor (GDI). GDI presents the complex to the specific organelles through an interaction catalyzed by GDI displacement factor (GDF) in conjunction with guanine nucleotide exchange factor (GEF). GTP-bound Rab proteins are recruited onto nascent

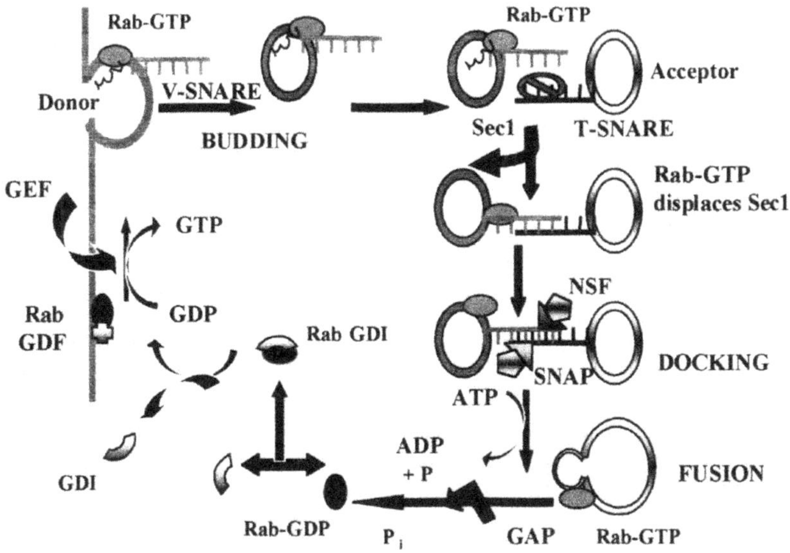

Fig. 3. Schematic representation of the membrane trafficking cycle. Redrawn based on [48, 49]

transport vesicles and catalyze the association of V SNARE [soluble NSF attachment protein (SNAP) receptor on vesicle] with T SNARE (SNAP receptor on target) to accomplish docking of the vesicles. Each V SNARE and T SNARE complex binds between 3 to 6 SNAP proteins and then NSF binds to this complex. Subsequently NSF hydrolyzes the ATP and uses the energy made available to disrupt the docking complex and release the SNAP protein [46, 47]. After the fusion GTPase activating proteins (GAP) increase the GTPase rate of the Rab protein, converting it into its GDP-bound conformation. Unoccupied GDI then retrieves the Rab protein for another round of vesicular transport [48, 49]. This is a generalized model (Fig. 3) but most of the interacting proteins for a particular Rab have not been identified.

Knowledge about the intracellular trafficking of the internalized ligands and their regulation by different signal transduction molecules provided clues to divert the intracellular route of the ligands by modulating the expression of Rab proteins. Recently, it has been shown that various intracellular pathogens like *Salmonella*, *Mycobacterium* survive in an early compartment by recruiting some early acting Rab protein on their phagosomes [50–54]. It will be interesting to determine if overexpression of the signal responsible for the transport of the cargo to the lysosomal compartment mediates the killing of invading pathogens.

6 Receptor-Mediated Endocytosis: Characteristics Important for Drug Delivery

The physiological process of receptor-mediated endocytosis is eminently amenable for the purposes of intracellular drug delivery because: (a) receptors present exclusively on specific cell-types permit the design of cell-type specific delivery systems, (b) high affinity binding of ligands to receptors allows their efficient retrieval even at low concentrations outside the cells, (c) directing ligand traffic to specific locations inside the cells (such as the lysosomes, the nucleus, cytosol) or transport across the cells is possible, and (d) recycling of receptors permits multiple rounds of ligand internalization, building up relatively high intracellular ligand levels. However, detailed knowledge about the nature of the ligand, the relative distributions of the receptor on various cell types and the intracellular pathways followed by the receptor-ligand complexes are essential prerequisites for success in exploitation of the process of receptor-mediated endocytosis in site-specific drug delivery applications. At present, only hepatocytes and cells of the macrophage lineage seem to possess receptor systems exploitable for selective drug delivery [5, 55, 56].

7 Macrophages as Target Cells

Macrophages, polymorphonuclear cells of the reticuloendothelial system, mount multipronged defensive responses that protect the host against a variety of invading microorganisms and developing cancers [57]. However, in many instances, these defensive responses are overwhelmed, circumvented or even misdirected so that these protective cells become the focal points in a large number of diseases of infectious (e.g., tuberculosis, leprosy, leishmaniasis, trypanosomiasis, dengue, Japanese encephalitis, HIV-AIDS), metabolic (rheumatoid arthritis, Gaucher's disease) or neoplastic (monocytic leukemia, histiocytosis) etiologies [58–60]. Furthermore, others and we have recently shown how invading pathogens successfully modulate the signal transduction pathways in the host cells to ensure survival in macrophages [50–54, 61]. Therefore, manipulation of the metabolic responses of macrophages in various disease states appears to be an attractive approach for the management of such diseases, success in which crucially depends on the ability to deliver appropriate drugs or biological response modifiers specifically to macrophages [5, 8, 55].

8 Types of Receptors Present on Macrophages

Receptors on the macrophage plasma membrane play an important role in controlling growth, differentiation, endocytosis, secretion and immune response. Macrophages express several members of the Ig superfamily [62] including distinct FcR as well as MHC class I and II molecules and CD4, which are involved in complex immune interactions. Macrophages also express several members of integrin family including the common β-chain of CR3/LFA and receptor for fibronectin [63, 64]. Macrophage receptors for a particular class of ligand are often heterogeneous, e.g., the distinct receptor molecules for different classes of IgG [65], and one receptor molecule may also interact with several ligands. Thus, most of the receptor systems reported so far fail to meet the criterion of adequate selectivity, being distributed ubiquitously or on many cell types. Nevertheless, attempts continue to be made to exploit quantitative differences in the content of such receptors between normal and, for instance, cancer cells for targeting antineoplastic drugs in cancer chemotherapy with limited success. Among the large variety of receptor systems expressed by macrophages, the mannosyl/fucosyl receptors [66] and the scavenger receptors [67–69] are the most well characterized.

8.1 Mannose Receptor

Predominantly found in liver and spleen macrophages, the mannose receptor was discovered as a carbohydrate-binding protein that recognizes mannose or fucose terminal glycoproteins such as lysosomal hydrolases and functions in their clearance from plasma [70]. Mannose receptor activity is also found in monocytes, macrophages from bone and liver Kupffer cells [66]. Neoglycoproteins [71] prepared by chemical attachment of mannose or fucose residues to bovine serum albumin (BSA) are effective ligands for the mannose receptor and have been used as a carrier of drugs to macrophages [72]. The human mannose receptor has been cloned and sequenced, revealing the important features of the receptor structure and function [73]. The primary structure of the mannose receptor revealed a type I transmembrane orientation with several distinct domains including a fibronectin type II-like repeats, suggesting a possible role in the interaction of differentiated macrophages with the extracellular matrix. The mannose receptor consists of eight segments resembling the carbohydrate-recognition domains (CRDs) of C-terminal lectins [74] which possibly explain the receptor's dependence on ligand multivalency for high affinity binding. This receptor has served as an excellent model for the study of endocytosis in macrophages. It is constitutively internalized through the coated pits and recycled from the early endosomes after dissociation from the ligand in acidic endosomes [75]. Ligand binding to the receptor was found to be highly sensitive to pH and Ca^{2+} concentration and the ligand is transported to the lysosome after dissociation from the receptor in the endosomes. Thus, being a recycling receptor permitting degradation of the ligand in the lysosomes, mannose receptor-mediated endocytosis could be effective for the delivery of drugs to macrophages. However, agents that activate macrophages like IFN-γ, lipopolysaccharides, phorbol esters as well as infection with intracellular pathogens like *Leishmania* selectively downregulate mannose receptor expression [76, 77] thereby limiting their use as generic drug carriers to macrophages for multiple purposes.

8.2 Scavenger Receptor

Scavenger receptors were discovered in 1979 during attempts to understand how cholesterol from LDL accumulates in macrophages in xanthomas and atherosclerotic plaques in patients with familial hypercholestolemia, as a set of binding sites present on macrophages that specifically bind and internalize acetylated low density lipoprotein [67, 69, 78]. Cross competition experiments revealed that these receptors bind a variety of polyanionic macromolecules. These include (a) chemically modified proteins such as acetylated LDL and maleylated albumin but not their unmodified counterparts, (b) anionic polysachharides and phospholipids such as fucoidin, carrageenan, bacterial LPS, lipoteichoic acid and phosphatidylserine, (c) other anionic substances like polyvinyl sulfate, dextran sulfate, asbestos, (d) certain polynucleotides like polyinosinic and polyguanylic acids, but not polycytidylic or polyadenylic acids suggesting that a base-quartet-stabilized four-stranded helical structure assumed by polyguanylic and -inosinic acids is recognized by the scavenger receptors [79]. A 220 kDa protein which specifically bind acetylated LDL was purified and characterized from bovine liver membrane [80, 81] and found to be a trimer of 77 kDa subunits with asparagine-linked carbohydrate chains.

Three classes of scavenger receptors are known, viz., classes A, B and C. The class A scavenger receptors are the most well characterized and are expressed primarily on monocytes, peritoneal and tissue macrophages. Some dendritic cells, specialized endothelial cells and smooth muscle cells in atherosclerotic lesions also express scavenger receptors [82, 83]. Two types of scavenger receptors were cloned from bovine liver, termed as SR-AI and SR-AII [84, 85]. The cDNA sequence and biochemical analysis have suggested that SR-AI are heterotrimeric integral membrane proteins consisting of six distinctive domains [80, 86]: domain I is the N-terminal cytoplasmic domain, domain II, a single transmembrane domain and four extracellular domains. Among the extracellular domains, it was shown that domain III is a spacer, domain IV is coiled coil consisting of 16 heptad repeats, domain V contains 23 or 24 uninterrupted Gly-X-Y triplet repeats with proline or lysine frequently present in the Y position which indicates the formation of a classic right-handed collagenous triple helix and domain VI comprises an 8 residue hinge followed by a 102-residue cysteine-rich domain. Work done by several groups [85–89] has demonstrated that a cluster of positive charges provided by the collagenous domain is responsible for the broad ligand specificity of the scavenger receptor. The type II scavenger receptor (SR-AII) differs from the SR-AI in that the domain VI of the SR-AII is highly truncated where the cysteine-rich domain is replaced by a short 6–17 amino acid C terminus [85, 87].

9
Receptor-Mediated Intracellular Drug Delivery to Macrophages

Scavenger receptors can be exploited for delivering drugs into macrophages because: (a) macrophages, and to a lesser extent, endothelial cells express the scavenger receptors, (b) the receptors are localized on the cell surface and mediate rapid transport of the ligands, (c) these receptors recycle and are not downregulated by the known ligands facilitating multiple rounds of ligand delivery, and (d) scavenger receptors recognize multiple ligands allowing flexibility for designing carriers for different purposes. The scavenger receptor-mediated endocytic pathway offers a facile means to deliver drugs to macrophages since drug molecules attached to the ligands of scavenger receptors are internalized by the macrophages with high efficiency and released inside these cells after the carrier is degraded in the lysosomes [90, 91]. Some examples of the application of scavenger receptor-mediated targeting approach are outlined below.

9.1
To Eliminate Intracellular Pathogens

Leishmania donovani, the etiological agent for visceral leishmaniasis or *kala-azar* in humans, reside and proliferate solely in the phagolysosomes of the macrophages of the infected host. All currently used anti-leishmanial drugs such as antimony compounds, diamidines and amphotericin, often lead to severe toxic side reactions. This probably happens because these drugs are taken up not only by the host macrophages but also by other cell types. We intended to design a cell-specific drug delivery system to macrophages that will increase the antiparasitc efficacy of the drug and simultaneously reduce its toxicity. Initially, we showed that methotrexate (MTX), a potent antifolate drug, conjugated with maleylated bovine serum albumin (MBSA), a specific ligand for macrophage scavenger receptor, was specifically recognized by scavenger receptors and killed the intracellular form of *Leishmania* [92] (Fig. 4). The power of this approach was established in an experiment in which hamsters were infected in the footpads with *Leishmania mexicana* which swelled to about 10-times the normal footpad size due to multiplication of the protozoa inside the macrophages. When free methotrexate (MTX) was given no substantial cure was achieved. Administering the drug in a form targeted to macrophages as MBSA-MTX conjugate brought the footpad size back to normal. All the animals remained healthy and no antibodies against the drug conjugate could be detected in these animals [90].

The scavenger-receptor-mediated mode of drug delivery was next extended to the treatment of an intracellular bacterial disease, tuberculosis. Tuberculosis continues to be a major public health concern and the incidence of tuberculosis is also increasing among the patients with AIDS in recent years. The pathogenicity is due to the multiplication of *Mycobacterium tuberculosis* within the mononuclear phagocytes. All currently used drugs are potent but toxic, thus intracellular delivery of antitubercular agent to macrophages may reduce the adverse side reactions. Moreover, drugs currently used in tuberculosis are quite expensive and difficult to afford by the common people in the developing world. Thus, a effective anti-tubercular drug like *para*-aminosalicylic acid (PAS) which is cheap but was discontinued mainly because of its toxicity will be welcome if drug can be delivered only to the host cells and thereby reducing the toxicity. We showed that the efficacy of PAS was enhanced nearly 80-fold by conjugating it with MBSA [93]. Studies in an animal model in which guinea pigs were infected with a virulent strain of *M. tuberculosis* showed that compared to the untreated control group, MBSA-PAS treatment reduced the bacterial load in the lung by six log units and increased the survival rate of the infected animals to 87% while the survival rate of animals treated with a three-times higher dose of free PAS was only 13% (Fig. 5). High and persistent drug activity was observed in the target organs

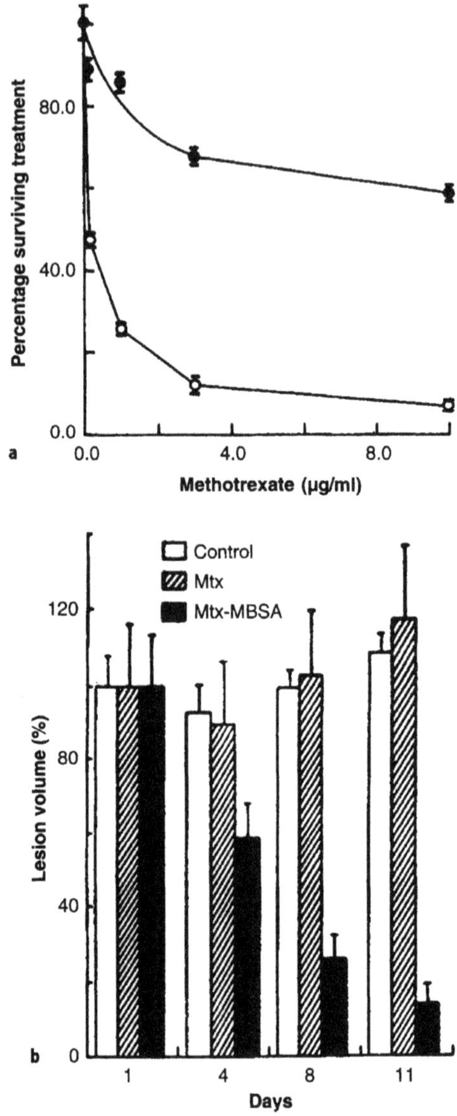

Fig. 4. Scavenger receptor-mediated drug delivery in chemotherapy of leishmaniasis. *A* The *Leishmania mexicana amazonensis* infected macrophages were incubated with indicated concentrations of MBSA-MTX or MTX at 37 °C for 3 h, then washed. After 20 h incubation, cells were fixed and stained with Giemsa and the number of amastigotes per 100 macrophages was determined. The number of amastigotes present in untreated culture was taken as 100%. *B* Hamsters were infected with *Leishmania mexicana amazonensis* amastigotes isolated from donor animals. Eight weeks postinfection, hamsters were treated with MBSA-MTX or MTX (1 mg/kg) on days 1 to 4. Control animals were injected with PBS alone. Size of the footpad lesion was measured on the indicated days. Reprinted from [90]

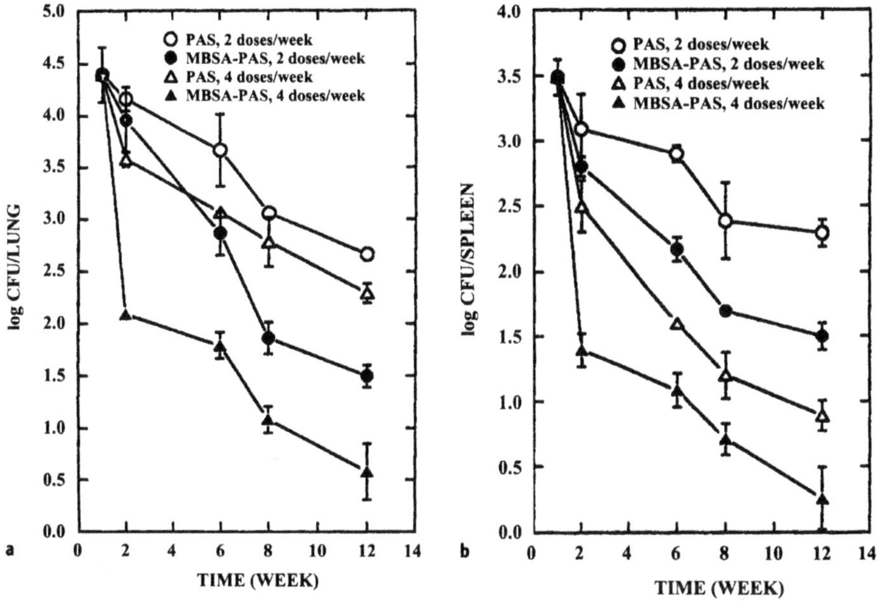

Fig. 5. Effect of MBSA-PAS against tuberculosis infection in guinea pigs. *Mycobacterium tuberculosis* infected guinea pigs (12 weeks postinfection) were treated with two or four doses of either free PAS (150 mg/kg) or MBSA-PAS (50 mg/kg) intravenously. Animals were sacrificed on the first day of the indicated weeks and CFU were determined from the homogenates of indicated tissues. Reprinted from [94]

when the drug was administered as an MBSA-PAS conjugate [94]. These studies demonstrated the feasibility of using scavenger receptor-mediated endocytosis to deliver the appropriate therapeutic agent to the intracellular infection associated with macrophages which will build up a high intracellular concentration of the drug at a low plasma concentration and thus reduce its toxicity.

Oligonucleotides complementary to the cognate mRNA sequences can bind to it and block its translation into a functional protein. Such antisense oligonucleotides have demonstrated, in vitro and in vivo, the ability to reduce or prevent the production of target proteins, and several such molecules are in clinical trials. Antisense oligonucleotides are particularly attractive as antiviral agents as they can be designed to block viral replication within infected cells without affecting the metabolism of the host cells. The therapeutic use of antisense oligonucleotides is hampered because of their susceptibility to degradation by nucleases, incomplete specificity for the target cells and poor cellular uptake. The selective and efficient intracellular delivery of antisense molecules to the target cells is essential for the success of this approach and better methods are needed [95].

We had established earlier that polyguanylic acid chains as small as 10 nucleotides are efficiently taken up by macrophages through scavenger receptor-mediated endocytosis. We therefore made various constructs consisting of a 15mer oligonucleotide sequence antisense to the translation initiation site of the N protein gene of vesicular stomatitis virus, ANS (CAT TTT GAT TAC TGT), with a 10mer PolyG tail at the 3′ end as follows:

- **ANS-G:** as native phosphodiester (CAT TTT GAT TAC TGT GGG GGG GGG G),
- **sANS-G:** as phosphorothioate (***CAT TTT GAT TAC TGT*** GGG GGG GGG G),
- **cANS-G:** as a chimeric molecule in which the first and last internucleotide linkages in the antisense portion of the oligonucleotide sequence were phosphorothioate (***CA***T TTT GAT TAC T***GT*** GGG GGG GGG G),
- **Sense-G:** complementary to ANS (ACA GTA ATC AAA ATG GGG GGG GGG G),

and two control sequences,

- **Scram-G:** a scrambled 15mer sequence (ATC GTA TTA CTT GTT GGG GGG GGG G),
- **ANS-C:** antisense sequence above with a 10mer polycytidilic acid (PolyC) tail at the 3′-end (CAT TTT GAT TAC TGT CCC CCC CCC C) since PolyC is not recognized by the SCR.

In order to determine if the scavenger receptors on macrophages recognized oligonucleotides with a 10mer PolyG tail at their 3′-ends, we tested the abilities of these oligonucleotide constructs to compete for the degradation of ^{125}I-MBSA in a murine macrophage cell line J774E. Data presented in Fig. 6 show that the 10mer PolyG and oligonucleotide constructs with 10mer PolyG tails at their 3′

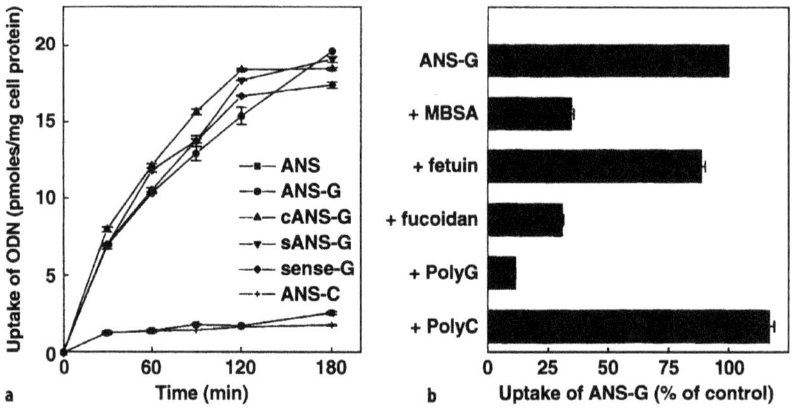

Fig. 6. Scavenger receptor-mediated enhanced uptake of PolyG constructs by J774E macrophages. **a** J774E cells in RPMI with BSA (1 mg/mL) and ^{32}P-labeled oligonucleotide (1 μM) were incubated at 37 °C for the indicated periods of time and processed to determine cell-associated radioactivity. **b** Cells were incubated as above with ^{32}P-labeled ANS-G (1 μM; specific activity 555 cpm/pmol) at 37 °C either alone or along with different polyanionic macromolecules (MBSA 500 μg/mL; fetuin 100 μg/mL; fucoidan 100 μg/mL; PolyG 7 μg/mL and PolyC 7 μg/mL. After 4 h, the contents of cell-associated radioactivity were determined. Results shown are mean±s.e. of three independent determinations. Reprinted from [95]

ends effectively compete for the degradation of ^{125}I-MBSA, while a 10mer PolyC and oligonucleotides containing 10mer PolyC tails did not interfere with the degradation of ^{125}I-MBSA. These results indicate that the scavenger receptors on J774E macrophages efficiently recognize oligonucleotides containing PolyG tails at their 3′ ends. Recognition of the PolyG constructs by the scavenger receptors was independent of both the sequences of the oligonucleotides and chemical modifications to the backbone of the oligonucleotides because Sense-G, Scram-G, cANS-G and sANS-G competed for the degradation of ^{125}I-MBSA as efficiently as ANS-G.

To assess the extent of scavenger receptor-mediated facilitated delivery to J774E cells, we examined the uptake of oligonucleotides with or without a PolyG or PolyC tail. Fig. 6 (Panel a) shows that uptake of the PolyG tail-bearing antisense N oligonucleotide construct by the macrophages was 8–12-fold higher compared to that of the antisense N oligonucleotide without the PolyG extension. We further found that ANS-G, cANS-G, sANS-G and Sense-G were taken up with comparable efficiencies whereas uptake of ANS-C was around the same levels as ANS. Uptake of the ANS-G construct was inhibited by the known scavenger receptor ligands MBSA, PolyG and fucoidan but not by fetuin or PolyC (Fig. 6, Panel b). These data indicate the feasibility of delivering antisense molecules to macrophages using a scavenger receptor-specific polynucleotide ligand.

Data in Fig. 7 demonstrate the bioefficacy of PolyG-mediated targeting of antisense oligonucleotides in arresting the replication of vesicular stomatitis virus (VSV) in J774E cells. The antisense oligonucleotides lacking a 10mer PolyG stretch were ineffective. The inhibition of VSV replication due to the ANS-G antisense oligonucleotide was completely abrogated in the presence of 10mer PolyG indicating that the antisense-effect of the ANS-G molecule was a consequence of scavenger receptor-mediated enhanced uptake. Importantly, antisense molecules linked by all natural phosphodiester bonds were as bioeffective as those synthesized with a mixed backbone of phosphodiester and phosphorothioate. The feasibility of delivering the antisense molecules to the macrophages by scavenger receptor-mediated endocytosis is further strengthened by the recent observation that PolyG-mediated delivery of DNAzyme specifically blocks HIV gene expression [96].

Selective targeting of therapeutic agents to macrophages using mannose receptors to combat intracellular infection has also been attempted. Several groups have generated mannose-coated liposomes for selective delivery of drug to macrophages exploiting the phagocytic properties of the mannose receptor. Approaches for targeting liposomes to macrophages include mannosylation of preformed liposomes [97, 98], formation of liposomes from mannosylated phospholipids [99] or grafting a mannose terminal protein on the liposome surface [100, 101]. Subsequent studies have demonstrated that mannose-bearing liposomes containing urea stibamine, an anti-leishmanial drug, have superior efficacy than the conventional drug in eliminating the parasites from experimental leishmaniasis in the hamster model [102, 103]. Similarly, tuftsin-bearing liposomes containing amphotericin B was shown to inhibit *Leishmania* and fungal infection in macrophages as well as in an appropriate animal model [104]. Another experimental design for macrophage-specific drug delivery exploiting mannose receptor-mediated endocytosis employs soluble macromolecular con-

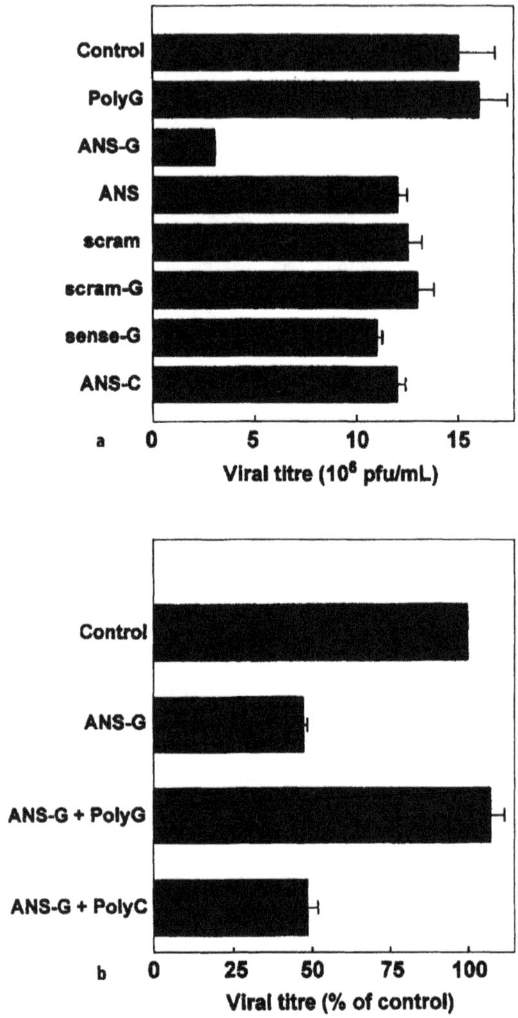

Fig. 7. Scavenger receptor-mediated antisense oligonucleotide delivery inhibits VSV replication in J774E macrophages. **a** Inhibition of VSV replication by PolyG-tailed ANS construct. **b** Reversal of antisense oligonucleotide inhibition by competing scavenger receptor ligands. Reprinted from [95]

jugates, e.g., mannose-BSA conjugated with therapeutic agents. Utilizing this approach, Chakraborty et al. [105] have shown that methotraxate-conjugated mannose-BSA significantly inhibited the growth of *Leishmania donovani* in macrophages in vitro as well as in infected animals.

9.2 To Eliminate Neoplastic Cells

The efficacy of the scavenger receptor-mediated drug delivery was next tested in a model for histiocytic malignancy in which cells of monocyte-macrophage origin turn malignant. We found that cancer cells of macrophage origin can efficiently take up and degrade a conjugate of MBSA with the highly toxic anticancer drug, daunomycin (DNM), while cancer cells of non-macrophage origin could not do so. Uptake of MBSA-DNM could stop the synthesis of DNA in macrophage cancer cells but not in non-macrophage cancer cells. Ex vivo treatment of the cells from murine tumors of macrophage origin abolished their ability to form tumors in BALB/c mice whereas such cells treated under identical conditions with free DNM were able to form tumors when transplanted back into mice as effectively as the untreated cells [106] (Fig. 8). When injected into mice, the drug conjugate suppressed the growth of J774A.1 tumors at much lower dosages of DNM relative to the free form of the drug. MBSA-DNM also effectively arrested the growth of pre-formed tumors whereas the tumors grew as rapidly in animals treated with free DNM or MBSA as in the animals that did not receive any drug or carrier [107]. As a prelude to determining the feasibility of extending this approach to human cancers of macrophage origin, we have shown the superior efficacy of MBSA coupled with the anticancer drug, doxorubicin (DXR) in arresting the growth of human histiocytic lymphoma cells in culture in contrast to the failure of free DXR to do so under similar conditions [91].

Similarly, mannose receptors were also used to deliver anticancer agents to macrophages using mannosylated vesicles loaded with macrophage activators [97, 99, 108]. In an interesting study, it was demonstrated that the A chain of the potent plant toxin, Ricin, is a ligand for the mannose receptor and has been shown to be toxic to macrophages in vitro [109, 110].

We have also examined the possible application of scavenger receptor-mediated drug delivery for circumvention of the insidious problem of multidrug resistance (MDR) in neoplastic cells, which is a major obstacle to success in cancer chemotherapy. For this purpose, a multidrug-resistant variant (JD100) of the murine macrophage tumor cells J774A.1 was developed. MDR in JD100 cells was shown to involve both P-glycoprotein-dependent and -independent mechanisms, resembling the complexity often encountered in clinical situations. We showed that the multidrug-resistant JD100 cells took up MBSA-DNM with high efficiency and their proliferation was arrested as measured by cessation of DNA synthesis [111]. It is likely that the high efficiency uptake of anticancer agents through scavenger receptor-mediated endocytosis serves to build up an adequate intracellular concentration of the drug even in the presence of a p-glycoprotein-like efflux pump in multidrug-resistant cells leading to selective killing of the multidrug-resistant tumor cells. It is tempting to speculate that in the receptor-mediated mode

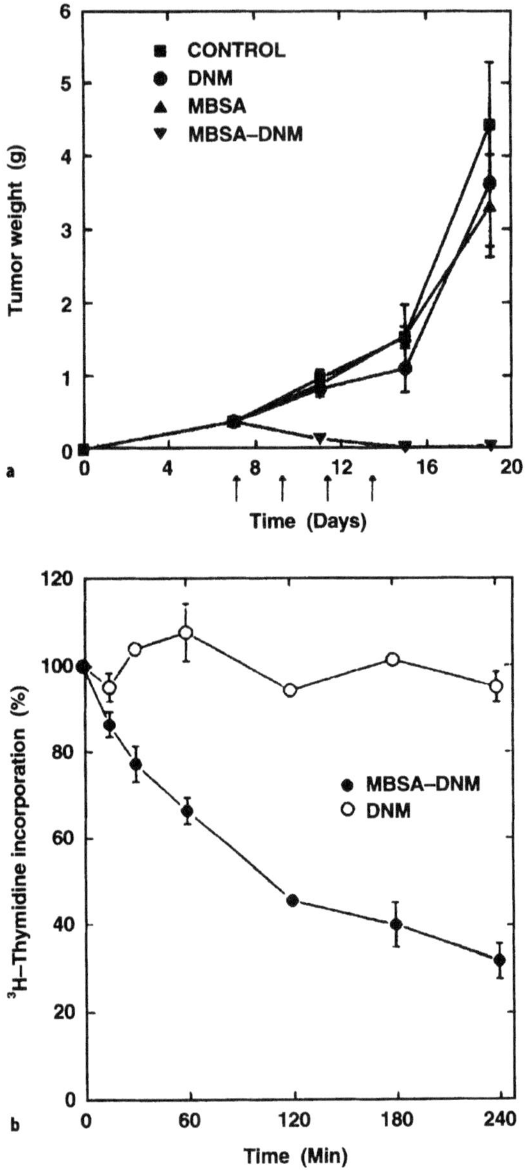

Fig. 8. Selective inhibition of macrophage cancer cells (**a**) and multidrug resistant cancer (**b**) cells by scavenger receptor-mediated delivery of daunomycin. **a** Effect of free or conjugated DNM (30 μg DNM) on the growth of preformed macrophage tumor in BALB/C mice. **b** Time course of killing of multidrug-resistant macrophage tumor cells (JD-100) by 0.3 μM of DNM either in free or conjugated form. Reprinted from [106, 111]

of drug delivery, the drug conjugate was delivered intracellularly to multidrug-resistant cells through vesicle fusion so that the drug was not accessible to the efflux pump.

9.3
To Modulate Immune Responses

As macrophages are known to migrate to solid tumor sites, we next explored if normal macrophages could be activated to a tumoricidal state by intracellular delivery of a biological response modifier, muramyl dipeptide [112, 113].

J774A.1 macrophages pretreated with MBSA-MDP brought about 50% lysis of the prelabeled B16F10 melanoma target cells at a concentration at which free MDP was ineffective. This cytolytic activity of the activated macrophages was found to be saturated at a low concentration of MBSA-MDP (Fig. 9). Moreover, treatment of the macrophages with BSA-MDP or MBSA at a dose equivalent to MBSA-MDP did not show any cytolytic activity towards the target cells. Data presented in Fig. 9 show that macrophages treated with MBSA-MDP displayed 7–10-fold higher cytotoxic activity against non-macrophage tumor cells compared to that elicited by free MDP-treated macrophages.

In order to elucidate the mechanism of antitumor activity, we explored whether the treatment of macrophages with MDP induces the secretion of various cytokines and other soluble mediators. MBSA-MDP treatment of murine peritoneal resident macrophages induced them to secrete significantly higher levels of IL-1, IL-6 and TNF-α in comparison with that released by macrophages on

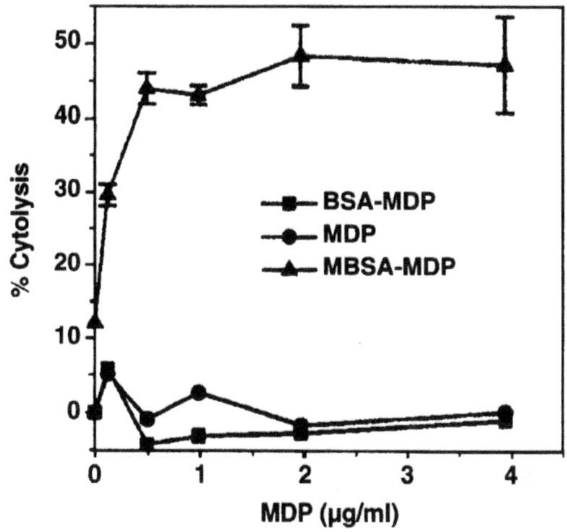

Fig. 9. Cytolysis of tumor cells by MBSA-MDP treated macrophages. J774A.1 macrophages treated with the indicated concentrations of MDP (free or conjugated with BSA or MBSA) for 24 h were washed and incubated with [^{3}H]-thymidine labeled B16F10 melanoma cells. Radioactivity released in the medium after 48 h was measured as an index of cytolysis. Reprinted from [112]

treatment with free MDP while PGE_2 levels were unaltered. Interestingly, MBSA-MDP-treated macrophages secreted higher, sustained levels of IL-12 while IL-10, a macrophage suppressor cytokine, could be detected only on free MDP treatment (Fig. 10).

To understand the role of various metabolites secreted by the activated macrophages in bringing about cytolysis of the tumor cells, cytotoxic assays were carried out in the presence of antibodies directed against one or more of the soluble mediators, individually or in combination. Addition of antibodies directed against IL-1, IL-6 or TNF-α singly to macrophage cultures partially inhibited the tumoricidal efficacy of MBSA-MDP-treated macrophages whereas the combined presence of anti-IL-6 and anti-TNF-α antibodies abrogated the tumoricidal response. The inhibition of bioefficacy due to neutralization of IL-6 and TNF-α was further modulated by indomethacin, indicating a regulatory role of PGE_2 (Fig. 11). These data suggest that the elevated secretion of specific soluble mediators, viz., IL-6 and TNF-α on MBSA-MDP treatment of macrophages augmented their antineoplastic activity. Importantly, the normal surface phenotype and functional integrity of macrophages was not altered by MBSA-MDP treatment indicating the immunotherapeutic potential of the receptor system and MDP, in combination, as an alternative approach to treatment of neoplasia.

In conclusion, our data demonstrate that it is possible to augment the antitumor efficacy of macrophages by delivering MDP via the scavenger receptor-mediated endocytic pathway. Intracellular delivery of MDP presumably modulates the signal transduction cascade in macrophages leading to secretion of various cytokines and other soluble mediators that are known to possess antitumor activity. TNF and

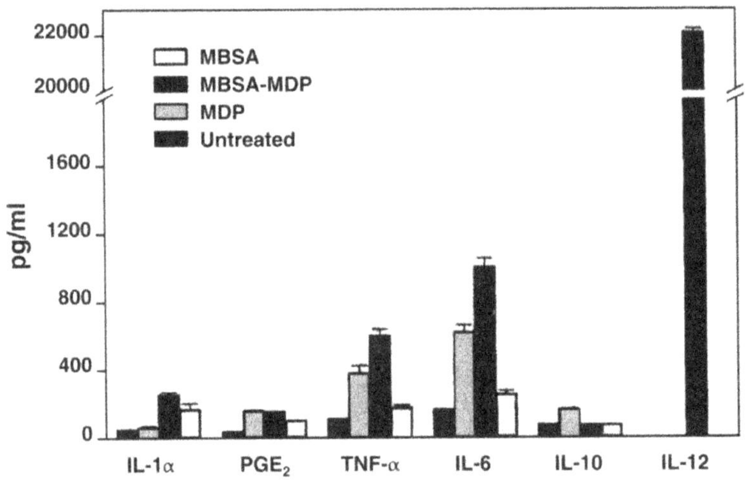

Fig. 10. Secretion of active metabolites by murine peritoneal resident macrophages. Peritoneal macrophages were treated in culture with 0.25 μg/mL MDP (free or conjugated with MBSA) for 24 h. The contents of the indicated metabolites secreted into the culture medium were measured by ELISA. Reprinted from [112]

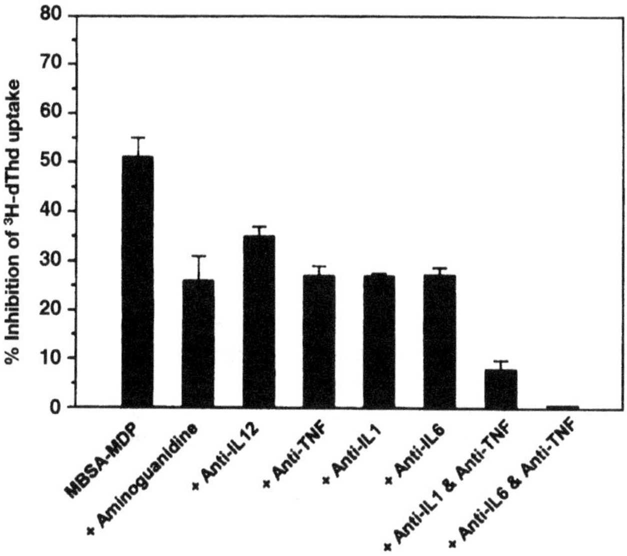

Fig. 11. Abrogation of antitumor activity by antibodies against secretory metabolites. Peritoneal macrophages were treated in culture with MBSA-MDP (0.25 μg/mL MDP equivalent) for 24 h, washed and incubated with L929 target cells along with the indicated neutralizing antibodies alone or in various in combinations (20 μg/mL). After 24 h, the cells were pulsed with [^{3}H]-dThd for a further 24 h period and the amounts of radioactivity in the target cells were expressed as % of radioactivity incorporated by the target cells in the absence of macrophages. Reprinted from [112]

IL-6 in conjunction with IL-1 and reactive nitrogen intermediates presumably mediate the tumoricidal ability of the macrophages treated with maleylated conjugate. Low immunogenicity and normal function of the MBSA-MDP treated macrophages indicate the potential of this general modality in cancer treatment.

Although the need for development of new and better chemotherapeutic agents is undeniable, vaccination appears to be the most cost-effective means for prevention and eventual eradication of infectious diseases. Success of anti-viral vaccines in eradicating small pox and effectively controlling measles, polio, mumps etc. has been spectacular. However, it has not yet been possible to make satisfactory vaccines against diseases such as malaria, kala azar, tuberculosis, etc. caused by more complex intracellular organisms. Recent studies carried out by Rath and his colleagues have shown that targeting of antigens to the scavenger receptors generates the Th1 type of immune response deemed necessary for protection against intracellular organisms [114–117]. It has also been shown recently that scavenger receptor-mediated delivery of allergens diverts an ongoing allergic immune response to a non-allergic route, brightening the prospects of immunotherapeutic measures against allergy [118]. These observations may provide a lead for manipulating immune responses to generate new generation vaccines and immunotherapeutics. Furthermore, they recently showed that the delivery of autologous serum albumin to scavenger receptors leads to a breakdown of T cell tolerance in mice to self-albumin, offering a new tool to dissect mechanisms of immune tolerance and the etiopathogenesis of autoimmunity [115, 119].

Diseases such as cancer, atherosclerosis, rheumatoid arthritis, multiple sclerosis etc. result from malfunctioning of signaling pathways. With the advent of the knowledge of the molecular basis of intercellular and intracellular communication, which is regulated by different signal transduction molecules, attempts have been made to exploit signaling pathway for rational drug design [120]. Inhibition of signaling pathways can be achieved by a variety of reagents, small molecules, antibodies and DNA encoding dominant negative proteins or antisense RNA. Usually, the activity of a particular signal transduction pathway is enhanced in disease cells compared to normal cells. For example, different oncogenic proteins like Ras and kinase activity are enhanced in most tumor cells. Recent studies have shown that inhibitor of EGFR kinase activity as well as JAK-2 kinase blocker significantly blocks the progression of tumor in mice [121, 122]. Macrophages have the ability to migrate toward the sites of inflammation, thus macrophages loaded with inhibitors of signal transduction intermediate exploiting receptor-mediated endocytosis may find an application in the near future.

10 Conclusion

Exploitation of the principles of scavenger receptor-mediated endocytosis for site-specific drug delivery is maturing into a rational approach which merits serious consideration both in designing new chemotherapeutic agents, and in resurrecting effective but highly toxic molecules for chemotherapy. In particular, scavenger receptor-mediated endocytosis has been demonstrated to offer a mechanism to deliver a wide variety of agents to macrophages including chemotherapeutic agents for intracellular infections and neoplasia, antisense oligonucleotides for viral infection as well as the activation/modulation of macrophage function. However, despite these promising results, there are many hurdles to cross before drugs based on these elegant targeting principles would find their way into the marketplace of reality. Thus, Paul Ehrlich's quest is still on; magic bullets still elude us.

Acknowledgement. The authors gratefully acknowledge the participation of Drs. S. K. Arora, Sunanda Basu, Gautam Chaudhuri, Shehla Hashim, Sadhana Majumdar, Bratati Mukhopadhyay, Konark Mukherjee, Vikram Prasad, Shantanu Sengupta, Shadab Siddiqi, S. Srividya, J. Tripathi; financial support from the Department of Biotechnology, the Department of Science and Technology, the Council of Scientific and Industrial Research, Indian Council of Medical Research; and collaboration with Drs. V. Bal, R. P. Roy, S. Rath and S. Sehgal at various stages of the work described are also greatly appreciated.

11 References

1. Pearse BMF (1976) Proc Natl Acad Sci USA 73:1255
2. Ehrlich P (1906) In: Collected Studies on Immunity, vol 2. John Wiley, New York, pp 442
3. Fiani ML, Stahl PD (1989) Trends Biotechnol 7:57
4. Poznansky MJ, Juliano RL (1984) Pharmacol Rev 36:277
5. Basu SK (1990) Biochem Pharmacol 40:1941
6. Pastan I, Willingham MC, FitzGerald DJP (1986) Cell 47:641
7. Goldstein JL, Basu SK, Brown MS (1983) Meth Enzymol 8:241
8. Basu SK (1984) J Biosci 6:535
9. Pearse BMF (1976) Proc Natl Acad Sci USA 73:1255
10. Pearse BMF (1988) EMBO J 7:3331
11. Lehrmann MA, Goldstein JL, Brown MS, Russell DW, Schneider WJ (1985) Cell 41:737
12. Davis CG, Lehrmann MA, Russell DW, Anderson RGW, Brown MS, Goldstein JL (1986) Cell 45:15
13. Anderson RGW, Brown MS, Goldstein JL (1977) Cell 10:351
14. Rothenberger S, Iacopetta BJ, Kuhn LC (1987) Cell 49:423
15. Miettinen .M, Rose JK, Mellman I (1989) Cell 58:317
16. Kosaka T, Ikeda K (1983) J Cell Biol 97:499
17. Van der Bliek AM, Meyerowitz EM (1991) Nature 351:411
18. Scaife R, Margolis RL (1990) J Cell Biol 111:3023
19. DeLuca-Flaherty C, Mckay DB, Parham P, Hill BL (1990) Cell 62:875
20. Goud B, Zahraoui A, Tavitan A, Saraste J (1990) Nature 345:553
21. Hopkins CR, Trowbridge IS (1983) J Cell Biol 97:508
22. Anderson RGW, Falck JR, Goldstein JL, Brown MS (1984) Proc Natl Acad Sci USA 81:4838
23. Stahl P, Schwartz AL (1986) J Clin Invest 77:657
24. Basu SK, Goldstein JL, Anderson RGW, Brown MS (1981) Cell 24:493
25. Ezekowitz RAB, Stahl PD (1988) J Cell Sci Suppl 9:121
26. Hanover CJ, Dickson RB (1985) In: Pastan I, Willingham MC (eds) Endocytosis. Plenum Press, New York, pp 131–162
27. Unanue ER (1984) Annu Rev Immunol 2:395
28. Stoschek CM, Carpenter G (1984) J Cell Biol 91:1048
29. Mostov KE, Friedlander M, Blobel G (1984) Nature 308:37
30. Blum JS, Diaz R, Mayorga LS, Stahl, PD (1993) Subcellular Biochemistry 9:69
31. Balch WE (1990) Trends Biochem Sci 15:473
32. Rothman JE, Sollner TH (1997) Science 276:1212
33. Schimmoller F, Simon I, Pfeffer SR (1998) J Biol Chem 273:22161
34. Chavrier P, Parton RG, Hauri HP, Simons K, Zerial M (1990) Cell 62:317
35. Goud B, Zahraoui A, Tavitan A, Saraste J (1990) Nature 345:553
36. Sluijs V, Hull PM, Zahraoui A, Tavitan A, Goud B, Mellman I (1991) Proc Natl Acad Sci USA 88:6313
37. Bucci C, Parton RG, Mather IH, Stunnenberg H, Simons K, Hoflack B, Zerial M (1992) Cell 70:715
38. Li G, Stahl PD (1993) J Biol Chem 268:24475
39. Barbieri MA, Li G, Colombo MI, Stahl PD (1994) J Biol Chem 269:18720
40. Lombardi D, Soldati T, Reiderer MA, Goda Y, Zerial M, Pfeffer SR (1993) EMBO J 12:677
41. Mukhopadhyay A, Barbieri AM, Funato K, Roberts R, Stahl PD (1997a) J Cell Biol 136:1227
42. Mukhopadhyay A, Funato F, Stahl PD (1997b) J Biol Chem 272:13055
43. Diaz R, Mayorga LS, Weidman PJ, Rothman JE, Stahl PD (1989) Nature 339:398
44. Mayorga LS, Colombo MI, Lennartz M, Brown EJ, Rahman KH, Weiss R, Lennon PJ, Stahl PD (1993) Proc Natl Acad Sci USA 90:20255
45. Li G, D'Souza-Schorey C, Barbieri MA, Roberts RL, Klippel A, Williams LT, Stahl PD (1995) Proc Natl Acad Sci USA 92:10207

46. Sollner T, Whiteheart SW, Brunner M, Erdjument-Bromage H, Geromanos S, Tempst P, Rothman JE (1993) Nature 362:318
47. Soggard M, Tani K, Ye R, Geromanos S, Tempst P, Kirchhausen T, Rothman JE, Sollner T (1994) Cell 78:937
48. Rothman JE, Wieland FT (1996) Science 272:227
49. Pfeffer SR (1999) Nature Cell Biol 1:E17
50. Via LE, Deretic D, Ulmer RJ, Hibler NS, Huber LA, Deretic V (1997) J Biol Chem 272:13326
51. Garcia-del Portillo F (1999) Trends Microbiol 6:467
52. Uchiya K, Barbieri MA, Funato F, Shah AH, Stahl PD, Groisman EA (1999) EMBO J 18:3924
53. Mukherjee K, Siddiqui S, Hashim S, Raje M, Basu SK, Mukhopadhyay A (2000). J Cell Biol 148:741
54. Hashim S, Mukherjee K, Raje M, Basu SK, Mukhopadhyay A (2000) J Biol Chem 275:16281
55. Gordon S, Rabinowitz S (1989) Adv Drug Deliv Rev 4:27
56. Fiume L, Busi G, Di Stefano A, Mattioli A (1994) Adv Drug Deliv Rev 14:51
57. Lewis CE, McGee JO (eds) (1992) The Macrophage. Oxford University Press, Oxford
58. Kleinerman ES, Schroit AJ, Fogler WE, Fidler IJ (1983) J Clin Invest 72:1
59. Mogensen SC (1979) Microbiol Rev 43:1
60. Schroit AJ, Hart IB, Madsen J, Fidler I (1983) J Biol Resp Modifiers 2:97
61. Mukherjee K, Parashuraman S, Raje M, Mukhopadhyay A (2001) J Biol Chem 276:23607
62. Williams AF (1987) Immunol Today 8:298
63. Rettenmier CW, Roussel MF, Sherr CJ (1988) J Cell Sci Suppl 9:27
64. Wright SD, Detmers PA (1988) J Cell Sci Suppl 9:99
65. Mellman I, Koch T, Healey G, Hunzeker W, Lewis V, Plutner H, Miettienen H, Vaux D, Moore K, Stuart S (1988) J Cell Sci Suppl 9:45
66. Pontow SE, Kery V, Stahl PD (1993) Int Rev Cytol B137:221
67. Basu SK, Brown MS, Ho YK, Goldstein JL (1979) J Biol Chem 254:7141
68. Goldstein JL, Ho YK, Basu SK, Brown MS (1979) Proc Natl Acad Sci USA 76:333
69. Brown MS, Basu SK, Falck JR, Ho YK, Goldstein JL (1980) J Supramol Struct 13:67
70. Schlesinger PH, Rodman JS, Frey M, Lang S, Stahl PD (1976) Arch Biochem Biophys 177:606
71. Lee YC, Stowell C, Krantz MJ (1976) Biochemistry 15:3956
72. Stahl PD (1992) Curr Opinion Immunol 4:49
73. Ezekowitz RAB, Sastry K, Bailly P, Warner A (1990) J Exp Med 172:1785
74. Drickamer K (1988) J Biol Chem 263:9557
75. Stahl PD, Schlesinger PH, Sigardson E, Rodman JS, Lee YC (1980) Cell 19:207
76. Shepherd VL, Abdolrasulnia R, Garrett M, Cowan HB (1990) J Immunol 145:1530
77. Basu N, Sett R, Das PK (1991) Biochem J 277:451
78. Goldstein JL, Brown MS, Anderson RGW, Russell DW, Schneider WJ (1985) Annu Rev Cell Biol 1:1
79. Pearson A, Rich A, Krieger M (1993) J Biol Chem 268:3546
80. Kodama T, Reddy P, Kishimoto C, Krieger M (1988) Proc Natl Acad Sci USA 85:9238
81. Krieger M, Herz J (1994) Annu Rev Biochem 63:601
82. Geng Y-J, Hansson GK (1995) Scand J Immunol 42:289
83. Li H, Freeman MW, Libby P (1995) J Clin Invest 95:122
84. Kodama T, Freeman M, Rohrer L, Zabrecky J, Matsudaira P, Krieger M (1990) Nature 343:531
85. Rohrer L, Freeman M, Kodama T, Penman M, Krieger M (1990) Nature 343:570
86. Penman M, Lux A, Freeman M, Rohrer L, Ekkel Y, McKinstry H, Resnick D, Krieger M (1991) J Biol Chem 266:23985
87. Ashkenas J, Penman M, Vasile E, Acton S, Freeman M, Krieger M (1993) J Lipid Res 34:984
88. Acton SL, Scherer PE, Lodish HF, Krieger M (1994) J Biol Chem 269:21003
89. Doi T, Higashino K, Kurihara Y, Wada Y, Miyazaki T, Nakamura H, Uesugi S, Imanishi T, Kawabe Y, Itakura H, Yazaki Y, Matsumoto A, Kodama T (1993) J Biol Chem 268:2126
90. Mukhopadhyay A, Chaudhuri G, Arora SK, Sehgal S, Basu SK (1989) Science 244:705
91. Basu S, Mukhopadhyay b, Basu SK, Mukhopadhyay A (1994) FEBS Lett 342:249

92. Chaudhuri G, Mukhopadhyay A, Basu S K (1989) Biochem Pharmacol 38:2995
93. Majumdar S, Basu SK (1991) Antimicrob Agents Chemother 335:135
94. Majumdar S, Basu SK (1995) Drug Delivery 2:144
95. Prasad V, Hashim S, Mukhopadhyay A, Basu SK, Roy RP (1999) Antimicrob Agents Chemother 43:2689
96. Unwalla H, Banerjea AC (2001) Biochem J 357:97
97. Yamashita C, Sone S, Ogura T, Kiwada H (1991) Jpn J Cancer Res 82:569
98. Muller CD, Schuber F(1989) Biochem Biophys Acta 986:97
99. Barratt GM, Nolibe D, Yapo A, Petit JF, Tenu JP (1987) Ann Inst Pasteur Immunol 138:437
100. Surolia A, Bachhawat BK, Poddar SK (1975) Nature 257:802
101. Garcon N, Gregoriadis G, Taylor M, Summerfield J (1988) Immunology 64:743
102. Das N, Mahato SB, Naskar K, Ghosh DK, Basu MK (1990) Biochem Med Metabolic Biology 43:133
103. Basu MK (1994) Biotech Genetic Eng Rev 12:383
104. Agrawal AK, Gupta CM (2000) Adv Drug Del Rev 41:135
105. Chakraborty P, Bhaduri AN, Das PK (1990) Biochem Biophys Res Commun 166:404
106. Mukhopadhyay A, Mukhopadhyay B, Srivastava RK, Basu SK (1992) Biochem J 284:237
107. Mukhopadhyay B, Mukhopadhyay A, Basu SK (1993) Biochem Pharmacol 46:919
108. Dumont S, Muller CD, Schuber F, Bartholeyns J (1990) Anticancer Res 10:155–160
109. Simmons BM, Stahl PD, Russell JH (1986) J Biol Chem 261:7912
110. Skilleter DN, Foxwell BMJ (1986) FEBS Lett 196:344
111. Mukhopadhyay A, Mukhopadhyay B, Basu SK (1995) FEBS Lett 376:95
112. Srividya S, Roy RP, Basu SK, Mukhopadhyay A (2000) J Leucocyte Biol 67:683
113. Srividya S, Roy RP, Basu SK, Mukhopadhyay A (2000) Biochem Biophys Res Commun 268:772
114. Abraham R, Singh N, Mukhopadhyay A, Basu SK, Bal V, Rath S (1995) J Immunol 154:1
115. Abraham R, Choudhury A, Basu SK, Bal V, Rath S (1997) J Immunol 158:4029
116. Singh N, Bhatia S, Abraham R, Basu SK, George A, Bal V, Rath S (1998) J Immunol 160:4869
117. Bansal P, Mukherjee P, Basu SK, George A, Bal V, Rath S (1999) J Immunol 162:4430
118. Bhatia S, Mukhopadhyay S, Jarman E, Hall G, George A, Basu SK, Rath S, Lamb JR, Bal V (2002) J Allergy Clin Immunol 109:321
119. Choudhury A, Mukherjee P, Basu SK, George A, Bal V, Rath S (2000) J Immunol 164:1713
120. Levitzki A (1996) Curr Opin Cell Biol 8:240
121. Fry DW, Kraker AJ, McMichael A, Ambroso A, Nelson JM, Leopold WR, Connors RW, Bridges AJ (1994) Science 9:1093
122. Mayden N, Gruenberger T, Dadi H, Shahar M, Arpaia E, Lapidot Z, Leader S, Freedman M, Cohan A, Gazit A (1996) Nature 379:645

Received: April 2002

Adv Biochem Engin/Biotechnol (2003) 84: 211–273
DOI 10.1007/b11042

Recent Advances in Tuberculosis Research in India

Anil K. Tyagi[1] · Neeraj Dhar[2]

[1] Department of Biochemistry, University of Delhi South Campus, Benito Juarez Road, New Delhi – 110021, India. *E-mail: AKT@dusc.ernet.in*, AKT1000@hotmail.com
[2] Department of Biochemistry, University of Delhi South Campus, Benito Juarez Road, New Delhi – 110021, India. *E-mail: needhar@yahoo.com*

Tuberculosis (TB) continues to be the leading killer of mankind among all infectious diseases, especially in the developing countries. Since the discovery of tubercle bacillus more than 100 years ago, TB has been the subject of research in an attempt to develop tools and strategies to combat this disease. Research in Indian laboratories has contributed significantly towards developing the DOTS strategy employed worldwide in tuberculosis control programmes and elucidating the biological properties of its etiologic agent, *M. tuberculosis*. In recent times, the development of tools for manipulation of mycobacteria has given a boost to researchers working in this field. New strategies are being employed towards understanding the mechanisms of protection and pathogenesis of this disease. Molecular methods are being applied to develop new tools and reagents for prevention, diagnosis and treatment of tuberculosis. With the sequencing of the genome of *M. tuberculosis*, molecules are being identified for the development of new drugs and vaccines. In this chapter, the advances made in these areas by Indian researchers mainly during the last five years are reviewed.

Keywords. Tuberculosis, Mycobacteria, India

1 **Introduction** 213

1.1 History of Tuberculosis Research in India 213
1.1.1 Ancient History 213
1.1.2 Modern Tuberculosis Research 214
1.2 Global Burden of Tuberculosis 215
1.3 Tuberculosis in India 215
1.4 Goals for Tuberculosis Research 216
1.5 Difficulties in Working with Mycobacteria 217

2 **Genetics of *Mycobacterium tuberculosis*** 217

2.1 Construction and Characterization of Plasmids 218
2.2 Mycobacteriophages 221
2.3 Transposition in Mycobacteria 222
2.4 DNA Replication, Homologous Recombination and DNA Repair . 223
2.5 Synthesis of Proteins and Growth Rate of Mycobacteria 228
2.6 Transcription in Mycobacteria 230
2.7 Expression of Genes in Mycobacteria 233

© Springer-Verlag Berlin Heidelberg 2003

3 Physiology of *Mycobacterium tuberculosis* 234

3.1 Cell Division in Mycobacteria 234
3.2 Proteins and Antigens of *M. tuberculosis* 235
3.3 Membrane Permeability and Transport in *M. tuberculosis* 239

4 Immunology of Tuberculosis 242

4.1 Immune Studies in Humans 244

5 Mechanisms of Pathogenesis of *M. tuberculosis* 248

5.1 Modulation of Host Cell Signaling 251

6 Approaches for Prevention and Treatment of Tuberculosis 253

6.1 Molecular Approaches for the Diagnosis of Tuberculosis 253
6.2 New Candidate Vaccines Against Tuberculosis 255
6.3 Development of Anti-Tubercular Drugs 258
6.3.1 X-Ray Crystallography of Novel Drug Targets 258
6.3.2 Development of Genetic Screens 260
6.3.3 Strategies for Enhancing Drug Efficacy 261

7 Host Genetics in Susceptibility/Resistance to Tuberculosis 263

8 Future Trends 265

9 References 267

Abbreviations

ATP Adenosine 5′-triphosphate
BCG Bacille Calmette Guerin
bp Base pair
CAT Chloramphenicol acetyl transferase
CMI Cell mediated immunity
DOTS Directly observed treatment, short-course
DNA Deoxyribonucleic acid
DTH Delayed Type Hypersensitivity
ELISA Enzyme linked immunosorbent assay
GFP Green fluorescent protein
h hour(s)
HIV Human immunodeficiency virus
HLA Human leukocyte antigen
IFN-γ Interferon γ
Ig Immunoglobulin
IL- Interleukin
INH Isoniazid

IS element	Insertion sequence element
kDa	Kilodalton
LAM	Lipoarabinomannan
mAb	Monoclonal antibody
MCS	Multiple cloning site(s)
MHC	Major histocompatibility complex
MIC	Minimal inhibitory concentration
mRNA	Messenger RNA
nM	nanomolar
NKT	Natural killer T-cells
NTP	National Tuberculosis Programme
ORF	Open reading frame(s)
PBMC	Peripheral blood mononuclear cell
PCR	Polymerase chain reaction
PPD	Purified protein derivative
rBCG	Recombinant BCG
RBS	Ribosome binding site
RNA	Ribonucleic acid
RNTCP	Revised National Tuberculosis Control Programme
rRNA	Ribosomal RNA
SDS-PAGE	Sodium dodecyl sulfate-polyacrylamide gel electrophoresis
SSB	Single stranded DNA binding protein
ssDNA	Single stranded DNA
TB	Tuberculosis
TCA	Tricarboxylic acid
tRNA	Transfer RNA
TSP	Transcription start point
UDG	Uracil DNA glycosylase
WHO	World Health Organization
X-gal	5-Bromo-4-chloro-3-indolyl-β-D-galactopyranoside

1 Introduction

1.1 History of Tuberculosis Research in India

1.1.1 *Ancient History*

Sushruta, considered as the Father of Surgery in India, described tuberculosis (TB) in *Sushruta samhita*, a compendium of works on ancient medicine and surgery written in 600 BC. TB was referred to as *Kshaya* (wasting away) or *Raja Yakshama* (King of Diseases) [1]. Several historians believe that the description of this disease passed orally from generation to generation long before the Vedas were written [2, 3]. Ancient Indian drugs were known to the Egyptians long before Hippocrates and mention of ancient Indian remedies has been found in the Old Testament. Several Indian scriptures have described methods to treat this disease. Treatment was also described on the principles of Ayurveda. Besides medicines, milk (especially from a lactating woman), a moderate quantity of alcohol and the flesh of birds and animals, which inhabit dry areas, were recommended [4]. The Yajur Veda recommended that a consumptive should go and live in high altitudes [3]. Even today in the field of Ayurvedic medicine, several concoctions are known and prescribed for the treatment of TB. Ayurveda also describes the anti-tubercular potency of several medicinal plants available in India.

1.1.2
Modern Tuberculosis Research

During the period 1960–1980, with lower incidences and decreasing trends, TB ceased to be pursued actively as a research subject in several western countries. However, it continued to be a focus of research in developing countries, especially in India, where TB was and continues to be a major health burden. The late Prof. M. Sirsi is acknowledged to have initiated TB research at the Indian Institute of Science, Bangalore, one of the premier institutions of India. He set up facilities for testing of compounds, synthesized in the Department of Organic Chemistry by Prof. B. H. Iyer and his group, on tubercle bacilli grown both in vitro and in experimental mice and guinea pigs [1]. Initially, the work focused on the metabolism of mycobacteria and later the emphasis shifted to the molecular biology of mycobacteria. Two major groups in India are credited with carrying out pioneering work in the field of TB – Prof. T. Ramakrishnan's group at the Indian Institute of Science, Bangalore and Prof. T. A. Venkitasubramanian's group at the V. P. Chest Institute, Delhi. They made significant contributions to the understanding of tubercle bacilli. Initially, the research in Prof. Ramakrishnan's laboratory was directed towards understanding the metabolism of mycobacteria with studies focusing on the mechanism of isoniazid action and resistance, the role of catalase peroxidase in the action of INH, the biosynthesis of isoleucine, valine, purines, enzymes of TCA cycle etc. Later, with the advent of molecular biology, several pioneering contributions were also made by this group, which included measurement of rates of DNA and RNA biosynthesis in mycobacteria, characterization of tRNA, purification of DNA and RNA polymerases, DNA methylation, protein synthesis and transduction etc. Prof. T. A. Venkitasubramanian's group made major contributions to the understanding of the metabolic processes in mycobacteria. Elucidation of pathways, involved in carbohydrate, acetate and malate metabolism, oxidative phosphorylation, synthesis of lipids, phospholipids, arginine, peptidoglycan, characterization of various mycobacterial enzymes such as phosphatases, glycerol kinase, fatty acid synthetases, aldolases, isocitrate lyase, glutamate dehydrogenase, malate dehydrogenase, adenylate and guanylate cyclases, pyruvate kinase, ATPase, ornithine transcarbamylase, isocitrate dehydrogenase and identification of calmodulin-like proteins from mycobacteria etc., all were carried out in the laboratory of Prof. T. A. Venkitasubramanian.

Recognizing the relevance of TB research to India, the Indian funding agencies have always categorized TB as a thrust area and agencies such as Council for

Scientific and Industrial Research, Department of Biotechnology, Department of Science and Technology and Indian Council of Medical Research have provided generous support for TB research. Investigators at several premier institutes such as All India Institute of Medical Sciences, New Delhi, Bose Institute, Calcutta, Central Drug Research Institute, Lucknow, Central JALMA Institute for Leprosy, Agra, Indian Institute of Science, Bangalore, Institute of Microbial Technology, Chandigarh, National Institute of Immunology, New Delhi, Postgraduate Institute for Medical Education and Research, Chandigarh, University of Delhi South Campus, New Delhi and several other centers are concentrating their efforts towards better understanding of this disease and its causative agent. Two national institutions – Tuberculosis Research Centre, Chennai and National Tuberculosis Institute, Bangalore have special distinction for focusing their efforts exclusively on TB and have made remarkable contributions in the field of TB research, prevention and control. In the following sections, we shall attempt to review the progress made by various research groups in India in the field of TB research mainly during the last five years.

1.2
Global Burden of Tuberculosis

Tuberculosis, which appeared to have been brought under control through vaccination and use of chemotherapy, has made a remarkable comeback in several parts of the world. This resurgence of TB is being attributed mainly to the advent of HIV and growing incidences of multidrug-resistant TB. With increasing globalization and world travel, the geographical barriers to the spread of the disease no longer exist and one country's scourge can very soon become another country's epidemic. About one-third of the world's population is infected with *M. tuberculosis*. Every year about eight million new cases develop and three million lives are lost due to this dreaded disease [5]. TB, which was a major problem in economically backward countries, is also rearing its head in developed countries, with outbreaks reported in USA and UK. Co-infection with HIV increases the probability of succumbing to TB by 30-fold [5]. The impact of the HIV pandemic is expected to be more serious in developing countries where a significant portion of the adult population is already infected with the tubercle bacilli. Presently 90% of the HIV-infected individuals are estimated to be residing in Asia and Africa. The problem is exacerbated by the emergence of *M. tuberculosis* strains that are resistant against most of the antibiotics used in the treatment of TB. Thus, in spite of rapid progress made in the last decade in our understanding of mycobacteria at the molecular level, TB continues to maintain its status as the number one killer among infectious diseases.

1.3
Tuberculosis in India

Despite the availability of a vaccine for the last 80 years and effective drugs against TB, the morbidity and mortality associated with this disease have not declined significantly. India accounts for 30% of the global incidence of TB [6] with an estimated 13–14 million TB cases at any point of time and one fourth of these cases being infectious in nature. In comparison to developed countries, the annual risk of infection in India is estimated to be 50–100 times higher [7]. Overall ~44% of our country's population are infected with the tubercle bacilli [8]. Every year about 5,000,000 people die of TB in India, which amounts to more than 1,000 patients dying of TB every day [9]. The majority of these deaths occur among people in the age group of 30 to 40 years, thus creating a burden on the country's resources.

BCG vaccine is still administered to all newborns within three months after birth, but in a major trial carried out in South India, its protective efficacy was observed to be 0%. A fifteen-year follow-up of this trial also did not exhibit any protection and BCG was observed to be totally ineffective against the adult form of TB [10, 11].

The National Tuberculosis Programme (NTP) has been in existence in India since 1962. However, there has not been an appreciable change in the epidemiological situation [12]. Due to certain drawbacks in NTP, a Revised National TB Control Programme (RNTCP) was formulated. The programme placed emphasis on the political commitment to support the national programme, quality diagnosis by smear microscopy, uninterrupted supply of good quality anti-tubercular drugs, treatment with short-course chemotherapy under direct obsevation and accountability through proper recording, reporting and effective supervision [12, 13]. The programme envisions to achieve an 85% cure rate for the newly diagnosed sputum smear positive TB patients and to detect at least 70% of new sputum smear positive patients. The RNTCP was implemented in 1993 in a population of 2.35 million in 5 sites in different states as a pilot project. Successful implementation of the programme has led to its continuous expansion [9] and by May 2001 it had covered 40% of the country's population.

1.4 Goals for Tuberculosis Research

From the perspectives of advancement in TB research, the control of TB requires a perfect vaccine, a rapid diagnostic test and drugs that are effective in a very short period. A perfect vaccine is required to provide protection against all forms of TB among the people from all age groups and populations. Diagnosis of the disease represents another critical aspect. There is no single diagnostic test which can provide rapid diagnosis on a mass scale. The major drawbacks of existing diagnostic tests are lack of sensitivity and the lengthy procedures. Besides, a good diagnostic test should be able to define the status of TB, and be useful for the diagnosis of extrapulmonary TB, especially TB meningitis. The test should also be able to differentiate between a dormant and an active infection. Moreover, it should not be affected by the BCG vaccination status of the individual. Drugs are most important in reducing the existing burden of the disease. However, drug therapy against TB involves the intake of multiple drugs for extended periods. This leads to several side effects and non-compliance, resulting in the development of drug resistance. Moreover, in a large country like India, it has been practically difficult to ensure the timely availability of drugs and compliance to therapy in remote areas. Although drugs effective against the TB bacilli are available, the long periods of intake and the development of resistance in bacilli and their being effective only against actively replicating bacteria but not against dormant bacteria, justify the need for the discovery of new drugs that can effectively shorten the duration of chemotherapy.

1.5 Difficulties in Working with Mycobacteria

The intrinsic properties of *M. tuberculosis* make it a challenging organism for laboratory investigations. It has a slow growth rate with a generation time of 22–24 h. It has a very thick cell wall, which makes isolation of DNA and RNA difficult. Selectable markers for mycobacteria were not known and transformation of this pathogen was very cumbersome. With the advent of electroporation, application of markers such as kanamycin and hygromycin and the development of transformation-efficient strains of mycobacteria, transformation of these pathogens has now become feasible. New techniques have also been developed for isolation of DNA and RNA from this pathogen [14–17]. However, there is a lack of knowledge about the mechanistics of growth of *M. tuberculosis* in the host. This pathogen has the capability of persisting in the host for decades, however, little is known about the mechanisms of persistence or the location of the pathogen in the host during this phase. The nature of protective immune responses against TB infection and the correlates of protection are also not known. However, as India has been continuously and severely affected by this disease, despite these daunting challenges, investigators in India have always shown keen interest in pursuing several critical areas towards the understanding of mycobacteria, tuberculosis and host-pathogen relationships.

2 Genetics of *Mycobacterium tuberculosis*

M. tuberculosis is a complex pathogen. It is a Gram-positive bacilli, with a genome having a very high G+C content (65.6%), which is reflected in its codon bias. The genome codes for ~4,000 putative ORFs, which account for ~91% of the potential coding capacity [18]. Consistent with the high G+C content of the genome, GTG initiation codon is used more frequently in *M. tuberculosis* (35%) than in *Bacillus subtilis* (9%) and *Escherichia coli* (14%), although ATG is the most common translation start codon (61%). There is also a slight bias in the orientation of genes with respect to the direction of DNA replication, as only 59% genes are transcribed with the same polarity as DNA replication, compared with 75% in *B. subtilis*. This has been suggested as one of the reasons for the slow growth of *M. tuberculosis* [18]. However, even ~120 years after its discovery, we still do not understand the crucial mechanisms employed by this organism to cause disease, to remain dormant within the host for long periods of time and to evade the host-immune mechanisms etc. As a prerequisite to unravel some of these mysteries, it is imperative to understand the basic biology of this organism.

2.1 Construction and Characterization of Plasmids

Since their discovery in mycobacteria, several investigators have attempted to isolate plasmids from various species of mycobacteria. Apart from the possibility that they might encode virulence factors, drug resistance etc., plasmids are important tools in recombinant DNA technology. As plasmids replicate independently from the bacterial genome, plasmid origins can be employed along with some selectable markers for the construction of vectors for molecular cloning. Numerous mycobacterial plasmid vectors have been developed for a variety of purposes. Most of them have employed the origin of DNA replication from plasmid pAL5000.

A *Mycobacterium-E. coli* shuttle promoter probe vector, pSD7, has been constructed [19]. The vector carries extrachromosomal origins of DNA replication from *E. coli* as well as pAL5000, which facilitate replication both in *E. coli* as well as in mycobacteria. Apart from these, it carries the kanamycin resistance gene for positive selection in mycobacteria and a promoterless chloramphenicol acetyl transferase (CAT) reporter gene to detect mycobacterial promoter elements in a homologous environment and to quantify their relative strengths. This vector was employed for the isolation and characterization of a large number of promoters from fast-growing as well as slow-growing mycobacteria [19]. These studies demonstrated that the majority of the mycobacterial promoters functioned very poorly in *E. coli*, thereby stressing the importance of developing tools for use in homologous systems. Using an *E. coli* promoter probe vector would have limited the isolation only to those promoters which are active in *E. coli* [19]. Another *Mycobacterium-E. coli* shuttle vector, pSD5, has been constructed for the expression of genes in mycobacteria [20]. This expression vector carries a kanamycin resistance gene as the selectable marker and origins of DNA replication from mycobacteria (ORI M) and *E. coli* (p15A). In order to generate a versatile expression system that would permit modulation of gene expression in mycobacteria, a modular expression cassette was designed. This expression cassette contained three different compartments (i) cloning sites for promoters, (ii) translational signals comprising of ribosome binding site and an ATG codon, and (iii) multiple cloning sites (MCS). The first two compartments were separated by translational termination codons in all three reading frames to uncouple transcriptional initiation from translational initiation. The RBS was derived from the *M. leprae hsp65* gene with suitable modifications to create an *Nde* I site at the ATG codon which helps in maintaining an optimal distance between the initiation codon and the vector-derived RBS, on cloning of the gene in the *Nde* I site. In order to stabilize the transcripts derived from strong promoters and to prevent any read-through transcription originating from ORI M, transcriptional terminators were cloned downstream of the gene cloning sites and upstream of the promoter cloning sites, respectively [20]. Such a system facilitates cloning of genes under a battery of promoters, and achieving varied levels of expression. This is especially important in the case of expression of candidate genes for development of rBCG vaccines. The level of expression of the antigen often determines the immunogenicity and also in some cases overexpression of the antigen to very high levels is known to cause lethality. Development of a modular expression vector such as pSD5 thus not only facilitates the expression of such lethal genes but also ensures the desired level of expression of the candidate antigens by using a single parental vector. The versatility of pSD5 to provide a range of expression was assessed by using mycobacterial promoters of various strengths to drive expression of the gene encoding *E. coli* *S*-adenosylmethionine decarboxylase (SAM decarboxylase). Expression levels varying by more than 100-fold were achieved by using different mycobacterial promoters [20].

For stable expression of genes in *M. bovis* BCG, to develop an rBCG system, vectors have to be stably maintained in the absence of any antibiotic or selection pressure. To facilitate this, integration-proficient vectors were constructed [21]. The expression vector pSD5 was modified by replacing ORI M with the integration signals from mycobacteriophage L5. The resulting vector, pDK20, replicates in *E. coli* extrachromosomally and has been demonstrated to integrate into the mycobacterial genome in a site-specific manner. By cloning different mycobacterial promoters in pDK20 and using *lacZ* as the reporter gene, an about 450-fold variation in the levels of β-galactosidase expression was demonstrated in *M. bovis* BCG [21].

The nodal expression vector, pSD5, was also used to construct two novel shuttle vectors, pSD5B and pSD5C [21]. Plasmid pSD5B is a promoter-selection vector containing a promoterless *lac*Z gene and allows the identification of mycobacterial promoters by the blue color of the colonies on solid media containing X-gal. Using pSD5B, mycobacterial promoters were isolated with a frequency of 10% [21]. Plasmid pSD5C has been designed to construct mycobacterial genomic libraries and express the cloned DNA inserts as fusion proteins with maltose binding protein (MBP) in mycobacteria. Such a vector system facilitates post-translational modifications such as glycosylation and acylation and also secretion of proteins in the native mycobacterial environment, which is not achievable if these genes were to be expressed in *E. coli*. pSD5C provides the possibility of expressing such genes in mycobacteria and easy purification of appropriately modified proteins directly from mycobacteria on affinity columns using amylose resins. The presence of a Factor Xa cleavage site between MBP and the expressed protein facilitates cleavage with Factor Xa to release pure mycobacterial protein synthesized in its native environment. Thus, the development of these vectors combines the ease of working in *E. coli* with the advantage of directly propagating them in mycobacteria without further manipulation [21].

Both replicative and integration-proficient vectors are being used in mycobacteria for the expression of antigens, for the development of candidate vaccines. In a study to investigate the stability of a gene in a replicative vector versus its integration in the chromosome, the *lac*Z gene was cloned in a replicative as well as in an integration-proficient vector [22]. Occurrence of white colonies on X-gal plates indicated the loss of expression of *lac*Z. In the case of integration-proficient vectors, the loss of lacZ phenotype was found to be due to the insertion of an IS element in the *lacZ* gene, whereas in the case of replicative vectors, the loss of LacZ phenotype occurred due to deletions in different regions of the *lacZ* gene and phsp60 promoter. The frequency of such events was calculated to be 1.7×10^{-5} in the case of an integration-proficient vector and 2×10^{-3} in the case of replicative vectors, thus implying that integration-proficient vectors could be a better choice for the development of vaccines [22].

To identify promoters whose transcription is under the control of a regulatory protein, a novel strategy has been described for mycobacteria using *lacZ* as the reporter gene [23]. In this strategy, the gene encoding the regulatory protein is cloned downstream of the inducible mycobacterial acetamidase promoter and the transcription of the gene is regulated by the presence or absence of acetamide in the culture medium. In addition, the vector contains a promoterless *lacZ* gene. By using this vector a promoter library can be generated by cloning the DNA fragments upstream to the *lacZ* gene. Subsequently, *M. smegmatis* can be transformed with this library and the transformants selected on LB agar plates containing the appropriate antibiotic. The colonies can then be replica plated onto minimal agar plates containing X-gal along with either succinate or acetamide. Promoters either down-regulated or up-regulated by the regulatory protein can be identified by the blue-to-white or white-to-blue transition of colonies, respectively. However, this strategy can be especially useful for genes, which are present only in the slow-growing species of mycobacteria and are absent in the fast-growing species [23].

In order to analyze the in vivo expression of mycobacterial promoters, a vector containing promoterless green fluorescent protein (GFP) has been constructed. Promoters can be cloned upstream and the constructs used to transform *M. tuberculosis*. The level of GFP expression can be monitored both in vitro and in vivo (after infection of animals or a cell line) to determine if the promoter is induced under such conditions [24].

A vector has also been developed for investigating the *M. tuberculosis*-phagocytic cell interaction [25]. In this vector, mycobacterial genomic fragments are cloned upstream of the *cre* gene. The vector carries an antibiotic resistance switch, which signals the interaction of mycobacteria and the macrophage. The antibiotic resistance marker is switched on using the Cre-LoxP system when the mycobacterial promoter activates the *cre* gene. The pathogen's interaction with the host cells activates the promoter, which in turn stimulates the Cre-LoxP-dependent recombination [25].

Plasmid pAL5000 was the first plasmid to have been characterized from mycobacteria. The plasmid is ~4.8 kb in size and it is believed that a ~1.8 kb fragment spanning two open reading frames (ORFs), ORF1 and ORF2 is necessary and sufficient for replication. Studying the replication mechanism of this plasmid can provide further insight into the replication mechanism of mycobacteria. The ORF1-ORF2 region can potentially produce two polypeptides RepA and RepB, although there is no direct evidence indicating that they are expressed in mycobacterial cells harboring pAL5000 vectors. The role of RepA and RepB in replication of the plasmid has been characterized [26]. It was observed that when the *repA-repB* genes were expressed in *E. coli*, a strong *ori* binding activity was generated in the host cell. When the gene encoding RepB was inactivated, a complete loss of activity was observed. However, inactivation of *repA* had a partial effect indicating that while *repB* plays an important role in the process, its activity is stimulated through the co-expression of *repA* [26]. Moreover, this stimulatory effect was observed only when the expression of *repA* and *repB* are coupled. At very high concentrations (1 μM), the purified RepB protein formed an *ori* complex of low specificity, which was sensitive to high salt and challenge by non-specific competitor. In contrast to this, the complex formed by the extract of cells expressing *repA-repB* was highly specific and resistant to both types of challenge. At a ten-fold lower concentration, however, it was observed that RepB failed to demonstrate any *ori* binding activity, but was able to form salt resistant complexes in vitro, provided host factors were present [26]. Supershift experiments using antiserum specific to RepB also demonstrated that RepB was a key component of the specific complex formed by extracts prepared from *E. coli* cells expressing the *repA-repB* genes and also mycobacterial cells harboring the pAL5000-derived vectors. Hence, these results indicate that in vivo RepB interacts with host factors and forms an *ori* complex and that this activity is maximal only when RepA and RepB are co-expressed [26]. Thus, for the first time direct involvement of a mycobacterial replication protein in *ori* complex formation has been demonstrated in mycobacteria, using RepB as a model [26].

While studying the replication of *E. coli*-mycobacteria shuttle vectors in *M. smegmatis*, a unique phenomenon of transposition was observed [27]. IS1096, an *M. smegmatis* transposon, was found to frequently transpose at a cluster of sites located within a small region of 60 bp immediately upstream of a kanamycin resistance gene present in these vectors. Because of these transpositions, the vectors, which were all derived from pAL5000 and pUC19, were frequently found to undergo structural alterations. Characterization of these structural alterations revealed deletions of large regions of vector which, in several cases, were found to extend into the ORF2 (RepB) coding sequences of the pAL5000 replication region, without affecting its replication capability [27]. This suggests that the entire coding sequences of the pAL5000 replication region might not be essential for replication of pAL5000-derived vectors. These deletion derivatives were found to be structurally stable and therefore can be potentially used for the construction of stable vector systems for use in mycobacteria [27].

2.2 Mycobacteriophages

Mycobacteriophages have played a key role in the development of mycobacterial genetics and have been helpful in evolving methods for the introduction of DNA into the mycobacteria and the generation of stable recombinants as well as for the construction of vectors and a variety of other genetic tools. Mycobacteriophage L1 is a lysogenic phage of mycobacteria which, along with mycobacteriophages L5 and D29, constitutes a closely linked family of mycobacteriophages. Mycobacteriophages have gained prominence due to their potential for use in genetic engineering of mycobacteria and diagnosis of mycobacterial infection. However, the effectiveness of phages depends on the efficiency with which they infect and grow within target cells. While working with L1clts, a temperature-sensitive mutant of mycobacteriophage L1, it was observed that it initially gave rise to low titer phage stocks [28]. Repeated passage through the host, *M. smegmatis*, generated high yielding stocks that were 10–100 times more efficient in producing phage particles. A high-yielding mutant, L1-P2 obtained from one such high yielding stock, was analyzed further [28]. The DNA obtained from L1-P2, when examined by restriction digestion, was observed to have undergone spontaneous loss of a DNA fragment from the right arm, which encodes early regulatory factors [28]. These high yielding mutants are currently being used as the starting material for developing diagnostic systems. The mechanism of adsorption by phages, as well as the events that take place during the early phase of growth following infection are also being investigated using proteomic analysis tools [29].

The structural proteins of mycobacteriophage I3 have been analyzed by SDS-PAGE, radioiodination and immunoblotting. Based on their abundance the 34-kDa and 70-kDa bands appeared to represent the major structural proteins. The gene encoding the 70-kDa protein was cloned and expressed in *E. coli*. The complete nucleotide sequence of this gene has also been determined [30].

2.3 Transposition in Mycobacteria

Transposons are genetic entities encoding functions that facilitate their mobility from one site on a DNA molecule to another, independent of host recombination functions. Transposons have been exploited as genetic tools for a variety of purposes such as for creating mutants by insertion, thereby enabling the identification of gene function or for the identification of genes involved in virulence by signature tagged mutagenesis. Transposons can be employed to identify vegetative as well as regulatory promoters for study of gene expression. Exported proteins can be identified by employing a reporter gene encoding an enzyme that is functional only in an extracytoplasmic compartment. Several insertion elements isolated from mycobacteria have been found to be species-specific. Hence, transposons have also been used for diagnosis as well as typing of mycobacterial strains by PCR tests.

A novel insertion sequence, IS219, was isolated from *M. fortuitum* by using a trap vector [31]. The trap vector is a temperature-sensitive (*ts*) *E. coli-Mycobacterium* shuttle plasmid and carries a *ts* ORI M, the kanamycin resistance gene as a selection marker and a *lacZ* expression cassette. The *ts* mutation present in the vector functions in mycobacteria and facilitates identification of transposable elements that disrupt the *lacZ* gene (by screening for white colonies on X-gal plates). IS219 was characterized and was observed to have inverted repeats at its ends. It contained two ORFs on one strand and duplicated host DNA at the target site. Southern blot analysis failed to reveal the presence of this insertion element in several slow- and fast-growing mycobacterial species analyzed. In the case of *M. fortuitum*, two copies of this insertion element were detected [31].

An insertion element-like repetitive sequence has also been described from *M. tuberculosis* [32]. The repeat sequence contained three putative ORFs, several direct repeats and seven inverted repeats. This repeat element had a low degree of polymorphism, was specific for the *M. tuberculosis* complex and was conserved among the various clinical isolates of *M. tuberculosis*. This repeat element can thus serve as an ideal target for PCR analysis to detect *M. tuberculosis* in clinical specimens, especially to detect strains that lack a copy of IS6110 [32]. Studies have been carried out for the evaluation of this element as a target for detection of tuberculous lymphadenitis by PCR analysis (described later in this chapter).

2.4 DNA Replication, Homologous Recombination and DNA Repair

Pathogenic mycobacteria are characterized by extremely slow growth rates. Comparative studies of metabolism in *M. tuberculosis*, *M. smegmatis* and *E. coli* have shown nucleic acid biosynthesis (replication and/or transcription) to be a strong candidate for regulation of the growth rate. However, the differences in the rate of transcription cannot account for the 5–7-fold slower growth rate of *M. tuberculosis* compared to *M. smegmatis*. Thus chromosomal DNA replication must be intimately associated with the growth rate of mycobacteria. In a classical study it was shown that the replication rate of 3,200 nucelotides/min in the case of *M. tuberculosis* was 11 times slower than in *M. smegmatis* and 13 to 18 times slower than in *E. coli* [33]. Hence, studying the mechanism of replication and the proteins involved can provide novel insights as well as unravel potentially novel targets for anti-tubercular drugs.

Topoisomerases represent a class of enzymes that function in several important cellular processes related to nucleic acids such as DNA replication, transcription, recombination, chromosomal segregation etc. DNA topoisomerases are essential for survival in all organisms. Characterization of this group of enzymes will help in understanding their role in the basic biology of the organism, in virulence gene expression and for their development as molecular drug targets. The DNA topoisomerase I and DNA gyrase enzymes from mycobacteria have been extensively characterized using various biochemical studies. DNA topoisomerase I was purified from *M. smegmatis* to near homogeneity using a series of chromatographic steps [34]. Although the topoisomerase was similar to the one from *E. coli*, the enzyme was considerably larger (110 kDa). The N-terminal sequence of the two enzymes was also different. On analysis of the biochemical properties and reactions of the enzyme, it was found to be similar to the *E. coli* counterpart in many respects such as (i) DNA relaxation, catenation and knotting reactions, (ii) absolute requirement for Mg^{2+} ion, (iii) covalent linkage to the 5′-end of the nicked DNA, (iv) presence of tyrosine at the active site etc. However, the mycobacterial enzyme also displayed some different characteristics with respect to the *E. coli* enzyme such as (i) pH optima, (ii) thermal stability, (iii) effect of spermidine on the enzyme activity and iv) role of cysteines in catalysis [34]. The mycobacterial enzyme was found to be active even after long periods of storage and over a broad pH range. The enzyme was not found to contain any bound Zn^{2+} and retained the activity even on modification of cysteine residues, suggesting a different mechanism of DNA binding for this enzyme [34]. Using a two-substrate assay system it was demonstrated that the enzyme acted processively at low ionic strength and switched over to distributive mode at high Mg^{2+} concentration. The enzyme activity was stimulated by single-stranded DNA binding (SSB) protein [34]. By employing a novel approach, the cleavage sites of *M. smegmatis* topoisomerase I were mapped [35]. An electrophoretic mobility shift assay of enzyme-DNA covalent complexes revealed that the enzyme bound to double-stranded DNA at regions having strong topoisomerase I sites with high affinity. Mg^{2+}, which is essential for the DNA relaxation activity of the enzyme, was not required for its binding to DNA [35]. The cleavage site within the binding region was determined to be CG/TCTT, which was surprisingly similar to the cleavage sites of some of the well-characterized eukaryotic topoisomerases [35]. Using DNA binding, oligonucleotide competition and covalent complex assays, the length of the substrate required for interaction was determined to be much longer (~20 nucleotides) in contrast to *E. coli* topoisomerase I and III (~8 nucleotides). P1 nuclease and $KMnO_4$ footprinting studies indicated the region spanning 20 nucleotides upstream and 2–3 nucleotides downstream of the cleavage site to be protected [36]. Both single-stranded and double-stranded DNA substrates containing the strong topoisomerase I sites interacted efficiently with the enzyme. In addition, oligonucleotides containing strong topoisomerase I sites were found to effectively inhibit the *M. smegmatis* topoisomerase I DNA relaxation activity [36]. Strong topoisomerase I binding sites of the enzyme were also mapped on the genomic DNA sequences from *M. smegmatis* and *M. tuberculosis* [37]. Mapping of the sites revealed conservation of a pentanucleotide motif CG/TCT/T at the cleavage site. The protein exhibited a very high preference for C or G residue at the +2 position with respect to the cleavage site. Based on the DNase I footprinting, methylation protection and interference analysis and use of conformation specific probes, a model of topoisomerase I binding to its substrate has been proposed [37, 38]. Within the large protected region of 30 bp, the enzyme makes direct contact at two locations in the scissile strand, one around the cleavage site and the other 8–12 bases upstream. Thus, the recognition of DNA is asymmetric in nature and the enzyme could carry out DNA relaxation by either of the two proposed mechanisms: enzyme-bridged and restricted rotation [37, 38]. Besides, it was demonstrated that SSB could specifically stimulate DNA topoisomerase I activity without direct protein-protein interactions. The stimulation was specific to topoisomerase I as DNA gyrase activity was unaffected by SSB protein, suggesting that such cases of functional collaboration between DNA transaction proteins play important roles in vivo [39].

DNA gyrase is a bacterial type II DNA topoisomerase which induces negative supercoils in DNA. The prototype enzyme from *E. coli* is a tetrameric protein (A_2B_2) having supercoiling-relaxation, catenation-decatenation and knotting-unknotting activities. It is encoded by two genes, *gyrA* and *gyrB*, producing the A and B subunits. The genes encoding the subunits of DNA gyrase, GyrA and GyrB have been cloned from *M. tuberculosis* and *M. smegmatis* [40, 41] and their organization and regulation from *M. smegmatis* has been studied. The two genes were part of a single operon, and the *gyr* promoter, which was specific to mycobacteria was non-functional in *E. coli*. In keeping with its involvement in increasing the supercoiling of DNA, the expression of the *gyr* genes was observed to be induced by the relaxation of the DNA template [42]. Relaxation-stimulated transcriptional activity was also observed in *M. smegmatis gyr* genes although the mechanism was different from the one reported for *E. coli*. It was observed that DNA elements that are present 600 bp downstream of the promoter are necessary for relaxation-stimulated transcription to occur in the plasmid context [42]. The *gyr* genes were also induced when the cells were treated with novobiocin (a known inhibitor of DNA gyrase) [42]. The purified recombinant GyrA and GyrB on combination exhibited strong ATP-dependent DNA supercoiling activity [41]. A GyrB-specific monoclonal antibody (mAb) from *M. smegmatis* did not cross react with GyrB of either *E. coli* or other mycobacterial species [43]. Hence, it has been suggested that monoclonal antibodies (mAbs) specific to *M. tuberculosis* or *M. leprae* GyrB proteins could be used for rapid diagnosis of mycobacterial infections. The epitope for the mAb was demonstrated to be located in the N-terminal region of *M. smegmatis* GyrB [43]. The binding site of the mAb did not overlap with the binding site of novobiocin which also interacts with the N-terminus of GyrB in *E. coli*. Immunoprecipitation studies with cell-free extracts revealed interaction of GyrA and GyrB subunits, which was inhibited by sodium chloride suggesting the nature of interactions to be ionic [43]. Monoclonal antibodies were also generated against GyrA of *M. smegmatis*, which showed a high degree of cross-reactivity with both fast-growing and slow-growing mycobacteria but failed to recognize *E. coli* GyrA [44]. The epitopes for these mAbs were mapped to different regions of the protein. Moreover, the two mAbs were shown to inhibit DNA supercoiling activity of the enzyme [44]. Thus, apart from serving as useful reagents for structure-function analysis of mycobacterial DNA gyrase, these will also be useful for designing peptide inhibitors against DNA gyrase.

Compounds that inhibit *E. coli* DNA gyrase include quinolones, coumarins, microcin B17, CcdB and cyclic peptide cyclothialidines. Detailed analyses of the effect of various groups of inhibitors on *M. smegmatis* gyrase demonstrated the latter to be refractory to the plasmid-borne proteinaceous inhibitors CcdB and microcin B17 [45]. Ciprofloxacin, a fluoroquinolone, exhibited a 10-fold decreased inhibition of *M. smegmatis* DNA gyrase as compared to the *E. coli* enzyme. However, DNA gyrases from both *E. coli* and *M. smegmatis* were equally susceptible to etoposide, an antineoplastic drug [45].

Mycobacteria being G+C-rich organisms might employ unique enzymes/proteins for their DNA replication. Characterization of these activities can be useful in the identification of new drug targets. Since many of the mechanisms involved in macrophage killing result in the production of DNA-damaging agents such as peroxide and nitric oxide, DNA repair mechanisms are likely to be important for intracellular survival of mycobacteria. Some of the important enzymes involved in DNA repair in mycobacteria were characterized. Uracil, a promutagenic base in DNA can be formed by spontaneous deamination of cytosine or incorporation of dUMP by DNA polymerase. Deamination of cytosine results in the appearance of uracil residues in DNA which, unless removed, lead to G:C → A:T mutations. Uracil DNA glycosylase (UDG), that initiates uracil excision repair to safeguard the genomic integrity, was purified from *M. smegmatis* by more than 3000-fold [46]. Unlike *E. coli* UDG which formed a physiologically irreversible complex with the inhibitor protein Ugi, *M. smegmatis* UDG formed a dissociable complex with Ugi. Studies to determine the substrate specificity of *M. smegmatis* UDG showed that it excised uracil from the 5′-terminal position of the 5′-phosphorylated substrates [46]. However, its excision from the 3′-penultimate position was extremely poor. pd(UN)p was found to be the minimal substrate of *M. smegmatis* UDG, as in the case of UDG from *E. coli*. Based on several kinetic parameters it was suggested that *M. smegmatis* UDG was an efficient enzyme and better suited for applications in molecular biology work [46].

Relatively inefficient excision of uracil from looped structures compared with excision from single-stranded 'unstructured' substrates suggested that destabilization of the loop structures may be required for efficient repair. It appears that SSB proteins may be involved in melting such structures. The effect of SSB proteins from *E. coli* and *M. tuberculosis* on uracil excision by UDG enzymes from *E. coli, M. smegmatis* and *M. tuberculosis* showed that the presence of SSBs with all three UDGs resulted in decreased efficiency of uracil excision from a single stranded "unstructured" oligonucleotide [47]. However, when a hairpin oligonucleotide containing a U residue at the 2nd position in a tetraloop was used as the substrate, addition of SSB protein to their respective UDG counterparts resulted in increased efficiency of uracil excision. Heterologous combinations did not exhibit this increase in efficiency. Using the surface plasmon resonance technique, the interactions between homologous combinations of SSBs and UDGs and a weaker or non-detectable interaction between heterologous combinations were demonstrated [47].

The gene encoding the SSB protein which plays an important role in DNA replication, DNA repair and recombination has been cloned from *M. tuberculosis*, overexpressed and characterized [48]. Sequence comparisons of *M. tuberculosis* SSB with *E. coli* SSB revealed that the former lacked or differed with respect to several highly conserved amino acids crucial for the *E. coli* SSB structure-function relationship. *M. tuberculosis* SSB also has a distinct C-terminal domain. However, despite these non-conserved changes, the SSB proteins from *E. coli* and *M. tuberculosis* have been found to be similar in their biochemical properties. Like *E. coli* SSB, the protein from mycobacteria also binds to DNA as a tetramer [48]. Using yeast two-hybrid assays, it was demonstrated that *E. coli* SSB and *M. tuberculosis* SSB were capable of hetero-oligomerization. Chimeric constructs were made between the N- and C-terminal domains of the *M. tuberculosis* SSB and *E. coli* SSB [49]. These existed as homotetramers and demonstrated DNA binding properties similar to their wild-type counterparts. However, neither the *M. tuberculosis* SSB nor the chimeric constructs were able to complement an *ssb* deletion construct of *E. coli*. The inability of the chimeric constructs to complement a *E. coli* mutant highlights the need for the N- and C-terminal domains to communicate with each other as well as with the other cellular proteins [49]. The chimeric proteins were also used to investigate the interaction between SSBs and UDGs. Using far Western analysis, surface plasmon resonance and the electrophoretic mobility shift assay, it was shown that the *E. coli* SSB or chimeras containing the C-terminal domain interacted with *E. coli* UDG in a binary (SSB-UDG) or ternary (DNA-SSB-UDG) manner [50]. However, the chimeras containing the N-terminal domain from *E. coli* SSB showed no interactions with *E. coli* UDG, implicating the C-terminal domain to be necessary and sufficient for interaction with UDG. Similarly, the interaction of *M. tuberculosis* SSB with *M. tuberculosis* UDG was through the C-terminal domain of SSB [50]. Recently, SSB from the fast-growing mycobacteria, *M. smegmatis* has been characterized and it has been shown that the C-terminal domain of SSB interacts within the DNA binding groove of UDG [51]. These observations thus suggest that SSB may facilitate recruitment of UDG to the site of the lesion. These studies have highlighted the significance of SSB-UDG interaction in uracil excision repair in the transiently occurring single stranded regions of the genome, such as the transcription bubble or replication fork.

The SOS bacterial response to DNA damage involves the coordinate induced expression of more than 20 genes. RecA protein plays a central role in the regulation of the SOS response, homologous recombination and DNA repair. Extensive studies have been carried out on the RecA protein of *M. tuberculosis*. The ability to achieve a high frequency of homologous recombination can significantly contribute towards the construction of mutants of *M. tuberculosis*, genetic analysis of mycobacteria and in understanding the mechanisms of pathogenesis. Allele replacement has been achieved in *M. smegmatis*, *M. intracellulare* and in the cases of certain genes in *M. tuberculosis*. However, in slow-growing species such as *M. tuberculosis* and *M. bovis* BCG it has been rather difficult to disrupt genes, perhaps due to their ability to promote random integration by non-homologous recombination. The *recA* genes of pathogenic mycobacteria are distinct from the *recA* genes of other organisms in that the genetic locus specifying the functional protein comprises a single ORF that is interrupted by an intein coding sequence. The active form of RecA protein is generated by excision of a portion of the RecA precursor and ligation of the N-terminal and C-terminal domains. By the use of in vivo assays, it was shown that *M. tuberculosis recA* partially complements *E. coli recA* mutants for recombination and in the repair of UV-damaged DNA [52]. However, splicing of the 85 kDa precursor to the 38 kDa RecA protein was necessary for manifestation of in vivo activity. Therefore, a strategy to purify both the 85 kDa precursor and *M. tuberculosis* RecA proteins to homogeneity was devised in order to gain an insight into the partial or total lack of complementation by *M. tuberculosis* RecA or 85 kDa protein, respectively [52]. Using a combination of assays, it was shown that the spliced form of *M. tuberculosis* RecA, but not 85 kDa protein, interacts with single-stranded DNA (ssDNA) to generate protein-DNA complexes. With regard to DNA-DNA pairing and strand exchange activities *M. tuberculosis* RecA exhibited unique properties. *E. coli* RecA protein binds to ssDNA in the presence or absence of ATP. In contrast, *M. tuberculosis* RecA protein required the presence of ATP for efficient binding, however, this binding was independent of ATP hydrolysis [52]. *M. tuberculosis* RecA also showed reduced affinity for ATP, and its hydrolysis, compared to that by *E. coli* RecA, but was proficient in homologous recombination. This along with the fact that the *M. tuberculosis* RecA protein is defective in displacing SSB from ssDNA, may account for the partial complementation of *E. coli* RecA mutants [52]. The 85-kDa protein showed a complete lack of DNA-binding, ATP-binding and ATPase activity. Molecular modeling studies suggest that the decreased affinity of *M. tuberculosis* RecA protein for ATP and its reduced efficiency for ATP hydrolysis might be due to the widening of the cleft. This alters the hydrogen bonds and the contact area between the enzyme and substrate and disposition of the amino acid residues around the magnesium ion and γ-phosphate [52].

The homologous DNA pairing and strand exchange activities of RecA of *M. tuberculosis* was demonstrated to be pH dependent [53]. The formation of joint molecules promoted by the concerted action of *M. tuberculosis* RecA and SSB was maximal around neutral pH. However, optimal strand transfer occurred only at higher pH. The order of addition experiments suggested the essential requirement for SSB when added after *M. tuberculosis* RecA. Studies using short or long duplex DNA to promote efficient strand exchange have led to observations that are consistent with a model in which the networks are formed by the invasion of hybrid DNA by the displaced linear single-stranded DNA. Titration of strand exchange reactions with ATP or salt distinguished a condition under which the formation of networks was prohibited without significantly affecting the strand exchange [53]. The X-ray structure of *M. tuberculosis* RecA protein has also been determined [54]. DNA binding by SSB proteins of *M. tuberculosis* and *M. smegmatis* was associated with enhanced sensitivity to proteases at the carboxyl-terminal domain, which was also the site of interaction of its cognate RecA [55]. The proteolytic signatures of free and bound forms of SSB proteins disclosed that DNA binding was associated with structural changes at the C-terminal domain. Under comparable experimental conditions the SSB protein from mycobacteria displayed high affinity for cognate RecA, but not for *E. coli* RecA. SSB and RecA were also co-immunoprecipitated from cell lysates, further supporting an interaction between these proteins in vivo [55]. These studies have improved our understanding of the molecular basis of intrinsic resistance of *M. tuberculosis* to allele exchange and the basis for higher frequency of illegitimate recombination.

Methylation of DNA plays an important role in several critical processes such as control of DNA replication and its repair. The status of Dam/Dcm methylation in the genus mycobacteria was not known. Hence, using isoschizomer restriction enzymes, the presence of 6-methyladenine and 5-methylcytosine at Dam (GATC) and Dcm (CCA/TGG) sites in DNA of mycobacterial species were investigated [56]. In all the species examined, the absence of methylation at the Dam and Dcm recognition sequences indicated the absence of these methylases. However, HPLC analysis of genomic DNA from *M. smegmatis* and *M. tuberculosis* exhibited significant levels of 6-methyladenine and 5-methylcytosine suggesting the presence of DNA methyltransferases other than Dam and Dcm. The presence of methylation in the genomic DNA was also confirmed by employing a sensitive genetic screen [56].

2.5 Synthesis of Proteins and Growth Rate of Mycobacteria

The genus *Mycobacterium* comprises fast growers such as *M. smegmatis* and *M. phlei* and slow growers such as *M. tuberculosis, M. bovis* and *M. leprae*. The growth rate of an organism is generally related to the rate of protein synthesis which, in turn, is dependent upon the abundance of ribosomes and the other cellular components associated with protein biosynthesis. In accordance with this premise, two ribosomal RNA operons are present in the fast grower *M. smegmatis* as opposed to only one in the case of the slow grower *M. tuberculosis*. Previously, qualitative differences in the total tRNA pools between *M. tuberculosis* and *M. smegmatis* have been reported [57]. Further, the initiator tRNA gene, *metA* and an initiator tRNA-like sequence, *metB* from *M. tuberculosis* $H_{37}Ra$ have been cloned and characterized [58]. The sequence of mature tRNA was found to contain all the three unique features of the eubacterial initiator tRNAs. However, RNA blot analysis of the total tRNA prepared from *M. tuberculosis* and *M. smegmatis* demonstrated that in vivo the tRNA exists in the formylated form. By using the CAT reporter gene, the promoter was demonstrated to be very strong with conserved sequence elements at the −10 and −35 regions [58]. A method to prepare total RNA from mycobacteria in order to analyze the in vivo status of tRNAs has also been described [58]. The initiator tRNA gene locus of the avirulent ($H_{37}Ra$) and virulent ($H_{37}Rv$) strains of *M. tuberculosis* have also been mapped, demonstrating that *metA* and *metB* are located on contiguous fragments, separated by an insertion element IS6110. Surprisingly, *M. smegmatis*, the fast-growing species showed the presence of a single initiator tRNA gene [59].

In eubacteria, once the translating ribosomes reach a termination codon, release factors in response to termination codons bind to the ribosome A-site triggering hydrolysis of the P-site-bound peptidyl-tRNA and culminating in the release of the nascent polypeptide chain. However, subsequent to the release of the polypeptide, 70S ribosome and the P-site-bound deacylated tRNA still remain attached to the mRNA. With the most important events in initiation of protein synthesis occurring on the P-site of the 30S ribosomal subunits, it is imperative that the terminating ribosomal complex dissociates before the start of a fresh round of polypeptide biosynthesis. The ribosome recycling factor, RRF, is involved in the dissociation of the termination complex. The gene encoding RRF of *M. tuberculosis* has been cloned [60]. However, the *M. tuberculosis* RRF failed to rescue an *E. coli* strain temperature sensitive for RRF. Interestingly, co-expression of *M. tuberculosis* elongation factor G (EFG) with *M. tuberculosis* RRF rescued the temperature sensitive RRF strain of *E. coli*. The simultaneous expression of *M. tuberculosis* EFG was also required to cause an enhanced release of peptidyl-tRNAs in *E. coli* by *M. tuberculosis* RRF [60]. Thus, the simultaneous expression of RRF and EFG of *M. tuberculosis* in *E. coli* provides the first genetic evidence for a functional interaction between RRF and elongation factor G [60]. The failure to detect an interaction between *M. tuberculosis* RRF and *M. tuberculosis* EFG by surface plasmon resonance and far-Western analysis indicated that the specific interactions between the two proteins might be effected on the ribosomes. In addition, the complementation of an *E. coli* strain, which carries a mutation in the EFG gene by *M. tuberculosis* EFG, suggests that the mechanism of RRF function is independent of the translocation activity of EFG [60]. Understanding the mechanism of action of the *M. tuberculosis* RRF would promote the possibility of its employment as a drug target to control mycobacterial growth.

M. tuberculosis contains only one *rrn* operon encoding rRNAs. Two promoters P1 and P3 driving the transcription of *rrn* operon were identified and the role of RNase III in pre-rRNA processing was demonstrated in mycobacteria for the first time [61]. In addition to demonstrating the high strength of the *rrn* promoter, the analysis of promoter using *M. smegmatis* as a host has demonstrated that *rrn* promoters are subject to regulation as a function of growth rate. These cross-species experiments indicate that the characteristically low content of RNA per unit of DNA of *M. tuberculosis* cannot be ascribed to the inherent structure or sequence of *rrn* promoter [62].

It has been shown that the morphology and staining pattern are different for mycobacteria grown in vitro and for persistors obtained from the lung lesions. It has also been shown that the morphology and staining pattern exhibited by persistors can be obtained in vitro by depriving the mycobacterial cultures of any nutrients. These observations thus suggest that the natural persistors may be physiologically similar to bacteria in nutritionally starved cultures. Hence, it has been hypothesized that studying the physiology and morphology of such starved cultures can help us in understanding the mechanism of latency.

Bacteria are known to adapt to nutritional stress for their survival through a mechanism termed as stringent response. Stringent response is characterized by the accumulation of guanosine tetraphosphate (ppGpp), also called the stringent factor, and downregulation of stable RNA (rRNA and tRNA) synthesis. Starvation in mycobacteria was studied using *M. smegmatis* as the model organism [63]. It was shown that ppGpp accumulation in mycobacteria was accompanied by morphological changes in *M. smegmatis* under conditions of carbon starvation. It has been proposed that the accumulation of ppGpp in *M. smegmatis* may be responsible for the morphological change from an elongated rod to a spherical coccus, as seen in the case of carbon-starved *M. smegmatis*. In order to correlate the morphological changes and accumulation of ppGpp, the *relA* gene from *E. coli* was expressed in *M. smegmatis*. The effect of ppGpp accumulation on the growth kinetics of *M. smegmatis* was also studied. As expected, a slower rate of growth was observed on expression of *relA*. Moreover, it was shown that *M. smegmatis* assumes a coccoid morphology (similar to persistors) when ppGpp is ectopically produced by overexpression of *E. coli relA* in an enriched nutritional medium [63]. Analysis of the genome sequences also revealed the presence of a single gene in *M. tuberculosis* and *M. leprae*, having almost 50% homology to both *relA* and *spoT* (encoding ppGpp hydrolase) of *E. coli*. This gene complemented the RelA phenotype in *E. coli* thus confirming it to be a RelA homologue. Using *M. smegmatis* as a surrogate host, the function of *M. tuberculosis relA/spoT* was characterized. The gene was cloned downstream of the inducible acetamidase promoter and expressed as an 89-kDa protein [63]. Induction of expression of *relA/spoT* was observed to be associated with elevated intracellular levels of ppGpp and reduced rates of growth. From these observations, an important role for ppGpp in the latency of the mycobacterium has been suggested [63]. Studies on the stringent pathways can thus answer important questions pertaining to the physiological transformation that occurs in this pathogen. A *relA*$^-$ mutant strain of *M. smegmatis* has also been constructed, which is currently being analyzed [64].

2.6 Transcription in Mycobacteria

The process of transcription represents an extremely important step in the regulation of gene expression in bacteria. Compared with other bacterial systems, mycobacteria have a low transcription rate, low content of RNA per unit of DNA and a lower rate of incorporation of ribonucleotides into RNA. Besides, the use of *E. coli* and *Streptomyces lividans* to study the expression of mycobacterial genes suggested that the efficiency of these heterologous expression systems was variable and did not permit the expression of the majority of mycobacterial genes. These observations pointed out that the transcriptional machinery of mycobacteria may not be comparable with the typical prokaryotic transcriptional apparatus.

To unravel the complexity of the transcriptional machinery of mycobacteria, the promoter probe vector pSD7 was used to isolate promoters from *M. smegmatis* and *M. tuberculosis*, based on the expression of CAT reporter gene [19]. It was found that the frequency of occurrence of promoter clones in *M. smegmatis* was 10–20% whereas in *M. tuberculosis* it was 1–2%. Moreover, it was also observed that the frequency of occurrence of strong promoters was higher in the case of *M. smegmatis* when compared to *M. tuberculosis*. About 350 promoter clones from *M. smegmatis* and 125 promoter clones from *M. tuberculosis* were isolated [19]. Subsequently, the nucleotide sequences of 35 mycobacterial promoter constructs, including 20 from *M. smegmatis* and 15 from *M. tuberculosis* were determined [65]. The promoter sequences were aligned on the basis of their transcription start point (TSP). Analysis of these promoter sequences from positions –1 to –50 showed that the promoters from *M. tuberculosis* had a higher G+C content (57%) compared to the promoters from *M. smegmatis* (43%). A highly conserved Pribnow-box like sequence was identified around the –10 region of all mycobacterial promoters analyzed in this study [65]. The conserved hexameric sequences TATAAT from *M. smegmatis* and TAYgAT from *M. tuberculosis* were similar to their counterparts in *E. coli*. The first, second and the sixth base in the hexameric sequence (T, A, and T, respectively), which represented the functionally more important positions in *E. coli* promoters were more strongly conserved in both mycobacterial species. However, on the basis of the limited number of promoter sequences analyzed in this study, it appeared that in *M. tuberculosis* the less important positions (the third and fourth) have become relatively more susceptible to GC intrusions in order to accommodate the higher G+C content of its promoters. More prominently however, the mycobacterial promoter sequences in the –35 region showed no homology with the TTGACA motif present in promoters of *E. coli* and several other bacteria [65]. In fact no sequence was found to be conserved in the –35 region of mycobacterial promoters. The comparison of sequences of the 2.4 and 4.2 regions from mycobacterial sigma factors, Sigma A and Sigma B with the corresponding regions of principal sigma factors from *Streptomyces aureofaciens* (HrdB) and *E. coli* (RpoD) demonstrated that the amino acid sequences of the 2.4 regions of mycobacterial Sigma A and HrdB were almost identical to the corresponding sequences in RpoD. However, a higher degree of variability in the 4.2 regions of the sigma factors was observed. Whereas the amino acid sequences of Sigma A and HrdB were nearly identical, Sigma A differed from RpoD at 14 out of 39 positions, of which 9 represented non-conserved changes [65]. Based on these observations it was proposed that the total cellular transcription in mycobacteria may represent a division of labor involving RNA polymerase with multiple sigma factors. This hypothesis was substantiated with the sequencing of *M. tuberculosis* genome, which revealed the presence of 13 sigma factors. Thus, the poor functioning of mycobacterial promoters in *E. coli* was ascribed to poor interaction of the 4.2 region of the *E. coli* sigma factor with the −35 region of the mycobacterial promoters [65]. A class of promoters, referred to as extended −10 promoters, has been described in other organisms that have a TGN motif immediately upstream of the −10 region. It has been observed that promoters having a TGN motif drastically lose their transcriptional activity if the TGN motif is mutated. Extended −10 promoters have been identified and shown to function in mycobacteria in a manner similar to *E. coli* [66]. In fact some reports have shown that, in presence of a TGN motif, the requirement of a −35 like sequence is reduced. However, this remains to be evaluated in the case of mycobacteria.

Currently, work is in progress to completely characterize the organization and functioning of mycobacterial promoters. Mycobacterial promoters constitutively active in mycobacteria have been cloned and are being characterized by electrophoretic mobility shift and footprinting assays using purified mycobacterial RNA polymerase [24].

In another study, by using chloramphenicol resistance as the marker, restriction fragments of mycobacteriophage I3 DNA capable of initiating transcription were identified [67]. CAT assays were carried out to determine the strength of these promoters. However, in contrast to the above-mentioned study, DNA sequence analysis revealed that these promoters had significant homology with *E. coli* promoters with respect to the −35 region but were less homologous to *E. coli* promoters with respect to the −10 region. Based on the sequence of phage I3 promoters and the reported sequences of mycobacterial promoters, a promoter consensus for mycobacteria was generated [67].

Recently, a spectrophotometric assay for quantitative analysis of promoters in mycobacteria, which is a modification of the XylE reporter assay, has been described. In this method, cells were grown by patching on agar plates and were used directly in the assay without lysis, making this assay system amenable for high-throughput analysis [68].

Isolation and analysis of mycobacterial promoters, which are under regulatory control, can be useful for the expression of cloned genes in mycobacteria. Studies have been carried out to map the transcriptional start site of the acetamidase gene from *M. smegmatis* and the mechanism of its regulation has been analyzed. This promoter is currently being employed to express important mycobacterial genes [69].

The phenomenon of transcriptional termination in mycobacteria has been investigated. Transcription terminators are classified into two groups based on the requirement of additional factors. Sequences capable of causing the release of a transcript in an in vitro system with purified RNA polymerase alone are referred to as intrinsic or simple or factor-independent terminators. Terminators that require the presence of additional factors are classified as complex or factor-dependent. Intrinsic terminators are characterized by the presence of a palindromic stretch followed by a trail of T (or U) residues in the coding strand. The palindromic region extrudes out as a hairpin in the nascent transcript. The stem-loop structure is believed to be responsible for pausing of the RNA polymerase and for weakening the interaction between the polymerase, nascent RNA and template DNA due to relatively weaker bonds between the trail of U and A residues. A highly efficient algorithm was developed to identify hairpin potential sequences in bacterial genomes in order to build a general model for intrinsic transcription termination [70]. Analysis of the *M. tuberculosis* genome with this algorithm revealed that hairpin potential sequences were concentrated in the regions immediately downstream to stop codons. While the structures were concentrated at 21 nucleotides downstream of the stop codons in *E. coli*, the structures peaked at 37 nucleotides downstream of the stop codon in *M. tuberculosis*. Also, while *E. coli* showed a marginal preference for L-shaped structures, *M. tuberculosis* almost exclusively used I-shaped structures. However, most of these structures either lacked the U trail entirely or had a mixed A/U trail [70]. The ability of these structures to bring about termination in mycobacteria was tested by employing a mycobacteria-specific terminator-selection vector, which was constructed by cloning *M. smegmatis* gyrase promoter upstream of a CAT reporter gene [70]. Thus, any fragment encoding a terminator sequence if cloned between the promoter and reporter gene would reduce transcriptional read-through and effect a quantitative decrease in specific CAT activity. Representative I-shaped terminators were also PCR-amplified from the *M. tuberculosis* genome and cloned into the termination vector. The terminators were observed to bring about transcription termination bidirectionally, demonstrating that the U trail was not essential for the functioning of the terminator in mycobacteria. The terminators were also observed to be functional only when cloned in the untranslated region. When cloned within the coding region, the structure had no detectable effect on transcription read-through in either orientation. Using purified RNA polymerase from *M. smegmatis* the ability of the terminators to bring about termination was demonstrated in vitro [70].

2.7
Expression of Genes in Mycobacteria

With the development of expression systems for mycobacteria, it became possible to express genes in mycobacteria. Initially, these expression systems were analyzed by expressing reporter genes such as *lacZ* and *cat*. It was observed that about ~400-fold variation in the level of expression could be obtained by using promoters of different strengths upstream of the reporter gene [20]. Moreover, it was observed that the basic determinants of transcription initiation in cases of slow-growing and fast-growing mycobacteria were similar. Hence, the fast-grower *M. smegmatis* can serve as a surrogate host for analysis of expression of genes. These expression systems have also been used to express several genes from both mycobacteria and other pathogens. Recombinant BCG can be used as a multivaccine vehicle for the delivery of antigens for immunization against a variety of pathogens. However, several studies have indicated that, for an optimal immune response, especially in case of cell-mediated immunity, maximum expression of the antigen does not always lead to elicitation of optimal protective efficacy. Hence, the expression systems developed using promoters of various strengths can be helpful in determining the optimal level of expression of an antigen required for elicitation of optimal immunogenicity. The genes encoding Hepatitis B virus surface antigen and Hepatitis E virus ORF2 and ORF3 proteins were cloned in the extrachromosomal replicative vector, pSD5 and in the integration-proficient vector, pDK20, under the transcriptional control of a battery of promoters. Expression was analyzed both in the fast-growing *M. smegmatis* as well as in the slow-growing *M. bovis* BCG [24]. In view of the fact that secretory antigens are known to be more immunogenic, the genes for these antigens were cloned by placing them downstream of the secretory signals derived from Antigen 85B gene. These rBCG constructs have been evaluated both in the mouse model as well as in the monkey model [24].

In an attempt to develop recombinant BCG vaccines against TB, the genes for six antigens, namely the members of the Antigen 85 Complex (85A, 85B and 85C), 19-kDa antigen, 38-kDa antigen and Esat-6 antigen were cloned downstream of a battery of mycobacterial promoters, in the extrachromosomal replicative vector pSD5 to obtain varied levels of expression. Analysis of expression in the cytosol as well as culture supernatant revealed that by using different promoters these antigens could be expressed at levels ranging from 2–13-fold when compared to native levels of the corresponding antigen. Overexpression of these antigens was demonstrated both in the fast-growing *M. smegmatis* as well as in the slow-growing *M. bovis* BCG strains [24, 71].

3 Physiology of *Mycobacterium tuberculosis*

3.1 Cell Division in Mycobacteria

The division of a bacterium into two daughter cells is accomplished by the completion of DNA replication and subsequent segregation of daughter nucleoids (karyokinesis), followed by formation of a septum (septation or cytokinesis) at the mid-site of the elongated cell. This process is regulated by a number of genes. Since proliferation of mycobacteria is a prerequisite for the establishment of an active infection, it has been proposed that the structural and functional characterization of the proteins involved in cell division would enable the design of new anti-tubercular drugs. Hence, the gene encoding FtsH protease from *M. tuberculosis* $H_{37}Rv$ was cloned and expressed in *E. coli* yielding a protein ~85-kDa in size, which localized in the membrane [72]. The mycobacterial FtsH carried the zinc-binding domain HEGGH as against HEAGH present in its counterpart from *E. coli*. Analysis of the FtsH protein sequence revealed the presence of two ATP-binding domains and a Second Region of Homology, which are characteristic features of the AAA family of proteases. The ATPase activity of the FtsH protein of *M. tuberculosis* has been characterized [72] and a unique domain specific to mycobacterial FtsH protein was identified. The gene for the FtsZ protein homologue of *M. tuberculosis* $H_{37}Rv$ has also been cloned and expressed in *E. coli* [73]. Amino acid sequence comparison with the *E. coli* homologue revealed the presence of several conserved domains and some specific differences. Expression of the *M. tuberculosis* FtsZ protein in certain *E. coli* strains was observed to cause filamentation of the host cell, possibly due to the interference with the septation process mediated by the endogenous *E. coli* FtsZ protein. *M. tuberculosis ftsZ* gene also failed to complement *E. coli ftsZ* temperature sensitive mutant [73]. FtsZ protein is considered to be the prokaryotic homologue of tubulins in mammalian cells and has a GTP-binding motif, GGGTGTG, identical to that found in tubulins. It binds to GTP and hydrolyzes it during polymerization. It is known to function as the guiding structural scaffold for the formation of the septum at the mid-site of the bacterial cell undergoing division. The level of FtsZ protein is also known to determine the frequency of cell division in bacterial cells. In vitro studies have demonstrated a functional link between the two proteins, FtsH and FtsZ, indicating a possible role for FtsH in the regulation of cytokinesis. Attempts are being made to crystallize these proteins for studies on their three-dimensional structure, with the aim of employing them as targets for the development of new drugs against TB [73].

3.2
Proteins and Antigens of *M. tuberculosis*

The failure of BCG as a perfect vaccine against TB has prompted research for the development of a subunit vaccine. This quest has resulted in the identification of an array of mycobacterial antigens. Since cell-mediated immune response is regarded as the principal protective mechanism against mycobacterial diseases, the focus has been to identify mycobacterial antigens capable of generating strong T-cell mediated immune responses. Mycobacterial antigens have been characterized from cytosol, cell wall, plasma membrane as well as cell supernatants.

The plasma membrane of mycobacteria is rich in protein content and houses several important enzymes, receptors and transporters. Several reports have suggested that membrane proteins, especially the "integral" proteins, could be highly immunogenic, and this immunogenicity has been attributed to their hydrophobicity as well as lipid modification. Analysis of the sequence of the *M. tuberculosis* genome predicted nearly 800 putative membrane proteins although available reports on such proteins, particularly on integral membrane proteins of *M. tuberculosis* or other mycobacteria, are scanty and preliminary. Attempts have been made to characterize antigens of the mycobacterial plasma membrane with respect to human T-cell proliferative responses. Extraction of membrane vesicles prepared from *M. tuberculosis*, *M. bovis* BCG and *M. fortuitum* yielded detergent-soluble "integral" and water-soluble "peripheral" membrane proteins (IMP and PMP) [74–76]. In *M. bovis* BCG and *M. tuberculosis* ($H_{37}Rv$ strain and a South Indian clinical isolate) the IMP pool was found to contain, three previously known antigens: the 19- and 38-kDa lipoproteins and 33/36 kDa "proline-rich" antigen, besides several unknown proteins. Similarly, the PMP pool showed the presence of 16 kDa α-crystallin heat-shock protein. The profile of IMP and PMP pools in *M. fortuitum* was, however, different, highlighting the differences between "slow-" and "fast"-growing species of mycobacteria. Proteins from the IMP pool but not from the PMP pool were found to be heavily glycosylated in all the three species. The membrane protein profiles of *M. tuberculosis* $H_{37}Rv$ and the South Indian isolate of *M. tuberculosis* were generally comparable not withstanding some differences in the levels of expression and extent of glycosylation with respect to a few of the proteins [74–76].

The evaluation of these proteins in in vitro assays for human T-cell stimulation showed that in the cases of all three mycobacterial species, IMP was significantly more potent than PMP or cytosol in inducing proliferation of either peripheral blood mononuclear cells or purified $CD4^+$ T-cells. $CD8^+$ T-cells were not induced significantly. IMP also induced a significantly higher secretion of IFN-γ and IL-10, compared to PMP [77]. Further purification of individual proteins from the IMP pool and their evaluation in human T-cell proliferation assays has demonstrated a greater potency of certain low molecular weight (<30 kDa) IMPs. The observation that a single IMP fraction (<20 kDa) produced positive responses in almost all the study subjects in a manner similar to the whole IMP indicates that these antigens may be stimulating T-cells in a manner uninhindered by the restrictions imposed by the heterogeneity in the MHC. Hydrophobicity and lipidation of the antigens, characteristics of IMP, may be responsible for this potent activation of T-cells [74, 75].

Currently, the comparative proteomic analysis of the integral membrane proteins of *M. tuberculosis* $H_{37}Rv$ and BCG is being carried out in order to identify differentially expressed proteins, which could form the basis for a new vaccine, diagnostic probe or drug targets [77]. By using 2-D gel electrophoresis and MALDI-TOF mass spectrometry, 18 proteins were identified. Four of these proteins (corresponding to genes Rv0822c, Rv0934, Rv1270c and Rv1368) were true "integral" membrane proteins identified as lipoproteins or proteins bearing putative trans-membrane domains. The remaining proteins included two heat-shock proteins (Rv0251c and Rv2031c) and several enzymes involved in energy or lipid metabolism. Three of these proteins (Rv0243, Rv1872c and Rv1368) were co- or post-translationally modified to appear in different "charge" forms with one of them (Rv0243, FadA2) also appearing in a different "mass" form [77].

Attempts have also been directed towards defining the major T-cell activating proteins of *M. habana*. A 23-kDa glycosylated protein, which was the most abundant constituent of cytosol and had a homodimeric configuration turned out to be a superoxide dismutase (SOD) bearing structural similarities with SODs of *M. leprae* and *M. tuberculosis* [78]. This required Fe/Mn as co-factor and revealed a structural relationship with human MnSOD but not with Cu or Zn SOD. In purified form, the protein induced in vitro T-cell responses in primed mice as well as in human volunteers comprising healthy subjects and leprosy patients. Induction of in vivo T-cell response in guinea pigs was also observed [78]. *M. leprae* SOD has been employed previously for vaccination of mice against footpad challenge with *M. leprae* with considerable success. Thus, mycobacterial SOD represents a highly antigenic and immunoprotective candidate antigen, in spite of bearing a prominent structural homology with the corresponding host proteins [78].

Most work on secretory antigens has been carried out using standard laboratory grown strains of *M. tuberculosis* ($H_{37}Rv$, $H_{37}Ra$ and Erdman strains). It was felt that the search for a potential vaccine candidate should be extended to contemporary clinical isolates of *M. tuberculosis*, which might be significantly different from the standard strains. Thus, the relative efficacies of culture filtrate proteins (CFPs) from two laboratory strains and four contemporary clinical isolates of *M. tuberculosis* to induce T-cell activation were evaluated [79]. CFPs from these six strains were evaluated for induction of T-cell proliferation, IFN-γ secretion and IL-12 secretion from PBMCs from 33 healthy donors. IL-12 was observed to be spontaneously secreted by the PBMC preparations. However, CFP preparations from two isolates were observed to be more potent in inducing T-cell proliferation and induction of IFN-γ secretion in human PBMCs [79]. Identification of the secretory antigens from these isolates, which are responsible for enhanced immune reactivity, is currently being pursued.

In previous studies, using malaria-infected erythrocytes and leishmania-infected macrophages as model systems, it has been shown that both polyclonal antiserum and monoclonal antibodies, generated by immunization with membranes of parasite-infected cells, reacted with the neo-antigenic determinants on infected cell surface both in vitro and in vivo [80, 81]. Such an immunization strategy was also observed to divert the immune response towards clinically relevant minor antigenic determinants in the absence of major parasite immunogenic components [82]. These observations have been subsequently extended to *Mycobacterium*. For the first time, it was shown that the membranes of *M. microti*-infected macrophages differ antigenically from those of the normal macrophages [83]. Identification of these antigenic differences can help in elucidating the mechanisms responsible for intracellular survival of mycobacteria. Antisera were generated against the bacterial extract, heat-killed bacteria and a crude preparation of membranes from *M. microti*-infected macrophages. Unlike anti-bacterial extract or anti-heat killed bacterial antiserum, anti-infected macrophage antiserum was observed to react specifically with the infected macrophage surface and this reactivity increased with the increase in post-infection time. However, this reactivity was not observed with serum raised against uninfected macrophages, PMA- or lipopolysaccharide-activated macrophages or those harboring *M. tuberculosis* $H_{37}Ra$ or heat-killed *M. microti* or *L. donovani* [83]. A 63-kDa molecule specifically present on *M. microti*-infected macrophage membranes is being currently characterized [83]. To understand the mechanism responsible for the intracellular survival of the bacteria, components commonly present on *M. microti* and infected macrophages and those common to *M. microti* and *M. tuberculosis* $H_{37}Ra$ are also being evaluated by using a panel of monoclonal antibodies [84].

In an attempt to identify new *M. tuberculosis* antigens to generate strong Th1 responses, the log-phase culture filtrate as well as the membrane component of a human isolate of *M. tuberculosis* were analyzed for their ability to stimulate lymphocytes isolated from TB patients. Mycobacterial proteins stimulating strong Th1 responses are being analyzed further for their vaccine potential [85].

The sequencing of the *M. tuberculosis* genome has led to the identification of two families of proteins – the PE and the PPE gene families. It was hypothesized that these proteins are the principal source of antigenic variation in *M. tuberculosis*. By in silico analysis, PE and PPE ORFs with regions of high antigenic index were short-listed. Two of these PPE family genes have been expressed as recombinant histidine-tagged proteins in *E. coli* [86]. Reactivity against these recombinant proteins was assayed by ELISA using serum isolated from fresh infection cases, patients with relapsed TB, extrapulmonary cases and multi-drug resistant cases. Preliminary results show that these recombinant proteins are recognized efficiently by the serum samples, thus indicating their antigenic potential [86].

Comparison of the protein profiles of *M. tuberculosis* and *M. bovis* BCG, has led to the identification of a 25-kDa protein, which was observed to be present in BCG strains but absent from *M. tuberculosis* suggesting its usefulness in distinguishing BCG strains from *M. tuberculosis* [87].

By using T-cell blot and immunosubtractive assays to identify antigens of *M. tuberculosis* associated with human immune responses in healthy contacts, a histone-like protein of *M. tuberculosis* (HLP_{Mt}, Rv2986c) was identified, which had homology with prokaryotic histone-like proteins (HU) and H1 histones from several eukaryotic species [88]. The sequence of a 16-amino acid long peptide of this protein exhibited 100% homology with an ORF, predicted to code for a protein of 214 amino acids, in the *M. tuberculosis* genome sequence. The gene encoding the predicted protein was PCR-amplified, cloned, sequenced and expressed in *E. coli* as a protein of 28 kDa [88]. A human immune response to the recombinant HLP_{Mt} protein has been demonstrated by its ability to induce lymphoproliferation in PBMCs and the detection of anti-HLP_{Mt} antibodies in pooled patient sera. The recombinant HLP_{Mt} protein elicited a vigorous lymphoproliferative response, especially in healthy tuberculin reactors compared to non-reactors and TB patients. The protein had unique dual domains with homology to both bacterial histone-like proteins (HU) and eukaryotic histone H1, which suggested that the protein could bind DNA. The DNA-binding properties of HLP_{Mt} were later confirmed by South-Western analysis [88]. Molecular dynamics simulations of interaction of the protein with 35 bp GC-rich U-bend DNA were also carried out [89]. The three dimensional structure of the first 92 residues of this protein was obtained by making use of secondary structure prediction programs, sequence search and HOMOLOGY-based modeling. Simulation of the protein and its complex with U^{35} DNA was carried out in vacuum by the molecular dynamics technique. These studies suggested the role of HLP_{Mt} in overwinding and packaging of DNA [89]. A strategy based on PCR amplification of the gene encoding HLP_{Mt} has recently been employed to differentiate between closely related mycobacterial species such as members of the *M. tuberculosis* complex and *M. paratuberculosis* and *M. avium* [90].

Two genes, encoding hemoglobin-like proteins, have been detected in the genome of *M. tuberculosis*. Mycobacterial hemoglobins, HbN and HbO, belong to the superfamily of truncated hemoglobins that are small heme proteins of 110–130 amino acid residues per chain carrying a substantial deletion within their N- or C-termini along with the replacement of the crucial heme binding F helix by an extended polypeptide loop. The gene encoding HbO of *M. tuberculosis* has been cloned and expressed in *E. coli* and *M. smegmatis* in an attempt to study the characteristics and potential function of this protein [91]. This 14.5 kDa homodimeric heme protein, which exhibited nearly 50-fold lower oxygen affinity than HbN interacts with the respiratory membranes of *E. coli* and *M. smegmatis* and participates in oxygen uptake and transfer during aerobic growth. HbO is produced in *M. bovis* BCG (0.2 to 0.5% of total cellular protein) throughout its aerobic growth and could play a vital role in the intracellular growth and survival of *M. tuberculosis* in the oxygen-deficient environment of its host by sequestering and competing for oxygen [91].

In an attempt to identify antigens that may be useful for immunodiagnosis of TB, the circulating antigen was isolated from bacteriologically confirmed tuberculous sera [92]. Its fractionation by SDS-PAGE and further analysis of various fractions resulted in the identification of a highly reactive antigen in a fraction. Further resolution of this fraction resulted in 4 antigenic fractions, of which the fraction CTA2-7D exhibited seroreactivity similar to a 31 kDa antigen (ESAS-7F) isolated from the in vitro culture medium. Biochemical characterization of CTA2-7D demonstrated it to be a lipoglycoprotein [92]. The 31 kDa antigen from *M. tuberculosis* $H_{37}Ra$ has been purified and shown to contain a metalloserine protease activity [93].

3.3 Membrane Permeability and Transport in *M. tuberculosis*

The emergence of drug-resistant strains of *M. tuberculosis* is considered as one of the main factors for the resurgence of tuberculosis. The elucidation of drug resistance at the molecular level is required in order to develop new intervention strategies.

Fluoroquinolines such as ciprofloxacin, ofloxacin and sparfloxacin are a group of drugs that have been found to be effective against mycobacteria. The primary intracellular target of fluoroquinolones has been shown to be the A subunit of DNA-gyrase (GyrA). Point mutations in this gene have been implicated in the resistance to this group of drugs in bacteria. However, utilizing *M. smegmatis* as a genetic model for *M. tuberculosis*, it was demonstrated for the first time that an active efflux of drugs was the predominant cause for high-resistance to fluoroquinolones exhibited by a laboratory-generated ciprofloxacin-resistant mutant [94]. Growth experiments indicated the mutant strain to be at least 64-fold less sensitive to ciprofloxacin predominantly due to the accelerated rate of drug efflux in the mutant strain compared to the wild-type strain. Incubation with verapamil, a calcium channel blocker, enhanced the drug accumulation in the mutant and thus reversed the resistant phenotype [94]. Although uncharacteristic for fluoroquinolone resistance in prokaryotes, similar findings were observed in studies of multidrug resistance in eukaryotes where ABC transporters were involved. Hence, the involvement of ABC transporters in the mutant was investigated. Using consensus sequences of the ABC family of transporters, the ATP binding region of the ABC transporter from *M. smegmatis* was PCR amplified, and was found to encode amino acid sequences that exhibited significant homology with different ABC transporters [95]. By analyzing its complete gene, this ABC protein has been identified as the B-subunit of the phosphate-specific transporter (PstB) [95]. In *E. coli*, PstB, a traffic ATPase, is a part of the binding protein-dependent phosphate uptake system. The phosphate-specific transport (Pst) system is known to scavenge phosphate and is active only when the phosphate concentration is limiting. By Southern analysis of the RT-PCR product, the mRNA of this transporter was demonstrated to be overexpressed in the ciprofloxacin-resistant mutant [95]. Southern hybridization also revealed chromosomal amplification of the *pstB* gene in the mutant [96]. These results provide a new insight into the mechanism of drug resistance in mycobacteria as the existing reports do not indicate the association of any ABC transporter in the process of efflux-based fluoroquinolone resistance in any prokaryote. The mutant was observed to exhibit enhanced phosphate uptake and, on inactivation of *pstB* in the parental strain of *M. smegmatis*, loss of high affinity phosphate uptake and hypersensitivity to fluoroquinolones and structurally unrelated drugs was observed [96]. To further investigate the role of this transporter in the process of efflux-mediated drug resistance, the *pst* operon in the mutant strain was inactivated [97]. Genotypic analysis revealed partial disruption of the *pst* operon due to transporter gene amplification in the mutant. However, phenotyopic characterization of the resulting strain (mutant strain with inactivated *pstB*) demonstrated a striking reduction in the minimal inhibitory concentration (MIC) of ciprofloxacin as well as in the drug extrusion profile [97]. These findings thus indicate, for the first time, the role of active efflux in drug-resistance in mycobacteria. It also establishes the involvement of an ABC transporter in the process and thus points towards the novel role of the mycobacterial Pst system besides its involvement in phosphate transport.

The *pstB* from *M. tuberculosis* has been overexpressed in *E. coli* and the purified protein was shown to exhibit ATP-binding and ATPase activity [98]. Interestingly, unlike other prokaryotic ABC proteins, the ATP hydrolyzing ability of PstB from *M. tuberculosis* strain $H_{37}Rv$ was observed to be magnesium-independent and resistant to known ATPase inhibitors. Mutation of the conserved aspartic residue to lysine in the Walker motif B abrogated the ATPase activity of PstB, confirming the recombinant protein to be an ATPase. Furthermore, these results convincingly established that the mycobacterial PstB is a thermostable ATPase and thus highlighted for the first time the presence of such an enzyme in any mesophillic bacteria [98].

The natural or intrinsic drug resistance of mycobacterial species has also been attributed to the organization and chemical nature of lipids in the outer layer of the cell wall. Understanding the steps and the enzymes involved in cell wall biosynthesis could thus lead to the identification of targets for new anti-tubercular drugs. A multipronged approach was initiated for understanding cell wall biosynthesis and antibiotic resistance in mycobacteria. The lack of efficacy of classical β-lactam antibiotics against mycobacteria has been associated with a combination of factors such as (i) capability of these drugs to permeate into mycobacteria, (ii) β-lactamase production, and (iii) affinity of the penicillin-binding proteins (PBPs) for the drugs, rather than the inertness of the drugs towards the peptidoglycan cross-linking machinery. A laboratory mutant of *M. smegmatis*, which was 7–78-fold more resistant than the parent strain to the β-lactams tested, was studied [99]. Characterization of this mutant revealed that the acquired β-lactam resistance in this strain was due to an interplay between the lowered affinity of at least one high-molecular-mass PBP for cephalosporins, as well as reduced permeability of the cell wall, rather than enhanced β-lactamase production. The efflux of antibiotics mediated by proton motive force and ATP-dependent efflux pumps also contribute to antibiotic resistance. Although several targets of INH have been identified, the mechanism of INH resistance remains incompletely understood. Using various inhibitors and a proton motive force (pmf) uncoupler, it was shown that both pmf- and ATP-dependent extrusion systems were involved in INH efflux in *M. smegmatis*, contributing to the high intrinsic resistance of *M. smegmatis* to INH. The possible existence of such pumps in pathogenic mycobacteria and their contribution to INH resistance are presently being evaluated [100]. The daunorubicin (Drr) efflux pump of *M. tuberculosis* has been expressed in *E. coli* and *M. smegmatis*. It was found to impart resistance to structurally unrelated drugs such as chloramphenicol, tetracycline and streptomycin. The Drr transporter and several other transporters of *M. tuberculosis* are currently being overexpressed and characterized in detail [101].

A set of membrane-bound proteins, known as PBPs, is involved in the final assembly of the bacterial cell wall peptidoglycan. The nine motifs characteristic of the Class A PBPs, which catalyze transglycosylation and transpeptidation reactions, are present in the PBPs designated as 1 and 1* in *M. tuberculosis* and *M. leprae*. These PBPs have low amino acid sequence similarity to their *E. coli* counterparts. Three of these PBPs have been characterized in detail [102–105]. The sequence of the *M. tuberculosis* genome has revealed the presence of two open reading frames, Rv3682 and Rv0050, encoding two putative Class A high-molecular mass PBPs. The PBP1* and its soluble derivative, encoded by the ORF, Rv0050, have been overexpressed and characterized. The C-terminal module of PBP1* of *M. tuberculosis* was observed to function independently of the N-terminal module as a penicillin-binding entity [105]. The *M. tuberculosis* genome has the largest repertoire of putative PBPs among all microbial genomes sequenced. A detailed analysis of these PBPs and their interactions with cell division machinery and with the components required for cell wall synthesis is currently in progress [101].

The outer membrane of Gram-negative bacteria serves as an efficient permeability barrier to the inward flow of antibiotics. Gram-positive bacteria, including mycobacteria, do not contain an outer membrane. In the case of mycobacteria, the mycolates form a highly hydrophobic lipid layer, which acts as a permeability barrier similar to the outer membrane of Gram-negative bacteria. Permeability coefficients of small molecules in mycobacteria are observed to be lower than in *E. coli*. Despite this, small hydrophilic molecules are known to penetrate through the water-filled channels of pore-forming proteins. A 40 kDa pore-forming protein from the cell wall of *M. smegmatis* has been purified. Although the N-terminal sequence did not show any similarity to the porins sequenced earlier, by employing liposome swelling assays, the permeation of small hydrophilic molecules such as sugars and amino acids has been demonstrated [106].

Twenty putative drug efflux pumps have been predicted in the published genome sequence of *M. tuberculosis* $H_{37}Rv$. Four pumps were selected and their expression levels were analyzed in different clinical isolates in the presence of drugs [86]. In one of the isolates, a direct correlation was observed between RNA transcript levels of one of the putative efflux pumps and different concentrations of drugs in the growth medium. However, the actual role of the efflux pump in imparting drug resistance in the clinical scenario has yet to be investigated [86].

4 Immunology of Tuberculosis

Intracellular pathogens such as *M. tuberculosis* exhibit an intense relationship with the immunological system of the host. These interactions are marked by a range of complexities. On one hand, the immunological response of the host can lead to the clearing of infection, on the other hand, tissue damage and necrosis prominent in TB results from these immunological perturbations. Hence, understanding the immunological responses in the context of host-pathogen interaction and their consequences in terms of onset of the disease and its progression are extremely vital for designing successful strategies for control of this disease.

Interferon-γ is known to play a key role in host defense against pulmonary mycobacterial infections. A variety of lymphocyte subsets might contribute towards this pulmonary IFN-γ response. To understand the relative contribution of the various subsets, a murine model of lung infection was standardized using *M. bovis* BCG and the kinetics of changes in lymphocytes sub-populations after murine pulmonary BCG infection was analyzed [107]. The BCG load in lungs peaked between 4–6 weeks post-infection. There was a marked increase (6-fold) in number of T-cells secreting IFN-γ, 5–6 weeks post-infection. Analysis of the cell populations responsible for IFN-γ responses revealed that a basal level of IFN-γ release continues to occur in lungs of uninfected mice, with T, NK as well as NKT cells in the lung interstitium being responsible for these responses. The NK cells are responsible for basal IFN-γ secretion in normal lungs, the peak of IFN-γ response post-infection is essentially due to T-cells [107].

Mitogens/antigens have the innate ability to trigger both proliferation and apoptosis simultaneously in what is known as the dual-signal hypothesis, dependent upon the dosage and duration of the trigger and various in vivo conditions. The dual signal hypothesis was tested in the murine model using PPD as the antigen [108]. The mice immunized with *E. coli* lipopolysaccharide served as the positive control. It was observed that PPD induced apoptosis only in the thymus whereas LPS induced apoptosis in both thymus and bone marrow. Plasma from PPD-treated animals also had higher levels of *M. tuberculosis*-specific antibodies. PPD, apart from triggering apoptosis in the thymus, also elicited a potent humoral immune response against the *M. tuberculosis* antigens. Thus, an in vivo defect in T-cell proliferation associated with an enhanced level of T-cell apoptosis was noted, which has been proposed as one of the components of the ineffective immune response against *M. tuberculosis*-induced immunopathology in susceptible hosts. It was hypothesized that high concentrations of *M. tuberculosis* antigens may incapacitate CMI in young susceptible mice by inducing thymic atrophy [108].

Monoclonal antibodies have been raised against secretory antigens of *M. tuberculosis* $H_{37}Rv$ by immunizing BALB/c mice with *M. tuberculosis* culture filtrate [109]. Intrasplenic and intraperitoneal routes of immunization were compared and the latter was found superior. Ten monoclonal antibodies (MAbs) were produced, which were characterized for their isotype, binding specificity, nature of binding epitope, reactivity in immunoassays etc. Two MAbs exhibited species specificity, one of which was found to be restricted in its reactivity to the *M. tuberculosis* complex and *M. leprae*, suggesting its potential in the immunodiagnosis of TB [109].

The fate of *M. tuberculosis* inside macrophages was monitored by infecting rat peritoneal macrophages with *M. tuberculosis*. Since the microbicidal nature of macrophages has been correlated with NO production, attempts were made to measure alterations in the levels of NO in terms of nitrite formed. Alterations in the levels of hydrogen peroxide, lysosomal enzymes such as acid phosphatase, cathepsin-D and β-glucuronidase in macrophages following infection with *M. tuberculosis* were also studied [110]. Elevation in the levels of nitrite was observed 72 h after *M. tuberculosis* infection. However, elevated levels of H_2O_2 were produced by *M. tuberculosis*-infected macrophages irrespective of the time point. A maximum increase in the level of acid phosphatase was observed at the 72 h time point whereas the maximum increase in the level of β-glucuronidase was observed 48 h after *M. tuberculosis* infection. Surprisingly, these microbicidal agents did not alter the intracellular viability of *M. tuberculosis*, even though total elimination of intracellular *S. aureus* by macrophages was seen [110]. The fate of *M. tuberculosis* was also analyzed in human peripheral blood monocytes [111]. In unstimulated PBMCs, the intracellular *M. tuberculosis* multiplied freely. Even after in vitro activation of monocytes with a cocktail of lipopolysaccharide, phorbol myristate acetate, IFN-γ and TNF-α, there was no killing of bacilli in the intracellular environment. The ability of ATP in reducing the intracellular viability of mycobacteria has also been investigated [111]. Infected monocytes upon ATP treatment underwent cell death, however, there was no loss in the intracellular viability of *M. tuberculosis* or *M. smegmatis*. It was hypothesized that the human monocytes inhibit the growth and multiplication of *M. tuberculosis* in vitro for up to 72 h, after which *M. tuberculosis* multiplies inside the intracellular environment of the monocytes. Activation with the various reagents stimulates the intracellular respiratory burst in macrophages but fails to inhibit the growth of intracellular mycobacteria [111].

TB lesions heal by fibrosis. However, an excess of fibrosis can lead to disability of the affected organ. This phenomenon on infection with *M. tuberculosis* has been characterized biochemically and pathologically in the guinea pig [112]. Establishment of the infection with *M. tuberculosis* in the guinea pig was assessed by determining the presence of viable organisms in the spleen and the presence of tuberculous granuloma in the lung, liver and spleen. The infection appeared to be self-limiting by the fourth week and fewer animals were observed to have viable organisms during the later periods of infection. The levels of collagen, elastin and hexosamines exhibited an initial decrease followed by an increase. The elastin and hexosamine levels returned to the basal levels in all three organs. However, the collagen levels were observed to rise in the lungs and were found to be comparable to those observed in the case of animals immunized with bleomycin as positive control. Thus, this model can be adopted for studying the mechanisms of fibrogenesis in tuberculous infection [112].

Attempts have also been made to define the relationship of intact tubercle bacilli and/or their antigenic fragments to a granuloma in the guinea pig in order to distinguish an active from a resolving granuloma. Granuloma was induced in the skin in one set of animals by injecting heat-killed *M. tuberculosis* intradermally, whereas in another set it was induced in the lungs and spleen by injecting live *M. tuberculosis* intramuscularly [113]. Histological examination for the presence of granuloma, bacilli and antigenic fragments revealed that, in the dermal lesion, the bacilli were cleared by day 42, followed by clearance of antigenic fragments by day 63 and finally, by day 84, the final resolution of the granuloma itself. However, in the case of the guinea pigs infected with live *M. tuberculosis*, while removal of bacilli followed by clearance of antigen was observed, the granuloma itself was not observed to subside completely even though a reduction in congestion and edema of the granulomatous area was observed. The results from these studies will be helpful in differentiating between a persisting and a resolving tuberculous granuloma [113].

To study the effect of environmental mycobacteria on the modulation of immune responses, *M. avium* complex (MAC) and *M. fortuitum* complex isolates (originating from different sources from the BCG trial area) were evaluated in the guinea pig model of TB infection [114]. Guinea pigs were sensitized with the various strains and the effects of prior exposure on BCG vaccination and on protection against *M. tuberculosis* were investigated. At 2 weeks after challenge, prior exposure to MAC or *M. fortuitum* complex was observed to have no modulatory effect on the protective immunity due to BCG. However, at 6 weeks after challenge, while the guinea pigs immunized with BCG were protected, modulation of the protective response resulting from BCG was observed in the guinea pigs sensitized with MAC or *M. fortuitum* [114].

4.1 Immune Studies in Humans

T-cell proliferation and IFN-γ production upon stimulation with *M. tuberculosis* has also been assessed in pulmonary TB patients and healthy contacts [115]. The studies were based on the lymphocyte transformation test and detection of intracellular IFN-γ production by $CD4^+$ T-cells. Patients exhibited lower levels of proliferation compared to the healthy contacts. Healthy contacts in general had elevated numbers of $CD4^+$ T-cells expressing IFN-γ compared to PBMC derived from patients. This feature was observed both in stimulated as well as in cultures devoid of antigen. The results of this study further demonstrated that PBMCs from healthy individuals exhibited enhanced frequency of $CD4^+$ T-cells expressing IFN-γ in response to *M. tuberculosis* antigen compared to control cultures [115].

Humoral and lymphocyte responses to *M. tuberculosis* culture filtrate antigens have also been analyzed in active pulmonary TB, inactive (cured/quiscent) TB and healthy control subjects [116]. High antibody titers were found in active TB patients compared to inactive TB patients or control subjects. On the contrary, low or suppressed lymphocyte responses to PHA, Con-A and *M. tuberculosis* culture filtrate antigens were observed in active TB patients compared to inactive TB patients and control subjects. This study suggested that when the disease is cured, the antibody response declines and the lymphocyte response to antigens and mitogens increases to the same level as found in controls, indicating the restoration of the normal immune status in cured patients [116].

Purified CFP 6 protein, a low molecular weight protein antigen, purified from the culture filtrate of *M. tuberculosis*, has been studied for its ability to induce proliferative responses of PBMCs isolated from different categories of human subjects [117]. CFP 6 was found to elicit highly proliferative responses in healthy contacts and patients recovering from disease and induced the release of significant amounts of IFN-γ in cell culture supernatants of healthy contacts compared to other groups. Moreover, the ability of CFP 6 to induce DTH in guinea pigs immunized with BCG or *M. tuberculosis* $H_{37}Rv$ pointed out the immunological importance of this protein [117].

Previous studies on the distribution of T-cell populations in TB patients have been carried out among adults. However, the distribution of T-cell populations during the initial response to infection cannot be assessed as the adults might have been infected in the distant past. In addition, several co-existing illnesses might affect the T-cell subpopulations in adults. To circumvent this problem, the T-lymphocyte subpopulations were investigated in children [118, 119]. Lymphocyte subpopulations were determined by flow cytometry in healthy tuberculin-positive children, in children newly diagnosed with pulmonary TB and in children after anti-tubercular therapy. It was found that the absolute numbers and percentages of $CD3^+$ and $CD4^+$ T-cells were reduced in children with TB, when compared to the controls. Following anti-tubercular therapy, the $CD4^+$ counts increased significantly compared to baseline values. The proportion of T-cells expressing the $\gamma\delta$ T-cell receptor was similar in TB patients as well as in the controls [118, 119].

In an attempt to understand the role of cytokines in susceptible and resistant subjects exposed to *M. tuberculosis* infection, intracellular IFN-γ and IL-4 in ex vivo peripheral blood-derived $CD4^+$ T-cells have been examined in TB patients and healthy subjects [120]. The patients had higher number of $CD4^+$ T-cells expressing IFN-γ and IL-4 in unstimulated cultures compared to healthy subjects, although the ratio of these two subsets was similar in both groups. This elevated response in patients, i.e., larger numbers of $CD4^+$ T-cells producing IFN-γ is indicative of an ongoing Th1 response. However, after in vitro stimulation of PBMC with heat-killed *M. tuberculosis* there was a significant reduction in the percentage of IFN-γ^+ $CD4^+$ T-cells in patients, which was also reflected in the IFN-γ ELISA assay using supernatants derived from stimulated cultures. In the case of a majority of patients, this reduction of IFN-γ^+ $CD4^+$ T-cells was seen to result in the dominance of IL-4^+ $CD4^+$ T-cells, which may contribute to the downregulation of IFN-γ expression and the crucial effector function of CD4 T-cells, leading to persistence of the disease and immunopathology characteristically seen in patients. The majority of high-risk healthy individuals examined differed in that, after mycobacterial-antigen stimulation, an enhancement in IFN-γ^+ $CD4^+$ T-cells was observed [120].

For a complete delineation of the protective response induced by BCG vaccination in South India, changes in cytokine secretion patterns were studied in addition to investigating skin test conversions and lymphocyte proliferation, which are usually considered as simple end-point tests [121]. Changes in cytokine secretion patterns before and after BCG vaccination in Mantoux-negative subjects were investigated and compared with the patterns in subjects who were Mantoux-positive in spite of never having received BCG or having any evidence of suffering from TB. The pre- and post-vaccination tuberculin skin tests carried out in this study indicated that all except one PPD-negative individual became positive following administration of BCG [121]. This suggested that, despite being largely ineffective in this population, BCG does induce some alterations in the T-cell repertoire. However, the in vitro proliferative responses to PHA, PPD and *M. tuberculosis* remained largely unaltered by vaccination, probably reflecting the fact that the response observed at the localized site of the PPD reaction reflects the recruitment of specific cells into a site of high antigen concentration. This is not mimicked in in vitro proliferation responses. In response to mycobacterial antigens, PPD-negative individuals exhibited very low production of IFN-γ at the protein level and only in some individuals was mRNA detected by using the very sensitive RT-PCR. BCG vaccination of these individuals did not alter the IFN-γ production, indicating that mycobacterial antigens might not have stimulated the appropriate subsets of T-cells [121]. The levels of IL-10 and IL-4 also did not significantly alter on vaccination with BCG. Compared to this group, the PPD-positive group produced significantly higher amounts of IFN-γ. This suggests induction of a Th1-type response to mycobacteria in unvaccinated PPD-positive individuals in the area, probably due to their exposure to *M. tuberculosis* or environmental mycobacteria. The results from this study indicated that BCG has little effect in driving the immune response towards a protective Th1-type response [121].

The effect of picroliv, a standardized fraction of root and rhizome of *Picrorhiza kurroa*, on proliferative T-cell response to the mycobacterial "Purified Protein Derivative (PPD)" antigen in subjects infected with or exposed to mycobacteria (tuberculoid leprosy patients and endemic normals), has also been analyzed [122]. Culturing of PBMCs isolated from low responders, with an optimal concentration of picroliv (0.5 μg/mL) significantly enhanced the proliferative response to PPD. However, the response of PBMCs isolated from high-responders, to PPD or PHA (a non-specific mitogen) in the presence of picroliv remained unaffected. Picroliv was also not observed to induce cell proliferation on its own. Thus, it has been proposed that picroliv could be useful as an adjunct to chemotherapy or as a short-term prophylactic agent [122].

Most of the studies conducted for identification and characterization of *M. tuberculosis* antigens have used monoclonal or polyclonal antibodies as reagents. During the course of an active infection, *M. tuberculosis* is known to modulate the immune response of the host. However, the exact mechanism of these modulations is still not clear. Expression of a battery of genes might be responsible for enabling *M. tuberculosis* to survive in the host. Thus, analysis of the host immune response towards these antigens will not only serve to enhance our knowledge about the antigens relevant for protection and for the altering host immune response, but will also help in the development of useful reagents. Towards this goal, genomic DNA from a human sputum isolate of *M. tuberculosis*, which showed extensive dissemination in the guinea pig model, was used to construct an expression library in the lambda ZAP vector [123]. The library was subsequently probed with pooled sera from chronically infected TB patients. The recombinants obtained were also probed with serum from patients who were clinically diagnosed to be in the early phase of infection. Interestingly, some recombinants that had given a strong signal in the original screen failed to react with the serum from these patients. On the other hand, some recombinants that had given very weak signals in the original screen reacted strongly with this serum, indicating a differential nature of either the expression of these antigens or the immune response elicited against these antigens with the progression of the disease [123]. Antisera raised in mice against killed antigen preparations of *M. tuberculosis* failed to recognize most of the recombinant antigens that had exhibited reactivity against TB patient sera. Similar results were obtained with serum from guinea pigs immunized with live and killed mycobacteria, suggesting that antigens derived from live or killed mycobacteria are presented to the host immune system in a differential manner [124]. Characterization of some of the TB patient sera-reactive antigens has led to the identification of one of them as FtsH and the other one as the aminoimidazole ribotide synthase of *M. tuberculosis*. Besides, one of these proteins, which was secretory in nature, was recognized by sera from the majority of TB patients but not by sera from BCG-vaccinated individuals, a property that could be useful in the development of immunodiagnostics of TB [125]. Using a novel technique to purify bacterial mRNA by polyadenylating mRNA on polysomes [17] and by employing differential display reverse transcription-polymerase chain reaction (DDRT-PCR), attempts have been made for isolating genes from *M. tuberculosis* expressed differentially within infected guinea pigs tissues [126].

Studies were conducted to analyze whether the presence of cold reactive lymphocytotoxic antibodies (LCA) in the system has any effect on immunity to TB [127]. The plasma samples of inactive-TB patients (cured patients) exhibited a higher percentage of positivity (36.7%) for lymphocytotoxic antibodies than the samples from active-TB patients (21.4%) and control subjects (18.8%). No significant difference was observed between LCA-positive and LCA-negative active-TB patients and normal healthy controls in antibody and lymphocyte response against *M. tuberculosis* culture filtrate antigens. Determination of the HLA-DR phenotype revealed that the percentage of individuals positive for LCA was higher among HLA-DR2-positive and DR7-positive active-TB patients and control subjects than among the non-DR2 and non-DR7 subjects, indicating that HLA-DR2 and DR7 may be associated with the occurrence of LCA activity. The presence of LCA had no immunoregulatory role on immunity to TB [127].

Several new assays have been described for in vitro measurement of cell-mediated immunity (CMI) such as determination of lymphocyte subsets, measurement of cytokines and soluble cytokine receptors etc., which have replaced the conventional lymphocyte blast transformation assay to some extent. However, these tests have one limitation or another. Moreover, the in vitro tests do not necessarily reflect the in vivo situation. Another approach to study CMI has been suggested, which involves the assay of a soluble marker such as neopterin, which is biologically stable and detectable in body fluids. A raised serum level of neopterin is generally associated with activation of CMI. Macrophages, particularly when stimulated with IFN-γ, are a major source of neopterin. The potential of neopterin in evaluating CMI in patients with pulmonary TB has been investigated [128]. Neopterin levels in serum and the culture supernatants of PBMCs after stimulation with PPD were measured in patients with pulmonary TB and in healthy individuals. Lymphocyte proliferative response assays to PPD were also carried out in these two groups. The mean concentration of serum neopterin was observed to be significantly higher in patients than in controls. The spontaneous release of neopterin in culture supernatants of PBMCs from patients was significantly higher when compared with those of healthy controls. However, on stimulation of PBMCs with PPD, the release of neopterin was similar in both the groups. Moreover, the release of neopterin and the proliferation of lymphocytes were not comparable, indicating that these two parameters did not run in parallel for measuring the state of CMI. Interestingly, the serum concentration of neopterin showed an apparent inverse relationship with lymphocyte proliferation [128]. Determination of serum neopterin levels was also carried out during chemotherapy in order to study whether elevated levels before treatment reflect the extent of tuberculous lesions and whether changes in their concentration during treatment correspond to a response to therapy [129]. Neopterin levels were also measured at the time of relapse and were compared with the levels observed in patients who did not relapse to ascertain whether these levels reflected the change in disease status. All patients were observed to have elevated levels of serum neopterin, which declined at 1 month and were near normal at the end of treatment. The levels continued to fall after completion of therapy in patients who did not relapse whereas an increase after completion of therapy was associated with relapse. Thus, the measurement of serum neopterin levels could be useful in evaluating the response to therapy [129].

5 Mechanisms of Pathogenesis of *M. tuberculosis*

All intracellular pathogens have an ability to grow inside the host cell. Pathogenic mycobacteria not only survive in the host cell but they are endowed with a unique ability to remain in the system for a very long time. However, our knowledge concerning the mechanisms that this pathogen employs for entry into the host cell, for its survival and multiplication, for its spread to neighboring cells, for circumventing the host defense mechanisms and for insulting the host to cause disease still remains rather poor. The pathogen has evolved several mechanisms to circumvent the hostile environment of macrophages, which include i) inhibition of phagosome-lysosome fusion, ii) inhibition of phagosome acidification, iii) recruitment and retention of tryptophan/aspartate-containing coat protein on phagosomes to prevent their delivery to lysosomes and iv) expression of members of the host-induced PE-PGRS family of proteins. For the identification of genes involved in virulence and pathogenesis two strategies have been primarily used – employment of recombinant DNA technology (for example, by generating transposon insertional mutants) and the use of suitable animal models.

The RNA subtractive hybridization strategy was employed to identify and clone genes that are differentially transcribed in the virulent strain of *M. tuberculosis* [130]. This approach was based on the premise that genes conferring virulence would be differentially expressed in *M. tuberculosis* $H_{37}Rv$ and RNA isolated from *M. tuberculosis* $H_{37}Rv$ would be enriched in transcripts specified by the differentially expressed genes. Differentially expressed genes in the virulent strain of *M. tuberculosis* (*dev* genes) were identified by screening a plasmid gene bank of *M. tuberculosis* $H_{37}Rv$ with a cDNA probe that was enriched in *dev* transcripts by subtraction of RNAs common to *M. tuberculosis* $H_{37}Ra$ [15, 130]. The genes identified using this assay included those likely involved in cell division, *ftsX* and *ftsE* [131], *smpB*, which is possibly involved in macrophage survival [131], a putative cysteine-rich protein *(mcp)* bearing homology with eukaryotic proteins implicated in cell-cell and cell-matrix interactions and *10Sa*, encoding a small stable RNA implicated in trans-translation [132] and a putative two-component regulatory system, *devR-devS (Rv3133c-Rv3132c)* [133]. This genetic system has been subjected to a detailed analysis. The *devR* and *devS* genes are co-transcribed with the upstream gene, *Rv3134c*, of unknown function. A *devR::kan* mutant of *M. tuberculosis* $H_{37}Rv$ was constructed by allelic exchange and characterized to delineate the function of DevR [134]. The *devR* mutant was impaired in growth in microaerophilic cultures but not under aerobic conditions. Both parental and mutant strains replicated at equivalent rates within human monocytes alone or supplemented with peripheral blood lymphocytes, suggesting the lack of a significant role for DevR in entry, survival and growth of tubercle bacilli within monocytes in vitro. However, the mutant strain was attenuated in virulence in guinea pigs as judged by the magnitude of gross lesions, pathological changes in the liver and lung and bacterial recovery from spleen. Experiments with the complemented strain are in progress to confirm these findings [135].

The expression of *devR*, *devS* and *Rv3134c* genes was up-regulated in wild-type cultures adapted to hypoxia. DevR-DevS has now been confirmed to be an authentic signal transduction system by demonstration of the phosphorylation activities characteristic of histidine kinase and response regulator proteins using purified proteins of *M. tuberculosis* [136]. DevR is a putative DNA-binding protein and recent studies have suggested *devR-devS* to be autoregulated [134]. Experiments are in progress to identify other target genes under the control of the DevR-DevS two-component system. It has been proposed that DevR-DevS mediates bacterial adaptation to hypoxia through a phosphorylation cascade and could serve as a regulatory link between oxygen limitation and the initiation and maintenance of the adaptive response to hypoxia. Since the environment within the granuloma is considered to be hypoxic, the identification of target genes under DevR control is likely to provide a better understanding of its role in bacterial adaptation to hypoxia and bacterial persistence in vivo. The DevR-DevS two component system was expressed in granulomas as determined by immunostaining of human and guinea pig tissue sections with antibodies raised against purified DevR and DevS proteins [135]. The identification, therefore, of candidate molecules that intercept DevR-DevS activity may provide an effective means of combating tubercle bacilli persisting within the recesses of granulomas in diseased individuals.

Accessibility to the unfinished DNA sequence of *M. smegmatis* has enabled its comparison in silico with the annotated genome of *M. tuberculosis* [137]. Due to conservation of the transcriptional machinery between *M. tuberculosis* and

M. smegmatis and genes for sigma factors including SigF, SigB and SigE and the DevR-DevS system, it is believed that *M. smegmatis* could serve as a useful system to study the mechanism of the adaptive response to hypoxia. Recent studies have indicated that the DevR-DevS system of *M. smegmatis* was upregulated in hypoxia at transcriptional as well as translational levels. A hypoxia-responsive pattern of expression of *M. tuberculosis devR* promoter has been observed in *M. smegmatis*, pointing towards the modulation of *M. tuberculosis* promoter by DevR protein of *M. smegmatis* [138].

Pathogenicity of a microorganism normally depends upon its ability to enter, survive and multiply within the host cell. *virS* gene was identified from *M. tuberculosis*, which encodes a 38 kDa protein (VirS) [139]. VirS protein exhibits close sequence and structural similarities with VirF protein of *Shigella*, VirFy protein of *Yersinia* and Cfad, Rns and FapR proteins from various enterotoxigenic strains of *E. coli*. All these proteins act as positive regulators of transcription and control the expression of structural genes required for establishment of infection and pathogenesis [139]. PCR and Southern blot analysis showed that *virS* gene is present exclusively in the organisms belonging to the *M. tuberculosis* complex [140, 141]. Another gene *mymA* has been identified from *M. tuberculosis*, which expresses a monooxygenase homologue and is located divergent to *virS*. Expression analysis of MymA in *M. tuberculosis* with antibodies against MymA showed that MymA is poorly expressed under in vitro culture conditions [24]. Deletion analysis of upstream region of *mymA* showed that its expression is regulated through possible involvement of trans-acting factor(s) specific to *M. tuberculosis* that are absent in *M. smegmatis*. This suggests that *mymA* expression could be induced optimally under specific conditions. The plausible role of *virS* and *mymA* in enhancing the survival of *M. tuberculosis* in macrophages has been suggested [24].

Since adhesion of pathogens to the host cell surface is the first and most crucial step in the entry of the pathogen into the host cell, this represents one of the steps which merits consideration as a target for developing new approaches for combating such infections. An extracellular lectin was characterized previously from *M. smegmatis* [142]. This 12–14 kDa lectin was specific for mannan and was demonstrated to be involved in mycobacteria-macrophage interactions [143]. Western blot analysis confirmed the presence of two mycotin-related proteins in *M. avium* (~67 kDa) and *M. tuberculosis* (~35 kDa), respectively. The specificity of lectin-mediated attachment was confirmed by observations that preincubation of the bacteria with mannan or anti-mycotin antibody considerably inhibited binding of the bacteria to the macrophages [143].

With the availability of the genome sequence of *M. tuberculosis* numerous techniques are being developed for the identification of genes involved in the virulence of *M. tuberculosis*. One such technique, fluorescent amplified fragment length polymorphism (FAFLP), is being exploited for differentiating the highly virulent, avirulent and drug-resistant strains of *M. tuberculosis*. The differential band patterns obtained provide insights into putative virulence markers. A large number of markers have been identified and their relevance to virulence is under investigation [86].

5.1 Modulation of Host Cell Signaling

The molecular basis of the pathogenicity of *M. tuberculosis* is poorly understood. A variety of mechanisms contribute to the survival of *M. tuberculosis* within macrophages, including inhibition of phagosome-lysosome fusion, inhibition of acidification of phagosomes and resistance to killing by reactive oxygen and reactive nitrogen intermediates. Macrophage apoptosis is known to contribute to host defense against *M. tuberculosis* infection. Human alveolar macrophages undergo apoptosis in response to *M. tuberculosis* infection. Recently, it has been demonstrated that bacillary control of host cell apoptosis is a virulence-associated phenotype of *M. tuberculosis*, with virulent strains having the ability to prevent apoptosis of infected macrophages. In the light of these observations, understanding the factors that may modulate host cell apoptosis may be crucial. Lipoarabinomannan (LAM), which is produced in large quantities by *M. tuberculosis*, can inhibit macrophage activation and triggering and represents a virulence factor contributing to the persistence of mycobacteria in macrophages. The ability of mannose-LAM (man-LAM) to impair responsiveness to IFN-γ and to attenuate TNF-α and IL-12 mRNA production through effects on the protein phosphatase SHP-1 has been suggested to be a major mechanism by which man-LAM promotes intracellular survival. A study was carried out to investigate whether man-LAM from a virulent strain of *M. tuberculosis* could modulate cell-signaling pathways known to control cell survival [144]. The phosphorylation of Bad is one of the mechanisms of protection of cells from programmed cell death. Through this study, it was demonstrated for the first time that man-LAM from the virulent Erdman strain of *M. tuberculosis* activates the serine/threonine kinase, Akt, in a phosphatidylinositol 3-kinase-sensitive manner leading to the phosphorylation of Bad at serine-136, suggesting that this may be one of the mechanisms by which LAM directly promotes macrophage cell survival [144]. This study has demonstrated for the first time the importance of a mycobacterial virulence factor, with the capability to up-regulate a macrophage survival signaling pathway, in the intracellular survival of the pathogen. The interplay of signaling pathways in macrophages following infection by *M. tuberculosis* and *M. avium*, an opportunistic pathogen, is also being investigated in detail [101].

Studies have also been initiated on bacterial signaling pathways. Phosphorylation of proteins serves as the molecular switch in signal transduction pathways, both in prokaryotes as well as in eukaryotes. These phosphorylations are mediated by protein kinases. The genome sequence of *M. tuberculosis* revealed the presence of eleven putative genes encoding eukaryotic-type Ser/Thr protein kinases. One of these kinases, PknA, located adjacent to the genes encoding bacterial morphogenic proteins, has been characterized [145]. The gene encoding PknA was PCR amplified and was expressed as a fusion protein (~97 kDa) with maltose binding protein in *E. coli*. PknA was found to have autophosphorylating activity which was strictly magnesium/manganese-dependent. Phosphoamino acid analysis revealed that PknA phosphorylates primarily at threonine residues. Protein-protein interaction studies indicated that PknA was capable of phosphorylating at least a ~56 kDa soluble protein from *E. coli* [145]. Furthermore, scanning electron micrography revealed that constitutive expression of this kinase resulted in elongation of *E. coli* cells. Thus, these studies in a heterologous system have established the involvement of PknA in regulating morphological changes associated with cell division [145].

Two other genes encoding Ser/Thr protein kinases, PknF and PknG of *M. tuberculosis* have also been characterized [146]. The genes encoding these proteins were cloned and expressed in *E. coli*. The ability of PknF and PknG to phosphorylate the peptide substrate myelin basic protein (MBP) was also examined. Purified PknF phosphorylated MBP at serine and threonine residues, while purified PknG phosphorylated only at serine residues. Both PknF and PknG were shown to have a lysine residue in the ATP binding site – a typical feature of Ser/Thr kinases. Substitution of this lysine residue with methionine resulted in the loss of activity in both PknF and PknG, suggesting the essential requirement of this lysine residue in catalysis [146]. Southern blot analysis revealed that homologues of the genes encoding the two kinases were present in *M. tuberculosis* $H_{37}Ra$ and *M. bovis* BCG, but not in *M. smegmatis*. Immunoblot analysis of various cellular fractions of *M. tuberculosis* $H_{37}Rv$ using polyclonal antibodies against these kinases revealed that, while PknF was a transmembrane protein, PknG was predominantly a cytosolic enzyme. The absence of *pknF* and *pknG* in non-pathogenic *M. smegmatis* suggests involvement of these proteins in the processes specific to pathogenic mycobacteria [146].

In some pathogenic bacteria secretion of nucleoside diphosphate kinase (Ndk) into the extracellular environment aids in survival of pathogen via P2Z receptor-mediated, ATP-induced death of infected macrophages. Ndk is a secretory protein in *M. tuberculosis*. The *ndk* gene from *M. tuberculosis* $H_{37}Rv$ has been cloned and expressed in *E. coli* [147]. Purified Ndk enhanced ATP-induced macrophage cell death, as assayed by the release of ^{14}C-adenine. Oxidized ATP, an antagonist of P2Z receptor, and a catalytic mutant of Ndk failed to enhance ATP-induced macrophage cell death. Purified Ndk undergoes autophosphorylation and conversion of a histidine residue (which is part of the nucleotide-binding site) to glutamine, which abolished this autophosphorylation. Purified Ndk also exhibited GTPase activity, which has been hypothesized to interfere with the intracellular signaling of host cells. The results from this study thus suggest that secretory Ndk of *M. tuberculosis* may be acting as a cytotoxic factor for macrophages, which may be helping in dissemination of the bacilli and evasion of the immune system [147].

Several other mycobacterial Ser/Thr kinases have been expressed. A detailed study of these kinases from *M. tuberculosis* could lead to the identification of new targets for the development of anti-tubercular drugs. Besides, an understanding of the phosphorylation activities of these kinases could provide means for the development of rapid assays for screening of compounds (kinase inhibitors) that could effectively block the signal transduction machinery necessary for intracellular survival of mycobacteria.

Partial purification and characterization of adenosine 3′,5′-monophosphate (cyclic AMP)-dependent protein kinase and Ca^{2+}/calmodulin-dependent protein kinase from *M. smegmatis* have also been reported [148, 149].

6
Approaches for Prevention and Treatment of Tuberculosis

6.1
Molecular Approaches for the Diagnosis of Tuberculosis

The global resurgence of tuberculosis in recent years has underscored the importance of rapid diagnosis of tuberculosis for its effective control. The main obstacles for the diagnosis of mycobacteria include the slow growth of the pathogen and the low numbers of bacilli present in clinical specimens. Effective chemotherapy is available for treatment, but rapid and early diagnostic methods are the most essential tools required to control the spread of the disease. Recent advances in molecular biology have greatly accelerated the development of more accurate and rapid tests. Nucleic acid amplification tests combine probe and genetic amplification technology in an assay system, which can be performed rapidly on clinical specimens. A new generation of antibody tests, using highly purified antigen derivatives and ELISA, has also been developed. Various approaches employed for the diagnosis of TB are summarized in Table 1.

Table 1. Summary of various approaches for diagnosis of tuberculosis

Test	Target	Disease	Sensitivity	Specificity	Ref.
PCR	IS6110	Intracranial tuberculoma	60%	100%	[150]
PCR	IS6110/ TRC4	Tubercular pleuritis (pleural fluid)	100%	85%	[151]
PCR	IS6110 TRC4	Tuberculous lymphadenitis (lymph node specimens)	69% 78%		[152]
PCR	*M. tuberculosis* 23S rRNA	Sputum and cerebrospinal fluid specimens	70.8%	Genus-specific assay	[153]
PCR+ histology	*M. tuber culosis* GC rich repetitive sequence	Tuberculous and non-tuberculous pleural fluid samples	73.3%	93.1%	[154]
PCR	*M. tuberculosis devR*	Lymphadenitis (Lymph node aspirates and tissues)	82%	86%	[155]
PCR	38 kDa antigen gene IS6110	Pulmonary TB (sputum)	45% 41%	96%	[156, 157]
Immuno-histochemical		Intracranial tuberculoma	80%	100%	[150]
RIA Antibody test	*M. tuber culosis* sonicate antigen	Pulmonary TB (serum)	81%	96%	[158]
Antibody test	*M. tuber culosis* sonicate antigen	Tuberculous meningitis (CSF)	76%	90%	[159]

Table 1 (continued)

Test	Target	Disease	Sensitivity	Specificity	Ref.
ELISA	*M. tuberculosis* 38 kDa antigen	Tuberculous meningitis (CSF)	60–80%	95%	[160]
ELISA	A 60 (IgM) A 60 (IgG) 38 kDa (IgG)	Pulmonary TB (serum)	71% 29% 45%	50% 86% 73%	[161]
ELISA	38 kDa (IgG) 38 kDa (IgA) 38 kDa (IgM) 38 kDa (IgG+IgA) 38 kDa (IgG+IgA+IgM))	Pulmonary TB (smear and culture positive cases)	61% 30% 10% 68% 71%	100% 96% 94% 96% 90%	[162]
ELISA	Mycobacterial antigens (PPD, Culture filtrate) r38 kDa Antigen 6		73% 61% 75%	79% 63% 72%	[163]
ELISA	*M. tuberculosis* 30 kDa antigen *M. tuberculosis* 16 kDa antigen	Childhood tuberculosis sera	84.6% 73%	96.7%	[164]
ELISA	A60 antigen	Pulmonary TB	60%	87.5%	[165]
ELISA	A60 antigen	Childhood Pulmonary TB Extrapulmonary TB	72% 76.6%	92%	[166]
Competitive ELISA		Tuberculous meningitis CSF	68.26%	100%	[167]
Sandwich ELISA	30 kDa antigen	Serum	68%	85%	[109]
Sandwich ELISA	*M. tuberculosis* $H_{37}Ra$ culture filtrate proteins	Sputum positive TB patients (urine)	75%		[168]
Slide agglutination test	*M. w* soluble antigens	Pulmonary TB Extra-pulmonary TB	90.2% 85.7%	92.7%	[169]
Stick Indirect ELISA	*M. tuberculosis* excretory-secretory antigen	Pulmonary TB Smear-positive Pulmonary TB Smear-negative	90% 71%	90%	[170]
	M. tuberculosis EST-6 antigen fraction	Pulmonary TB Smear-positive Pulmonary TB Smear-negative	94% 78%	88%	

6.2 New Candidate Vaccines Against Tuberculosis

Vaccines represent one of the most efficacious and potent defenses against infectious diseases. At present we have no vaccine to ensure protection against TB in adults and older people and efforts in this direction have not resulted in a TB vaccine that is better than BCG. Rapid developments in TB research, access to the *M. tuberculosis* genome sequence, availability of new technologies such as gene arrays, proteomics and availability of cytokine assays at the epidemiological level, however, have provided fresh input and hopes for the development of a more efficacious TB vaccine. Several approaches such as recombinant BCG, DNA vaccine, subunit vaccine, atypical mycobacterial strains and auxotrophic vaccine have been employed to develop an efficient vaccine against this deadly disease.

In a recombinant BCG approach for the development of candidate vaccines against TB [171], six candidate antigens, namely members of the Antigen 85 Complex – 85A, 85B, 85C, 19 kDa lipoprotein, 38 kDa protein involved in phosphate transport and early secretory antigenic target protein (Esat-6), have been expressed under the transcriptional control of a battery of promoters in BCG (Danish) [24, 71]. Initially, the immunogenicity of rBCG strains was evaluated in BALB/c mice by immunizing them intravenously with 1×10^6 cfu of rBCG/BCG strains. In the cases of each candidate antigen, three rBCG strains expressing the antigen to different levels were analyzed. Humoral responses were analyzed by measuring total antibody response and the level of IgG1 and IgG2a antibodies using ELISA. Cellular responses were analyzed by culturing isolated splenocytes in the presence of antigen and measuring the extent of proliferation as well as by measuring the level of cytokines, IFN-γ and IL-10, in the culture supernatants [24]. For ELISA as well as T-cell stimulation, both purified antigen as well as BCG extract were used as the coating/stimulating antigen. In the case of mice immunized with rBCG strains overexpressing the antigens belonging to the Antigen 85 Complex, an earlier and enhanced immune response was induced when compared to mice immunized with wild-type BCG [172]. Moreover, overexpression of these antigens did not always result in induction of Th1/increased Th1 responses. The immune responses were observed to range from Th1 through mixed Th1/Th2 to Th2, depending upon the level of expression of the recombinant antigen. The general patterns of immune responses elicited against rBCG overexpressing Ag 85A and Ag 85B were more or less similar. Despite the strong homology shared by all three antigens (85A, 85B and 85C), rBCG strains overexpressing Ag 85C elicited a different pattern of immune response. Some of these rBCG strains induced increased Th1 responses, manifested by increased secretion of IFN-γ and a stronger IgG2a antibody response [24, 172].

Wild-type BCG immunized mice exhibited a mixed type Th1-Th2 or Th2 type of T-cell response against purified 19 kDa antigen, 38 kDa antigen and BCG extract. Due to the deletion of the Esat-6 gene from the BCG strain, no native expression of Esat-6 in the wild-type BCG was observed and hence, no responses against this antigen were elicited in mice immunized with wild-type BCG. Overexpression of Esat-6 in BCG also induced only a mixed Th1-Th2 or a Th2 type of T-cell response against the purified antigen as well as against the BCG extract (as seen in the case of immunization with wild-type BCG) [24]. Overexpression of 19 kDa antigen in BCG induced a prominent Th1 type immune response against purified 19 kDa antigen and a predominant Th2 type immune response against BCG extract. However, overexpression of the 38 kDa antigen in BCG shifted the host immune response towards a Th1 phenotype against both the purified antigen as well as against BCG extract. In the cases of each of the six antigens, one rBCG strain was selected for evaluation of its protective efficacy in guinea pigs [24].

For evaluation of the protective efficacy of these candidate vaccines (carried out at Tuberculosis Research Centre, Chennai), guinea pigs immunized intradermally with 1×10^6 cfu of the various rBCG/BCG strains were challenged subcutaneously (after eight weeks) with 4 different doses of *M. tuberculosis* $H_{37}Rv$ in separate groups [24]. Protection was evaluated by analyzing the gross pathological damage to the various organs, bacterial load in the spleen and histological analysis of the liver and lung organs. Preliminary results from these studies suggest that two of the rBCG strains exhibited increased protection against challenge compared to wild-type BCG. In the cases of rBCG strains expressing two other candidate antigens, the protection imparted was found to be comparable to that imparted by wild-type BCG. The recombinant BCG strain overexpressing 19 kDa antigen, however, was found to be detrimental for protection and abrogated the protective efficacy of BCG itself. These results are being currently confirmed by using a larger number of guinea pigs [24].

The DNA vaccine approach represents one of the most recent and interesting strategies for the development of vaccines. By using this approach, significant advances have been made towards the development of vaccines against several diseases. By virtue of its stability, ease of production and cost and feasibility to elicit a Th1 immune response, especially crucial in the case of TB, this approach has gained significant importance for the development of vaccines against TB. Towards this, the genes for three immunodominant antigens, viz., Esat-6, α-crystallin and superoxide dismutase, have been cloned and expressed using a eukaryotic vector [24]. These candidate DNA vaccines are currently being evaluated for elicitation of immune responses in mice and protective efficacy in guinea pigs [24].

In another approach towards the development of a candidate TB vaccine, two proteins of *M. tuberculosis* (designated as 65 and 202) were observed to stimulate strong Th1 responses in the healthy, subclinically infected individuals [126]. A detailed characterization of these immune responses further demonstrated that the specific T-cell responses to these two antigens, which were robust in the healthy population were, however, depressed in the TB patients [126]. The protective efficacy of these two *M. tuberculosis* proteins was subsequently evaluated in guinea pigs (in collaboration with the National Tuberculosis Institute, Bangalore). The two antigens were tested in various forms – naked DNA constructs, recombinant BCG as well as recombinant pox virus carrying the genes encoding the two antigens. When tested in guinea pigs, protein 202 but not 65 was found to afford some protection, comparable to that imparted by BCG. Moreover, a BCG recombinant expressing protein 202 was observed to be the most effective, and resulted in sterilizing immunity in the vaccinated animals, with no challenge organisms detected in the spleen. A pox virus expressing protein 65 was also observed to afford dramatic protection, in contrast to the DNA vaccine expressing protein 65. In light of recent studies demonstrating that prior exposure to environmental mycobacteria reduces the efficacy of BCG, it has been proposed that live vaccine vectors such as pox virus are likely to be superior to BCG as carriers of T-cell antigens identified from *M. tuberculosis* [126].

Although BCG has not performed satisfactorily as a TB vaccine, the concept of a live vaccine against TB still draws tremendous support as experience with killed vaccines has demonstrated that live vaccines perform better for eliciting protective immune responses. Hence, several atypical mycobacteria have been employed to evaluate their ability to impart protection against this disease. *Mycobacterium w*, is an anti-leprosy vaccine and phase III immunotherapeutic trials of this vaccine have been completed. *M. w* as an immunotherapeutic vaccine against leprosy has been launched for public use.

M. w is a non-pathogenic strain, a fast-grower and its growth and metabolic properties resemble those listed in Runyons group IV. On the basis of data on both B- and T-cell epitopes, *M. w* was observed to share epitopes with both *M. leprae* and *M. tuberculosis*. It has also been shown to confer protection against challenge with *M. tuberculosis* in all the four strains of mice tested. However, in parallel experiments, BCG conferred protection only in two strains of mice [173]. In order to characterize the immune mechanisms responsible for this protection, mice were vaccinated with heat-killed *M. w* and subsequently T-cells were isolated and restimulated in vitro with mycobacterial antigens. On stimulation, IL-2 and IFN-γ were produced, but IL-4 and IL-5 were not detectable, indicating induction of a Th1-based immunity. Killed *M. w* was observed to induce stronger T-cell responses compared to killed *M. tuberculosis* and both vaccination regimens were improved by administration in a mild adjuvant, which consisted of trehalose dimycolate, monophosphoryllipid A and mycobacterial cell wall skeleton [174]. In vivo depletion of $CD4^+$ and $CD8^+$ T lymphocytes led to impairment of *M. w* vaccine-induced protection against *M. tuberculosis* in mice. Lower levels of IFN-γ production by both the depleted pools of T-cells suggested that both $CD4^+$ and $CD8^+$ T-cells might be participating in resistance against TB induced by vaccination with *M. w* [175]. On extending these studies to guinea pigs, it was demonstrated that immunization of guinea pigs with *M. w* prevented the formation of lesions and the growth of *M. tuberculosis*, on challenge with *M. tuberculosis* $H_{37}Rv$ [176]. Since *M. w* vaccine confers protection against TB in guinea pigs, the incidence of TB (in addition to leprosy) in the large number of subjects undergoing immunoprophylaxis trials, is being determined to evaluate whether the use of this vaccine can provide additional benefit of protection against TB.

In another attempt to develop an effective TB vaccine using atypical mycobacteria, the protective efficacy of *M. habana* was studied. It was shown that this mycobacteria can provide protection against TB in a mouse model and serves as a more effective vaccine than BCG. Currently, *M. habana* is being evaluated for its protective efficacy in a guinea pig model of TB. Besides, the evaluation of protective antigens of cell wall, cell membrane, cytosol and peripheral and integral compartments of the membrane fraction of *M. habana*, using an experimental TB model in the mouse has demonstrated that protective proteins were distributed both in peripheral and integral membrane compartments, with proteins from the peripheral compartment being more protective in nature [177].

In another subunit approach for the development of a candidate vaccine against TB, several antigens from the secretory fraction as well as from the cell wall peptidoglycan fraction of *M. tuberculosis* were evaluated for their ability to elicit cell-mediated and humoral immune responses. Among the various antigens, a 30 kDa antigen from the secretory fraction and a 71 kDa cell wall-associated antigen, which demonstrated maximum immuoreactivity, were evaluated for their protective efficacy in mice using various adjuvant systems, viz., Freund's incomplete adjuvant (FIA), liposomes, dimethyldioctadecylammonium bromide (DDA), polylactide-co-glycolide microparticles (PLG-MPs) and poly(DL-lactide-co-glycolide) (DL-PLG). In comparison to animals immunized with BCG, higher immune responses, manifested by increased IL-2 and IFN-γ production and increased protective efficacy, manifested by increased survival rates and decreased bacterial load in the target organs, have been demonstrated in the mouse model, by using these various preparations [178–184]. These mycobacterial proteins separately and in combination are being evaluated for their protective efficacy in a guinea pig model [185].

6.3 Development of Anti-Tubercular Drugs

With the availability of several effective anti-tubercular drugs such as streptomycin, rifampicin, isoniazid, pyrazinamide and ethambutol, and effective regimens of short-term chemotherapy, an urgent need for new drugs was not felt and no frontline new drug has been developed since the discovery of rifampicin in 1967. However, with the advent of drug resistance, and multi-drug resistant TB, the scenario has changed and new drugs are urgently required. Besides, to overcome the problem of non-compliance, it would be ideal to have a drug that would be effective in a single dose. Thus, there is an urgent need for the development of new drugs by identification and structural determination of new targets, development of high-throughput drug screens, identification of potential molecules and improving the efficacy and delivery of existing drugs.

6.3.1 *X-Ray Crystallography of Novel Drug Targets*

Several isoniazid-resistant strains of *M. tuberculosis* have been found to overexpress alkyl hydroperoxidase (AhpC). AhpC is a decameric protein, composed of five identical dimers held together by ionic interactions with dimerization occurring through an intersubunit disulfide linkage. Based on the UV absorption spectrum and three-dimensional modeling of AhpC, interesting conformational changes during oxidation and reduction of the intersubunit disulfide linkage have been predicted [186].

Chaperonin-10 (Cpn10) and chaperonin-60 (Cpn60) are essential proteins involved in ATP-dependent folding of several intracellular proteins in the bacterial cell. Folding of the nascent substrate polypeptide is known to take place in the large central cavity formed by each ring of the tetradecameric Cpn60, which is closed upon capping by the heptameric Cpn10. Cpn10s are known to interact with Cpn60s primarily through a 17-residue mobile loop and regulate the release and binding of the substrate polypeptide from Cpn60 surface. The Cpn10 protein of *M. tuberculosis*, which is involved in protein folding, has also been characterized to reveal that the plasticity of its oligomeric structure is due to a highly flexible segment of approximately 10 residues that covers the dome-like structure of Cpn10 [187]. By exploiting a Trp residue in the dome loop of this protein for intrinsic fluorescence measurements, it has been shown that, in the absence of metal ions and at neutral pH, the Trp is almost fully solvent and exposed. However, in the presence of metal ions or at low pH, the dome loop assumes a closed conformation [187].

The *M. tuberculosis* genome sequence has greatly aided efforts towards understanding the biology of this slow-growing bacterium. Recently, in addition to attempts by individual laboratories, coordinated structural genomics projects have been initiated to determine the structures of a large number of proteins from *M. tuberculosis* in order to provide deeper insights into the biology of this bacterium as well as to identify new targets for the development of anti-tubercular drugs.

The structure of *M. tuberculosis* Cpn10 has been recently determined at 3.5 Å resolution [188]. The Cpn10 monomer was comprised of a four-stranded beta-barrel and two long stretches of highly flexible segments: the dome loop and the mobile loop. The seven subunits in the heptamer exhibited very little conformational difference and had nearly perfect seven-fold geometry. Binding sites for metal ions in the dome loop of Cpn10 have been identified, which suggests the role of metal ions in the stabilization of the protein [188].

RecA, as described in the previous sections, plays an important role in homologous DNA recombination and DNA repair. Structural knowledge of *M. tuberculosis* RecA is imperative for a full understanding of these activities, especially since both ATP binding as well as hydrolysis catalyzed by it are considerably less efficient than those by the prototype *E. coli* RecA, although it is capable of binding ssDNA and is proficient in homologous pairing and strand exchange. The crystal structure of uncomplexed *M. tuberculosis* RecA and its complex with ADP-AlF_4, an ATP analogue, have been determined [54]. The crystal structure was observed to have six molecules in the unit cell forming a 6_1 helical filament with a deep groove capable of binding DNA. The filamentous structure of RecA further aggregates into bundles, mainly as a form for storage and as a mechanism to keep RecA inactive when not needed, which may have implications in the efficiency of recombination and random integration. The observed weakening in the higher order aggregation of filaments into bundles may also have implications in recombination in mycobacteria. The crystal structure of *M. tuberculosis* RecA complexed with ADP and AlF_4 revealed that each molecule of RecA binds one molecule of the ATP analogue in the P-loop region, as in the structure of the *E. coli* RecA-ADP complex [54]. These structures also help explain the reduced levels of interactions of *M. tuberculosis* RecA with ATP, despite sharing the same fold, topology and high sequence similarity with *E. coli* RecA. A histidine residue in loop L1 also appeared to be positioned appropriately for

DNA interaction. The crystal structure of *M. tuberculosis* RecA thus allows its exploitation as a drug target as well as providing insights into the biology of recombination in mycobacteria [54].

S-Adenosyl-L-methionine (AdoMet)-dependent methyltransferases are involved in a wide variety of cellular processes such as modification of bioactive amines, bulk metabolic transformations and participate in a variety of regulatory biological processes through methylation of nucleic acids, proteins, phospholipids and small molecules utilizing the ubiquitous methyl donor AdoMet. The crystal structure of Rv2118c, an AdoMet-dependent methyltransferase from *M. tuberculosis* has been determined [189]. The crystallographic asymmetric unit is comprised of a monomer, but symmetry-related subunits interact extensively, leading to a tetrameric structure. The monomer unit had two domains: the larger catalytic C-terminal domain that binds the cofactor AdoMet and is involved in the transfer of methyl group from AdoMet to the substrate and a smaller N-terminal domain, which was primarily a β-structure with a fold not found in other methyltransferases of known structure. The structure of the catalytic domain was observed to be very similar to that of other AdoMet-dependent methyltransferases [189].

As described earlier in this chapter, single-stranded DNA-binding proteins play an important role in DNA replication, repair and recombination. The SSB protein from *M. tuberculosis* has been crystallized under various conditions and, by using X-ray crystallographic parameters, attempts are being made to determine its structure [190].

1-L-myo-Inositol is essential for the biogenesis of mycothiol, phosphatidylinositol and glycosyl phosphatidylinositol anchors, which are linked to complex carbohydrates in *M. tuberculosis*. However, the inositol biosynthetic pathway in *M. tuberculosis* was not known. A unique combination of tertiary structure prediction and yeast-based functional assays was employed to identify the gene encoding inositol-1-phosphate synthase in *M. tuberculosis*, that was distinct from the eukaryotic analogs [191].

6.3.2
Development of Genetic Screens

With rise in the frequency of drug-resistant strains of *M. tuberculosis*, new drugs and drug targets need to be discovered. With advances in combinatorial chemistry and the potential of rich and natural biodiversity of the country, assays are required that can screen the potency of these new drugs in a relatively short time in a high-throughput fashion. Recombinant *M. smegmatis* and *M. aurum* expressing *E. coli* β-galactosidase have been employed for assaying anti-tubercular activity of compounds [192, 193]. *M. aurum* was employed as the test organism as it is non-infectious, fast grower and obviates the need for a special containment facility. Moreover, since *M. aurum* resembles *M. tuberculosis* in mycolic acid and antibiotic sensitivity, it was hypothesized that it could serve as a predictor of inhibitory activity of the test compounds. The screen was validated by using standard anti-tubercular drugs. This method can be easily and conveniently adopted in 96-well microtiter plate format to suit high throughput screening of natural and synthetic products [192, 193]. Green fluorescent protein and firefly luciferase have also been used as reporters for rapid screening of anti-tubercular compounds in vitro and in macrophages. A recombinant *M. aurum* strain expressing green fluorescent protein has been constructed. Fluorescence exhibited by this recombinant strain was observed to be inhibited in vitro within 8 to 24 h by frontline anti-mycobacterial drugs at their reported MICs whereas in the case of infected macrophages the inhibition occurred in 72 h [194]. By employing *M. aurum* expressing firefly luciferase, bioluminescence was determined using luciferin as the substrate. Inhibition of bioluminescence was observed as early as 2 h in vitro and within 24 h in infected macrophages, with frontline anti-mycobacterial drugs used at their reported MICs [195]. The bioluminescence assay of ATP has also been tested as an alternative, reliable and rapid method and has been compared with the broth dilution method for drug susceptibility testing of *M. tuberculosis* [196].

In contrast to the time-consuming conventional method or the very expensive radiometric method, a simple method to assess resistance to frontline drugs has been developed by using Alamar Blue, a resazurin-based oxidation-reduction indicator [197]. The sputum samples from TB patients, processed by Petroff's method are inoculated on LJ-medium for 4–8 weeks and cell viability in the presence of various drugs is determined by using Alamar Blue. The assay is based on the ability of viable cells to oxidize and thereby change the color of the dye from blue to pink [197].

6.3.3
Strategies for Enhancing Drug Efficacy

The antibiotic rifampicin inhibits transcription of DNA template exclusively by binding to the β-subunit of RNA polymerase. Moreover, all the rifampicin-resistant mutants have been mapped in three different clusters in the β-subunit of *E. coli* RNA polymerase. The impressive success of rifampicin to control mycobacterial infection has made it a subject of critical research. Attempts have been made to discover natural products that could mimic the action of rifampicin. Towards this goal, piperine, a black pepper extract, which exhibits diverse biological activities has been studied. Piperine alone did not exhibit any anti-mycobacterial activity. However, in combination with rifampicin (24:1 mixture of rifampicin and piperine), it augmented the anti-tubercular activity of rifampicin by several fold [198]. It was demonstrated that piperine acts with rifampicin in a synergistic manner to inhibit transcription by purified *M. smegmatis* RNA polymerase. Interestingly, the rifampicin augmentation activity of piperine was maintained even in a rifampicin-resistant strain of *M. smegmatis*. Studies based on molecular modeling of mycobacterial RNA polymerase have revealed that both rifampicin and piperine could be accommodated in the rifampicin pocket through stacking interactions between them, which may help to augment the transcription inhibitory activity of rifampicin [198].

New chemotherapeutic drugs and new treatment regimens are required to shorten the duration of treatment and to combat drug-resistant TB including MDR-TB. Several new compounds and drugs have shown significant anti-tubercular activity under in vitro conditions. However, for appropriate usage of these drugs in treatment regimens, in vitro determination of their minimal inhibitory concentrations (MIC) is an important first step. A study was undertaken to compare the susceptibility pattern of ofloxacin using different susceptibility testing methods [199]. The direct sensitivity method has been standardized by employing L-J medium for streptomycin, isoniazid and rifampicin leading to a reduction in the time taken for reporting of results to 3–5 weeks in smear-positive specimens. Besides, the time required for the conventional indirect sensitivity method used for smear-negative specimens has also been reduced from 4 to 2 weeks by advancing the measurements by 2 weeks [200]. A comparison of the BACTEC radiometric method with the conventional culture and drug susceptibility testing methods on isolates from clinical specimens of pulmonary and extrapulmonary TB, childhood TB and TB in HIV-infected individuals has also been carried out [201]. The rate of isolation of positive cultures was significantly faster with the BACTEC method, especially in smear-negative pulmonary TB, extrapulmonary TB, HIV-TB and childhood TB in comparison to conventional methods. Further, most of the drug susceptibility test results became available within 8 days by the BACTEC method. Thus, by facilitating early diagnosis, the BACTEC method may prove to be cost effective in a population with a high prevalence of TB, particularly in the extrapulmonary and paucibacillary forms of the disease [201].

The studies to assess bioavailability indices for rifampicin, isoniazid and pyrazinamide separately or in a fixed triple-drug formulation suggest that the pharmacokinetic properties of these drugs on individual or combined administration do not change when combined in a single pharmaceutical preparation [202]. The action of drugs such as metronidazole in combination with isoniazid and rifampicin on persisting organisms has been determined using an experimental murine TB model [203]. Metronidazole exhibited an additive effect with rifampicin in reducing the $\log_{10}$ cfu of *M. tuberculosis* in both the lung and spleen of mice by the end of 3 months of treatment. However, when metronidazole was combined with isoniazid this effect was observed in the spleen only [203]. Studies have also been conducted to determine the in vitro susceptibility of clinical isolates of *M. tuberculosis* to cefadroxil, a cephalosporin antibiotic, and trifluoperazine, an antipsychotic drug [204, 205]. Studies were also carried out to determine the MICs of sublactam/ampicillin, lomefloxacin and minocycline for drug-sensitive and drug-resistant *M. tuberculosis* isolates [206, 207]. In vitro, ex vivo and in vivo activities of ethambutol and sparfloxacin alone and in combination against mycobacteria and the biochemical mechanisms of their action have also been studied [208–210]. The therapeutic potential of human neutrophil peptide 1 against experimental TB in mice has also been investigated [211–213].

For the successful chemotherapy of tuberculosis, a lung-specific liposomal drug delivery system has been found to be a highly effective carrier for drugs such as isoniazid and rifampicin [214, 215]. A PLG polymer-based drug delivery system for effective treatment of TB, using higher doses of drugs has been developed, which provides sustained release of drugs. Hardened PLG microparticles have been recommended as the ideal carrier for a sustained level of drugs in plasma/different organs based on the observations in the mouse model

[216–219]. Development of this drug delivery system promises an improvement in patient's compliance to therapy.

7
Host Genetics in Susceptibility/Resistance to Tuberculosis

Susceptibility to TB is multifactorial. As in the case of many other diseases, apart from environmental and socio-economic factors, epidemiological and genetic factors have been implicated in the infection and spread of the disease. Therefore, the genetic make-up of an individual may influence susceptibility or resistance to TB through the involvement of genetic factors such as human leukocyte antigen (HLA) and non-HLA genes. Thus, identification of HLA and non-HLA genes that are associated with susceptibility or resistance to TB will serve towards the development of genetic markers for prognosis of the disease. Moreover, studying the role of HLA and non-HLA genes or their products on the immune mechanisms of susceptibility or resistance to TB will be useful to understand the immunopathogenesis of the disease.

Studies have been carried out on HLA and non-HLA gene polymorphism in (i) pulmonary TB patients and control subjects, (ii) relapse quiescent cases of pulmonary TB and (iii) spinal TB patients. HLA studies in pulmonary TB carried out by different groups [220–222] revealed the association of HLA-DR2 with susceptibility to pulmonary TB. Moreover, HLA-DQ1 was also found associated with susceptibility to TB [221]. Studies have also been carried out on non-HLA gene polymorphisms [mannose binding protein (MBP) or mannose binding lectin (MBL), and vitamin D receptor genes] and it has been shown that variant mutant genes are associated with susceptibility to pulmonary TB. MBP or MBL is an acute phase protein secreted by the liver, which binds mannose, and *N*-acetylglucosamine terminated glycoproteins and plays an important role in host defense against pathogens. Upon binding to certain carbohydrate moieties such as terminal *N*-acetylglucosamine or mannose on various pathogens, MBP activates complement via a specific protease and acts directly as an opsonin using the C1q receptor on macrophages. Mutations in the coding regions of MBP genes result in low or near absent serum MBP levels in heterozygotes and homozygotes, respectively. A low serum level of MBP is associated with a common opsonic defect that predisposes to severe and frequent infections during early life. In a recent study in Indian patients, increased genotype frequencies of MBP functional mutant homozygotes were found among pulmonary TB patients (10.9%) when compared to control subjects (1.8%) [223].

HLA-DR2 and its subtypes (DRB1*1501 and DRB1*1502) were found to be associated with far advanced and drug-resistant sputum-positive pulmonary TB patients [220, 222, 224]. A similar observation of DR2 association has also been made in other endemic areas of the world, like Russia and Indonesia etc. Moreover, DR2 association transcended ethnic barriers, suggesting a direct MHC mechanism for TB susceptibility rather than a simple linkage [222, 224].

The Lenzini's disease and immunological spectrum identified in pulmonary TB has also been identified in healthy, hospital contacts. The extent of Mantoux reactivity was observed to be inversely correlated to anti-PPD, anti-BCG and anti-

HSP70 serum antibodies of the IgG isotype [225, 226]. The existence of a similar immunological spectrum has been confirmed in healthy populations from Indonesia and Brazil. Recent studies using *M. tuberculosis* recombinant antigens and synthetic peptides demonstrated the existence of this spectrum at the cytokine level as well [227]. The investigation on chemotherapy in TB patients revealed the importance of immune status of the patient in the rate of recovery from the disease – better CMI in patients was correlated with early recovery [228].

TCRVβ plays an important role in the recognition of the peptide-MHC complex. In the endemic areas of South India, it was observed that more individuals in control group use TCRVβ 4, 6, 8–12 and 14, compared to patients and also individuals in the control groups show higher expression of TCRVβ 2, 3, 7, 13 and 17. Furthermore, majority of the patients and half of the individuals in the control group showed restricted TCRVβ usage (<5 TCRVβ families used). It was observed that half of the BCG scar-negative patients carrying high risk alleles for TB (HLA-DRB1*1501, DRB1*08 and DQB1*0601), all the BCG scar-negative, non-risk allele carriers and BCG scar-positive patients showed a restricted TCRVβ usage [227]. A restricted TCRVβ usage may thus lead to a skewed immune response and the disease in the absence of BCG vaccination. Infection by itself may be responsible for a good T-cell response and the high risk HLA alleles may be responsible for the disease.

Studies employing DRB1*1501 binding peptides from vaccine candidates of *M. tuberculosis* such as 19 kDa, antigen 85A, 85B, 85C, Esat-6, Gro-ES, HSP-14 kDa, MPT64, MPT-83 and 65 kDa in IFN-γ ELISPOT assays revealed that high Mantoux reactive, BCG scar-positive, controls responded to many peptide pools, irrespective of their HLA DRB1*15 status, while none of the BCG scar-negative controls from this endemic region responded to all the pools studied. The majority of the BCG scar-negative patients irrespective of their DRB1 status responded to many peptide pools [229].

A positive association of HLA-DRB1*15/16 in mycobacterial diseases (both TB and leprosy), was reported earlier, especially in patients with more severe disseminated forms of the disease [220, 230]. This study was further extended by including patients across the clinical spectrum. Sequence analysis of polymorphic residues in the peptide binding groove indicated that an arginine-rich motif (DRB1*15/16 and DRB1*1404), in the DRB1 peptide binding groove occurred with >90% frequency in patients with tuberculoid leprosy [231]. Similar observations were made in pulmonary TB where a predominant occurrence of DRB1*1501 (>80%) was observed among DR2+ve patients who developed more severe pulmonary infection with multi-drug resistant TB. No such association was observed in patients with the restricted form of the disease. These observations suggested that V/G dimorphism at position β86 in the "pocket 1" of the MHC class II peptide binding groove might play an important role in disease manifestation following *M. tuberculosis* infection. Additionally, a positive association was observed at the HLA-DQ locus whereby DQB1*05031 and DQB1*0603 were significantly increased in patients with multidrug-resistant TB. The observed significant increase of single or multiple HLA alleles in the various clinical forms suggested that there may be a common peptide binding motif or amino acid sequence involved in tripartite interaction during antigen presentation that might play a significant role in disease manifestation following *M. tuberculosis* infection.

The role of immunogenetics and various risk factors for the development of drug-induced hepatotoxicity (DIH) in patients with pulmonary TB has also been investigated. These studies revealed that, in addition to the clinical factors such as old age, low serum albumin and advanced disease, the presence/absence of particular HLA-DQ alleles play an important role in the development of hepatotoxicity. For example, it was observed that the presence of DRB1*0201 in patients with pulmonary TB puts them at a higher risk of developing DIH. On the other hand, the absence of DQA1*0102 conferred protection from such a side effect [232].

Earlier studies carried out using *M. leprae* heat-shock protein (HSP) antigens had revealed an HLA-linked control of the immune response in polar leprosy patients [233, 234]. To further elucidate the role of helper T-cell subsets and their cytokine profile in mycobacterial infections, a flow cytometry assay was utilized to investigate the expression of differentiation markers on polarized helper T-cells at the cellular level. Using this novel technique of intracellular cytokine staining, studies were carried out on the functionally polarized subsets of CD4 memory T-cells, identifiable on the basis of expression of CD11a, CD45RAand CD62L, viz., those that primarily produce IL-4 (MT_2) or IFN-γ (MT_1). Memory T-cells with the type 1 profile (IFN-γ producer) were categorized as $CD45RA^-$, $CD62L^-$ and $CD11a^{bright}$, while the type 2 cells (IL-4 producers) were categorized as $CD45RA^-$, $CD62L^+$ and $CD11a^{dim}$ [235].

In tuberculosis pleural effusion, IFN-γ producing $CD11a^{bright}$ helper T-cells were observed in the pleural fluid rather than in PBMC. On the contrary, low levels of IFN-γ producing $CD11a^{bright}$ helper T-cells were observed in lymph node-derived T-cells of patients with lymph node TB. Furthermore, in vitro T-cell responses stimulated with *M. tuberculosis* culture filtrate antigens in the tuberculosis disease spectrum were evaluated. These studies based on the antigen-specific proliferative response and cytokines measured in the supernatants, revealed a comparatively increased level of IFN-γ in patients with more extensive and multidrug-resistant disease [236]. These observations point towards an important paradox whereby a preferential production of IFN-γ and Th1 responses in multidrug-resistant patients rather than an expected Th2 immune responsiveness was observed. Exact mechanisms underlining this apparent lack of protective immunity in MDR-TB, despite the production of high levels of IFN-γ, are not clear. It has been hypothesized that this could be due to a consistently increased bacillary load and constant exposure of *M. tuberculosis* bacilli to the host immune system making it redundant [236]. Further investigations into the role of IFN-γ and its receptor gene polymorphism may be useful in understanding this paradox.

8 Future Trends

In spite of its tremendous relevance as a pathogen, *M. tuberculosis* was not a favorite organism among research investigators for a long time. However, due to spectacular advancements in technology over the last two decades of the 20th century, significant progress has been made towards understanding the molecular biology and immunology of this organism and mycobacteria no longer remain intractable for laboratory research. The elucidation of the genome sequence of *M. tuberculosis*, along with an explosion in bioinformatics tools, has enabled us to predict the amino acid sequence of every potential protein of this pathogen. It can be expected that application of new technologies in future will provide molecular insights into the mechanism(s) of pathogenesis and protective immunity and provide us with opportunities for developing new and more potent drugs, diagnostics and vaccines for the control of TB. The structures of several new drug targets are being determined. This knowledge along with the development of high-throughput screens would make the rational design of antitubercular drugs more feasible and significantly improve the chances of identification of new lead molecules. The identification and structure determination of proteins involved in the persistence of this pathogen inside the host will provide new approaches to chemotherapy effective against dormant bacilli and would help in reducing the duration of therapy. The availability of new systems for expression of genes in various species of mycobacteria will help in the analysis of mycobacterial proteins in their native environment and in the development of new recombinant vaccines. Development of tools for disruption of genes in slow-growing mycobacteria would aid the identification of virulence genes. However, the path leading towards these goals, in spite of all advances, is ridden with difficulties. There are hurdles that need to be overcome and questions that need to be answered. Identification of virulence genes, for example, is only the first step and sustained efforts will be required to investigate the detailed function of the identified virulence genes. Better models of infection will have to be developed, especially to mimic the latent state of TB. New surrogate markers have to be identified to help shorten the vaccine trials and provide early indications of protective efficacy. The development of new TB drugs, despite the identification of new target molecules, is going to be rather arduous. Most of the major pharmaceutical companies have not shown great enthusiasm in developing anti-mycobacterial drugs mainly because TB is considered a disease of the developing countries and anti-tubercular drugs are not considered profitable. However, it is hoped that initiatives taken by groups such as The Bill and Melinda Gates Foundation and The Wellcome Trust, will result in the follow up of promising lead candidates and culminate in the development of new, novel and better antitubercular drugs in the next decade. The study of intracellular metabolism of mycobacteria merits greater attention to answer questions such as, what nutrients are utilized by this pathogen in macrophages? Is there a requirement for induction of novel metabolic pathways during the growth of mycobacteria in host cells? Are the transport properties of this pathogen altered in the intracellular environment? A clear understanding of these functions would emerge by employing methods that permit the host and the pathogen to be studied simultaneously. The recent developments in the molecular biology and genetics of *M. tuberculosis* offer new approaches to provide answers to some of these questions. The significant convergence of microbiology, genetics, molecular biology, cell biology and immunology during the last few years, directed at understanding this remarkably successful pathogen, provides great hope to gain a more complete understanding of the TB bacillus and thereby develop effective disease control strategies.

Acknowledgement. The authors are thankful to many colleagues and friends, including Prof. N. K. Mehra, Prof. H. K. Prasad, Prof. J. S. Tyagi from All India Institute of Medical Sciences, New Delhi, Dr. G. V. Kadival from Bhabha Atomic Research Centre, Mumbai, Dr. Joyoti Basu, Dr. Parul Chakraborti, Dr. Sujoy DasGupta, Dr. M. Kundu from Bose Institute, Calcutta, Dr. Yogendra Singh from Centre for Biochemical Technology, Delhi, Dr. S. E. Hasnain, Director, Centre for DNA Fingerprinting and Diagnostics, Hyderabad, Dr. Sudhir Sinha from Central Drug Research Institute, Lucknow, Prof. Dipankar Chatterjee, Prof. K. Muniyappa, Prof. V. Nagaraja, Prof. Umesh Varshney, Prof. S. Vijaya from Indian Institute of Science, Bangalore, Dr. Pradip K. Chakraborti and Dr. Girish Varshney from Institute of Microbial Technology, Chandigarh, Prof. S. K. Kar and Prof. Rajeev Saxena from Jawahar Lal Nehru University, New Delhi, Prof. R. M. Pitchappan from Madurai Kamaraj University. Madurai, Dr. P. Jagota and Dr. V. K. Chadha from National Tuberculosis Institute, Bangalore, Dr. Sathish Mundayoor from Rajiv Gandhi Centre for Biotechnology, Trivandrum, Dr. P. R. Narayanan, Director, Tuberculosis Research Centre, Chennai, Prof. G. P. Talwar from The Talwar Research Foundation, New Delhi, for furnishing information and data pertaining to their work or institutions. We would also like to acknowledge Prof. J. S. Tyagi and Prof. H. K. Prasad, for providing expert comments on the manuscript and Prof. K. K. Kohli, PGIMER, Chandigarh and Vivek Rao and Nisheeth Agarwal, UDSC, New Delhi for critical reading of the manuscript. Mr. Rajiv Chawla is acknowledged for efficient secretarial help.

9 References

1. Ramakrishnan T, Chandrasekhar P (1999) J Biosci 24:143
2. Kiple K (ed) (1993) The Cambridge World History of Human Disease. Cambridge University Press, Cambridge
3. Webb G (ed) (1936) Tuberculosis. Clio Medica, Paul B. Hoeber Inc., Medical Book Department of Harper & Brothers, New York
4. Davis AL (2000) In: Hershfield ES, Reichman LB (eds), Tuberculosis: A Comprehensive International Approach (Lung Biology in Health and Disease), vol 144. Marcel Dekker, New York, p 3
5. WHO Fact Sheet No. 104, 2000
6. Pathania V, Almeida J, Kochi A (1997) WHO/TB/97:223
7. Chadha V (2000) In: Nayak R, Shaila M, Ramananda Rao G (eds), Status of Tuberculosis in India -2000. Society for Innovation and Development, Bangalore, p 19
8. Dye C, Scheele S, Pathania V, Raviglione M (1999) JAMA 282:677
9. Khatri G (2000) In: Gupta S, Sood OP (eds), Tuberculosis: Proceedings of the Seventh Round Table Conference held at New Delhi. Ranbaxy Science Foundation, Gurgaon, p 15
10. Baily GV (1980) Indian J Med Res 72 Suppl:1
11. Tuberculosis Research Centre, Chennai (1999) Indian J Med Res 110:56
12. Vijay S (2000) In: Nayak R, Shaila M, Ramananda Rao G (eds), Status of Tuberculosis in India -2000. Society for Innovation and Development, Bangalore, p 91
13. Khatri G (1999) Indian J Tuberc 46:157
14. Kumar RA, Indulakshmi R, Mundayoor S (2000) Anal Biochem 286:310
15. Kinger AK, Verma A, Tyagi JS (1993) Biotechniques 14:724
16. Bashyam MD, Tyagi AK (1994) Biotechniques 17:834
17. Amara RR, Satchidanandam V (1997) Nucleic Acids Res 25:3465
18. Cole ST, Brosch R, Parkhill J, Garnier T, Churcher C, Harris D, Gordon SV, Eiglmeier K, Gas S, Barry CE, 3rd, Tekaia F, Badcock K, Basham D, Brown D, Chillingworth T, Connor R, Davies R, Devlin K, Feltwell T, Gentles S, Hamlin N, Holroyd S, Hornsby T, Jagels K, Krogh A, McLean J, Moule S, Murphy L, Oliver K, Osborne J, Quail MA, Rajandream M-A, Rogers

J, Rutter S, Seeger K, Skelton J, Squares R, Sulston JE, Taylor K, Whitehead S, Barrell BG (1998) Nature 393:537
19. DasGupta SK, Bashyam MD, Tyagi AK (1993) J Bacteriol 175:5186
20. DasGupta SK, Jain S, Kaushal D, Tyagi AK (1998) Biochem Biophys Res Commun 246:797
21. Jain S, Kaushal D, DasGupta SK, Tyagi AK (1997) Gene 190:37
22. Kumar D, Srivastava BS, Srivastava R (1998) Vaccine 16:1212
23. Tyagi AK, DasGupta SK, Jain S (2000) In: Hatfull G, Jacobs WJ (eds), Molecular Genetics of Mycobacteria. ASM Press, Washington, DC, p 131
24. Tyagi AK and colleagues, unpublished observations
25. Mundayoor S, personal communication
26. Basu A, Chawla-Sarkar M, Chakrabarti S, Das Gupta SK (2002) J Bacteriol 184:2204
27. Chawla M, Das Gupta SK (1999) Plasmid 41:135
28. Chatterjee S, Mitra M, Das Gupta SK (2000) FEMS Microbiol Lett 188:47
29. Das Gupta SK, personal communication
30. Ramesh G, Gopinathan KP (1994) Gene 143:95
31. Waskar M, Kumar D, Kumar A, Srivastava R (2000) Microbiology 146 (Pt 5):1157
32. Narayanan S, Sahadevan R, Narayanan PR (1997) Curr Sci 73:259
33. Hiriyanna KT, Ramakrishnan T (1986) Arch Microbiol 144:105
34. Bhaduri T, Bagui TK, Sikder D, Nagaraja V (1998) J Biol Chem 273:13925
35. Bhaduri T, Sikder D, Nagaraja V (1998) Nucleic Acids Res 26:1668
36. Bhaduri T, Basak S, Sikder D, Nagaraja V (2000) FEBS Lett 486:126
37. Sikder D, Nagaraja V (2000) Nucleic Acids Res 28:1830
38. Sikder D, Nagaraja V (2001) J Mol Biol 312:347
39. Sikder D, Unniraman S, Bhaduri T, Nagaraja V (2001) J Mol Biol 306:669
40. Madhusudan K, Ramesh V, Nagaraja V (1994) Biochem Mol Biol Int 33:651
41. Madhusudan K, Nagaraja V (1995) Microbiology 141 (Pt 12):3029
42. Unniraman S, Nagaraja V (1999) Genes Cells 4:697
43. Manjunatha UH, Somesh BP, Nagaraja V, Visweswariah SS (2001) FEMS Microbiol Lett 194:87
44. Manjunatha UH, Mahadevan S, Visweswariah SS, Nagaraja V (2001) Eur J Biochem 268:2038
45. Chatterji M, Unniraman S, Mahadevan S, Nagaraja V (2001) J Antimicrob Chemother 48:479
46. Purnapatre K, Varshney U (1998) Eur J Biochem 256:580
47. Purnapatre K, Handa P, Venkatesh J, Varshney U (1999) Nucleic Acids Res 27:3487
48. Purnapatre K, Varshney U (1999) Eur J Biochem 264:591
49. Handa P, Acharya N, Thanedar S, Purnapatre K, Varshney U (2000) Nucleic Acids Res 28:3823
50. Handa P, Acharya N, Varshney U (2001) J Biol Chem 276:16992
51. Acharya N, Varshney U (2002) Manuscript submitted
52. Kumar RA, Vaze MB, Chandra NR, Vijayan M, Muniyappa K (1996) Biochemistry 35:1793
53. Vaze MB, Muniyappa K (1999) Biochemistry 38:3175
54. Datta S, Prabu MM, Vaze MB, Ganesh N, Chandra NR, Muniyappa K, Vijayan M (2000) Nucleic Acids Res 28:4964
55. Reddy MS, Guhan N, Muniyappa K (2001) J Biol Chem 276:45959
56. Hemavathy KC, Nagaraja V (1995) FEMS Immunol Med Microbiol 11:291
57. Bhargava S, Tyagi AK, Tyagi JS (1990) J Bacteriol 172:2930
58. Vasanthakrishna M, Kumar NV, Varshney U (1997) Microbiology 143 (Pt 11):3591
59. Vasanthakrishna M, Rumpal N, Varshney U (1998) J Biosci 23:101
60. Rao AR, Varshney U (2001) Embo J 20:2977
61. Verma A, Kinger AK, Tyagi JS (1994) Gene 148:113
62. Verma A, Sampla AK, Tyagi JS (1999) J Bacteriol 181:4326
63. Ojha AK, Mukherjee TK, Chatterji D (2000) Infect Immun 68:4084
64. Chatterji D, personal communication
65. Bashyam MD, Kaushal D, Dasgupta SK, Tyagi AK (1996) J Bacteriol 178:4847

66. Bashyam MD, Tyagi AK (1998) J Bacteriol 180:2568
67. Ramesh G, Gopinathan KP (1995) Indian J Biochem Biophys 32:361
68. Dastur A, Varshney U (2001) Tuberculosis (Edinb) 81:267
69. Narayanan S, Selvakumar S, Aarati R, Vasan SK, Narayanan PR (2000) FEMS Microbiol Lett 192:263
70. Unniraman S, Prakash R, Nagaraja V (2001) J Biol Chem 276:41850
71. Dhar N, Rao V, Tyagi AK (2000) FEMS Microbiol Lett 190:309
72. Anilkumar G, Chauhan MM, Ajitkumar P (1998) Gene 214:7
73. Ajitkumar P, Anand SP, Anilkumar G, Srinivasan R (2000) In: Nayak R, Shaila MS, Ramananda Rao G (eds), Status of Tuberculosis in India -2000. Society for Innovation and Development, Bangalore, p 149
74. Mehrotra J, Bisht D, Tiwari VD, Sinha S (1995) Clin Exp Immunol 102:626
75. Mehrotra J, Mittal A, Dhindsa MS, Sinha S (1997) Clin Exp Immunol 109:446
76. Mehrotra J, Mittal A, Rastogi AK, Jaiswal AK, Bhandari NK, Sinha S (1999) Scand J Immunol 50:411
77. Sinha S, personal communication
78. Bisht D, Mehrotra J, Dhindsa MS, Singh NB, Sinha S (1996) Microbiology 142 (Pt 6):1375
79. Siddiqui SM, Orme IM, Saxena RK (2000) Int J Tuberc Lung Dis 4:980
80. Owais M, Varshney GC, Choudhry A, Chandra S, Gupta CM (1995) Antimicrob Agents Chemother 39:180
81. Choudhury A, Goel A, Raje M, Vohra H, Varshney GC (1997) Parasitol Res 83:746
82. Goel A, Vohra H, Varshney GC (1999) Parasitol Res 85:19
83. Majumdar S, Kaur H, Vohra H, Varshney GC (2000) FEMS Immunol Med Microbiol 28:71
84. Varshney GC, personal communication
85. Kar SK (2000) In: Nayak R, Shaila MS, Ramananda Rao G (eds), Status of Tuberculosis in India -2000. Society for Innovation and Development, Bangalore, p 188
86. Hasnain SE, personal communication
87. Kumar D, Srivastava BS, Singh NB, Srivastava R (1996) J Clin Microbiol 34:224
88. Prabhakar S, Annapurna PS, Jain NK, Dey AB, Tyagi JS, Prasad HK (1998) Tuber Lung Dis 79:43
89. Kothekar V, Raha K, Prasad HK (1998) J Biomol Struct Dyn 16:223
90. Prasad HK, personal communication
91. Pathania R, Navani NK, Rajamohan G, Dikshit KL (2002) J Biol Chem (In Press)
92. Nair RE, Banerjee S, Kumar S, Harinath BC (2000) Scand J Infect Dis 32:551
93. Nair ER, Banerjee S, Kumar S, Reddy MV, Harinath BC (2001) Indian J Chest Dis Allied Sci 43:81
94. Banerjee SK, Bhatt K, Rana S, Misra P, Chakraborti PK (1996) Biochem Biophys Res Commun 226:362
95. Banerjee SK, Misra P, Bhatt K, Mande SC, Chakraborti PK (1998) FEBS Lett 425:151
96. Bhatt K, Banerjee SK, Chakraborti PK (2000) Eur J Biochem 267:4028
97. Banerjee SK, Bhatt K, Misra P, Chakraborti PK (2000) Mol Gen Genet 262:949
98. Sarin J, Aggarwal S, Chaba R, Varshney GC, Chakraborti PK (2001) J Biol Chem 276:44590
99. Mukhopadhyay S, Chakrabarti P (1997) Antimicrob Agents Chemother 41:1721
100. Saha Choudhuri B, Sen S, Chakrabarti P (1999) Biochem Biophys Res Commun 256:682
101. Basu J, Chakrabarti P, Kundu M, personal communication
102. Basu J, Mahapatra S, Kundu M, Mukhopadhyay S, Nguyen-Disteche M, Dubois P, Joris B, Van Beeumen J, Cole ST, Chakrabarti P, Ghuysen JM (1996) J Bacteriol 178:1707
103. Lepage S, Dubois P, Ghosh TK, Joris B, Mahapatra S, Kundu M, Basu J, Chakrabarti P, Cole ST, Nguyen-Disteche M, Ghuysen JM (1997) J Bacteriol 179:4627
104. Mahapatra S, Bhakta S, Ahamed J, Basu J (2000) Biochem J 350 Pt 1:75
105. Bhakta S, Basu J (2002) Biochem J 361:635
106. Mukhopadhyay S, Basu D, Chakrabarti P (1997) J Bacteriol 179
107. Saxena RK, Wiessman D, Saxena QB, Simpson J, Lewis DM (2002) Clin Exp Immunol (In Press)
108. Vivekanandhan A, Das S (2001) Curr Sci 81:301

109. Senthil Kumar KS, Raja A, Uma Devi KR, Paranjape RS (2000) Indian J Med Res 112:37
110. Vishwanath V, Meera R, Puvanakrishnan R, Narayanan PR (1997) Mol Cell Biochem 175:169
111. Vishwanath V, Narayanan S, Narayanan PR (1998) Curr Sci 75:942
112. Jayashankar K, Ramanathan VD (1999) Indian J Med Res 110:91
113. Shakila H, Jayashankar K, Ramanathan VD (1999) Indian J Med Res 110:4
114. Kamala T, Paramasivan CN, Herbert D, Venkatesan P, Prabhakar R (1996) Indian J Med Res 103:201
115. Bhattacharyya S, Das SN, Dey AB, Nagarkar K, Lobo J, Kapoor SK, Prasad HK (1997) J Biosci 22:99
116. Uma H, Selvaraj P, Reetha AM, Xavier T, Prabhakar R, Narayanan PR (1999) Tuber Lung Dis 79:199
117. Bhaskar S, Khanna SP, Mukherjee R (2000) Vaccine 18:2856
118. Swaminathan S, Gong J, Zhang M, Samten B, Hanna LE, Narayanan PR, Barnes PF (1999) Clin Infect Dis 28:1290
119. Swaminathan S, Nandini KS, Hanna LE, Somu N, Narayanan PR, Barnes PF (2000) Indian Pediatr 37:489
120. Bhattacharyya S, Singla R, Dey AB, Prasad HK (1999) Infect Immun 67:5597
121. Das SD, Narayanan PR, Kolappan C, Colston MJ (1998) Int J Tuberc Lung Dis 2:836
122. Sinha S, Mehrotra J, Bala L, Jaiswal AK, Dhawan BN (1998) Immunopharmacol Immunotoxicol 20:579
123. Amara RR, Satchidanandam V (1996) Infect Immun 64:3765
124. Amara RR, Satchidanandam V (1997) Infect Immun 65:4880
125. Amara RR, Shanti S, Satchidanandam V (1998) Microbiology 144 (Pt 5):1197
126. Satchidanandam V, personal communication
127. Uma H, Selvaraj P, Reetha AM, Xavier T, Prabhakar R, Narayanan PR (1999) Indian J Med Res 109:5
128. Immanuel C, Swamy R, Kannapiran M, Vijayalakshmi S, Sundaram V, Jagannath K, Paramasivan CN (1997) Int J Tuberc Lung Dis 1:175
129. Immanuel C, Rajeswari R, Rahman F, Paul Kumaran P, Chandrasekaran V, Swamy R (2001) Int J Tuberc Lung Dis 5:185
130. Kinger AK, Tyagi JS (1993) Gene 131:113
131. Tyagi JS, Das TK, Kinger AK (1996) Gene 177:59
132. Tyagi JS, Kinger AK (1992) Nucleic Acids Res 20:138
133. Dasgupta N, Kapur V, Singh KK, Das TK, Sachdeva S, Jyothisri K, Tyagi JS (2000) Tuber Lung Dis 80:141
134. Kapur V, Sharma D, Ramanathan VD, Shakila H, Das TK, Chakravorty S, Saini DK, Li Q, Silver RF, Narayanan PR, Tyagi JS, manuscript submitted
135. Tyagi JS, personal communication
136. Saini D, Tyagi JS, manuscript submitted
137. Tyagi JS, Sharma D (2002) Trends Microbiol 10:68
138. Mayuri, Bagchi G, Tyagi JS, unpublished observations
139. Gupta S, Tyagi AK (1993) Gene 126:157
140. Gupta S, Jain S, Tyagi AK (1999) FEMS Microbiol Lett 172:137
141. Tyagi AK (1998) Tuberculosis Research: Into the 21st Century. Tuberculosis Research Centre, Chennai, p 29
142. Kundu M, Basu J, Chakrabarti P (1989) FEBS Lett 256:207
143. Goswami S, Sarkar S, Basu J, Kundu M, Chakrabarti P (1994) FEBS Lett 355:183
144. Maiti D, Bhattacharyya A, Basu J (2001) J Biol Chem 276:329
145. Chaba R, Raje M, Chakraborti PK (2002) Eur J Biochem 269:1078
146. Koul A, Choidas A, Tyagi AK, Drlica K, Singh Y, Ullrich A (2001) Microbiology 147:2307
147. Chopra P, Singh A, Koul A, Ramachandran S, Drlica K, Tyagi AK, Singh Y (2002) manuscript submitted
148. Kaur H, Sharma S, Khuller GK (1996) Biochem Mol Biol Int 38:141

149. Sharma S, Giri S, Khuller GK (1998) Mol Cell Biochem 183:183
150. Sumi MG, Mathai A, Sheela R, Radhakrishnan NS, Radhakrishnan VV, Indulakshmi R, Mundayoor S (2001) Clin Neuropathol 20:176
151. Parandaman V, Narayanan S, Narayanan PR (2000) Indian J Med Res 112:47
152. Narayanan S, Parandaman V, Rehman F, Srinivasan C, Gomathy D, Kumaraswami V, Paramasivan CN, Ramanathan VD, Narayanan PR (2000) Curr Sci 78:1367
153. Verma A, Rattan A, Tyagi JS (1994) Indian J Biochem Biophys 31:288
154. Verma A, Dasgupta N, Aggrawal AN, Pande JN, Tyagi JS (1995) Ind J Biochem Biophys 32:429
155. Singh KK, Muralidhar M, Kumar A, Chattopadhyaya TK, Kapila K, Singh MK, Sharma SK, Jain NK, Tyagi JS (2000) J Clin Pathol 53:355
156. Kadival GV, D'Souza CD, Kolk AH, Samuel AM (1995) Zentralbl Bakteriol 282:353
157. Kadival GV, D'Souza CD, Kulkarni SP, Samuel AM (1996) Ind J Tub 43:151
158. Kadival GV, Kameswaran M, Ray MK (2000) Ind J Tub 47:97
159. Kadival GV, personal communication
160. Kadival GV, Kameswaran M, Doshi R, Todiwala SS, Samuel AM (1994) Zentralbl Bakteriol 281:95
161. Swaminathan S, Umadevi P, Shantha S, Radhakrishnan A, Datta M (1999) Ind J Pediatr 66:837
162. Uma Devi KR, Ramalingam B, Brennan PJ, Narayanan PR, Raja A (2001) Tuberculosis (Edinb) 81:249
163. Raja A, Acharyulu CS, Selvarj R, Khudoos A (2001) Biomedicine 21:63
164. Raja A, Ranganathan UD, Bethunaickan R, Dharmalingam V (2001) Pediatr Infect Dis J 20:1161
165. Mahajan M, Singh NP, Gadre DJ, Talwar V, Gupta HC, Agarwal DS (1996) J Commun Dis 28:176
166. Gupta S, Bhatia R, Datta KK (1997) Tuber Lung Dis 78:21
167. Naidu AK, Gogate A (1997) Ind J Pathol Microbiol 40:531
168. Choudhry V, Saxena RK (2002) Eur J Clin Microbiol Infec Dis 21:1
169. Bhaskar S, Jain NK, Mukherjee R (1996) Tuber Lung Dis 77:160
170. Lodam AN, Reddy MV, Narang P, Gupta OP, Harinath BC (1996) Phytochemistry 42:66
171. Tyagi AK, Jain S (1999) In: Mahajan RC, Therwath A (eds), Multidrug resistance in Emerging and Re-emerging diseases. Narosa Press, p 109
172. Dhar N, Rao V, Tyagi AK (2002) manuscript submitted
173. Singh IG, Mukherjee R, Talwar GP (1991) Vaccine 9:10
174. Singh IG, Mukherjee R, Talwar GP, Kaufmann SH (1992) Infect Immun 60:257
175. Guleria I, Mukherjee R, Kaufmann SH (1993) Med Microbiol Immunol (Berl) 182:129
176. Talwar GP (1999) Int Rev Immunol 18:229
177. Chaturvedi V, Srivastava A, Gupta HP, Srivastava BS (1999) Vaccine 17:2882
178. Sinha RK, Verma I, Khuller GK (1997) Vaccine 15:689
179. Sinha RK, Khuller GK (1997) Immunol Cell Biol 75:461
180. Sharma AK, Verma I, Tewari R, Khuller GK (1999) J Med Microbiol 48:757
181. Dhiman N, Khuller GK (1997) Med Microbiol Immunol (Berl) 186:45
182. Dhiman N, Khuller GK (1998) FEMS Immunol Med Microbiol 21:19
183. Dhiman N, Khuller GK (2000) Lett Appl Microbiol 30:345
184. Dhiman N, Sinha RK, Khuller GK (1998) Indian J Exp Biol 36:663
185. Chadha VK, personal communication
186. Chauhan R, Mande SC (2001) Biochem J 354:209
187. Taneja B, Mande SC (2001) Protein Eng 14:391
188. Taneja B, Mande SC (2002) Acta Crystallogr D Biol Crystallogr 58:260
189. Gupta A, Kumar PH, Dineshkumar TK, Varshney U, Subramanya HS (2001) J Mol Biol 312:381
190. Saikrishnan K, Jeyakanthan J, Venkatesh J, Acharya N, Purnapatre K, Sekar K, Varshney U, Vijayan M (2002) Acta Crystallogr D Biol Crystallogr 58:327
191. Bachhawat N, Mande SC (1999) J Mol Biol 291:531

192. Srivastava R, Kumar D, Subramaniam P, Srivastava BS (1997) Biochem Biophys Res Commun 235:602
193. Srivastava R, Kumar D, Srivastava BS (1997) Biochem Biophys Res Commun 240:536
194. Srivastava R, Deb DK, Srivastava KK, Locht C, Srivastava BS (1998) Biochem Biophys Res Commun 253:431
195. Deb DK, Srivastava KK, Srivastava R, Srivastava BS (2000) Biochem Biophys Res Commun 279:457
196. Vanaja Kumar N, Selvakumar P, Venkatesan P, Chandrasekaran V, Paramasivan CN, Prabhakar R (1998) Indian J Med Res 107:75
197. Ajay Kumar R, Laiza Paul K, Indulakshmi R, Manju YK, Vinod Kumar K, Ayyappan P, Joshi M, Mundayoor S (2001) Curr Sci 80:70
198. Balakrishnan V, Varma S, Chatterji D (2001) Curr Sci 80:1302
199. Sulochana S, Venkataraman P, Paramasivan CN (1999) Indian J Med Res 110:186
200. Mathew S, Narayanan A, Segaran R, Paramasivan CN (1999) Indian J Med Res 109:167
201. Venkataraman P, Herbert D, Paramasivan CN (1998) Indian J Med Res 108:120
202. Gurumurthy P, Ramachandran G, Vijayalakshmi S, Hemanth Kumar AK, Venkatesan P, Chandrasekaran V, Vijasekaran V, Kumaraswami V, Prabhakar R (1999) Int J Tuberc Lung Dis 3:119
203. Paramasivan CN, Kubendiran G, Herbert D (1998) Indian J Med Res 108:115
204. Selvakumar N, Kumar V, Paramasivan CN (1997) Indian J Med Res 105:58
205. Selvakumar N, Vanaja Kumar P, Krishnamurthy PV, Prabhakar R, Murthy PS (1997) Curr Sci 73:79
206. Paramasivan CN, Venkataraman P, Herbert D (1997) Microbios 89:135
207. Kamala T, Herbert D, Venkatesan P, Paramasivan CN (1997) Int J Lepr Other Mycobact Dis 65:375
208. Kaur D, Khuller GK (2001) Int J Antimicrob Agents 17:51
209. Kaur D, Verma I, Khuller GK (2001) Indian J Biochem Biophys 38:235
210. Kaur D, Verma I, Khuller GK (2001) Indian J Exp Biol 39:238
211. Sharma S, Verma I, Khuller GK (1999) Arch Microbiol 171:338
212. Sharma S, Verma I, Khuller GK (2000) Eur Respir J 16:112
213. Sharma S, Verma I, Khuller GK (2001) Antimicrob Agents Chemother 45:639
214. Deol P, Khuller GK (1997) Biochim Biophys Acta 1334:161
215. Deol P, Khuller GK, Joshi K (1997) Antimicrob Agents Chemother 41:1211
216. Dutt M, Khuller GK (2001) J Antimicrob Chemother 47:829
217. Dutt M, Khuller GK (2001) Antimicrob Agents Chemother 45:363
218. Dutt M, Khuller GK (2001) Int J Antimicrob Agents 17:115
219. Dutt M, Khuller GK (2001) Int J Antimicrob Agents 18:245
220. Rajalingam R, Mehra NK, Chopra GS, Puri MM, Jain RC (1996) Tuber Lung Dis 77:502
221. Selvaraj P, Uma H, Reetha AM, Xavier T, Prabhakar R, Narayanan PR (1998) Indian J Med Res 107:208
222. Ravikumar M, Dheenadhayalan V, Rajaram K, Lakshmi SS, Kumaran PP, Paramasivan CN, Balakrishnan K, Pitchappan RM (1999) Tuber Lung Dis 79:309
223. Selvaraj P, Narayanan PR, Reetha AM (1999) Tuber Lung Dis 79:221
224. Brahmajothi V, Pitchappan RM, Kakkanaiah VN, Sashidhar M, Rajaram K, Ramu S, Palanimurugan K, Paramasivan CN, Prabhakar R (1991) Tubercle 72:123
225. Brahmajothi V (1990) Ph.D. thesis, Madurai Kamaraj University, Madurai, India
226. Pitchappan R, Brahmajothi V, Rajaram K, Subramanyam PT, Balakrishnan K, Muthuveeralakshmi P (1991) Tubercle 72:133
227. Shanmugalakshmi S (2001) PhD thesis, Madurai Kamaraj University, Madurai, India
228. Brahmajothi V, Pitchappan RM, Paramasivan CN, Rajaram K, Sankakumar U, Prabhakar R (1995) The Lancet 346:790
229. Pitchappan RM, personal communication
230. Mehra NK (1990) Trop Med Parasitol 41:352
231. Zerva L, Cizman B, Mehra NK, Alahari SK, Murali R, Zmijewski CM, Kamoun M, Monos DS (1996) J Exp Med 183:829

232. Sharma SK, Balamurugan A, Saha PK, Pande RM, Mehra NK (2002) Am Res Cri Care Med (In press)
233. Mitra DK, Mehra NK, Maiti TK, Banerjee A, Taneja V, Rajalingam R, Ahuja RK, Bhattacharya BC (1995) Int J Leprosy 63:518
234. Mitra DK, Rajalingam R, Taneja V, Bhattacharya BC, Mehra NK (1996) Clin Immunol Immunopathol 82:60
235. Mitra DK, De Rosa SC, Luke A, Balamurugan A, Khaitan BK, Tung J, Mehra NK, Terr AF, O'Garra A, Herzenberg LA, Roederer M (1999) Internat Immunol 11:1801
236. Mehra NK, personal communication

Received: May 2002

Author Index Volumes 51 – 84

Author Index Volumes 1 – 50 see Volume 50

Ackermann, J.-U. see Babel, W.: Vol. 71, p. 125

Adam, W., Lazarus, M., Saha-Möller, C. R., Weichhold, O., Hoch, U., Häring, D., Schreier, Ü.: Biotransformations with Peroxidases. Vol. 63, p. 73

Ahring, B. K.: Perspectives for Anaerobic Digestion. Vol. 81, p. 1

Ahring, B. K. see Angelidaki, I.: Vol. 82, p. 1

Ahring, B. K. see Gavala, H. N.: Vol. 81, p. 57

Ahring, B. K. see Hofman-Bang, J.: Vol. 81, p. 151

Ahring, B. K. see Mogensen, A. S.: Vol. 82, p. 69

Ahring, B. K. see Pind, P. F.: Vol. 82, p. 135

Ahring, B. K. see Skiadas, I. V.: Vol. 82, p. 35

Akhtar, M., Blanchette, R. A., Kirk, T. K.: Fungal Delignification and Biochemical Pulping of Wood. Vol. 57, p. 159

Allan, J. V., Roberts, S. M., Williamson, N. M.: Polyamino Acids as Man-Made Catalysts. Vol. 63, p. 125

Allington, R. W. see Xie, S.: Vol. 76, p. 87

Al-Rubeai, M.: Apoptosis and Cell Culture Technology. Vol. 59, p. 225

Al-Rubeai, M. see Singh, R. P.: Vol. 62, p. 167

Alsberg, B. K. see Shaw, A. D.: Vol. 66, p. 83

Angelidaki, I., Ellegaard, L., Ahring, B. K.: Applications of the Anaerobic Digestion Process. Vol. 82, p. 1

Angelidaki, I. see Gavala, H. N.: Vol. 81, p. 57

Angelidaki, I. see Pind, P. F.: Vol. 82, p. 135

Antranikian, G. see Ladenstein, R.: Vol. 61, p. 37

Antranikian, G. see Müller, R.: Vol. 61, p. 155

Archelas, A. see Orru, R. V. A.: Vol. 63, p. 145

Argyropoulos, D. S.: Lignin. Vol. 57, p. 127

Arnold, F. H., Moore, J. C.: Optimizing Industrial Enzymes by Directed Evolution. Vol. 58, p. 1

Autuori, F., Farrace, M. G., Oliverio, S., Piredda, L., Piacentini, G.: "Tissie" Transglutaminase and Apoptosis. Vol. 62, p. 129

Azerad, R.: Microbial Models for Drug Metabolism. Vol. 63, p. 169

Babel, W., Ackermann, J.-U., Breuer, U.: Physiology, Regulation and Limits of the Synthesis of Poly(3HB). Vol. 71, p. 125

Bajpai, P., Bajpai, P. K.: Realities and Trends in Emzymatic Prebleaching of Kraft Pulp. Vol. 56, p. 1

Bajpai, P., Bajpai, P. K.: Reduction of Organochlorine Compounds in Bleach Plant Effluents. Vol. 57, p. 213

Bajpai, P. K. see Bajpai, P.: Vol. 56, p. 1

Bajpai, P. K. see Bajpai, P.: Vol. 57, p. 213

Banks, M. K., Schwab, P., Liu, B., Kulakow, P. A., Smith, J. S., Kim, R.: The Effect of Plants on the Degradation and Toxicity of Petroleum Contaminants in Soil: A Field Assessment. Vol. 78, p. 75

Barut, M. see Strancar, A.: Vol. 76, p. 49

Bárzana, E.: Gas Phase Biosensors. Vol. 53, p. 1
Basu, S. K. see Mukhopadhyay, A.: Vol. 84, p. 183
Bathe, B. see Pfefferle, W.: Vol. 79, p. 59
Bazin, M. J. see Markov, S. A.: Vol. 52, p. 59
Bellgardt, K.-H.: Process Models for Production of β-Lactam Antibiotics. Vol. 60, p. 153
Beppu, T.: Development of Applied Microbiology to Modern Biotechnology in Japan. Vol. 69, p. 41
Berovic, M. see Mitchell, D. A.: Vol. 68, p. 61
Beyeler, W., DaPra, E., Schneider, K.: Automation of Industrial Bioprocesses. Vol. 70, p. 139
Beyer, M. see Seidel, G.: Vol. 66, p. 115
Bhardwaj, D. see Chauhan, V.S.: Vol. 84, p. 143
Bhatia, P. K., Mukhopadhyay, A.: Protein Glycosylation: Implications for in vivo Functions and Thereapeutic Applications. Vol. 64, p. 155
Bisaria, V. S. see Ghose, T. K.: Vol. 69, p. 87
Blanchette R. A. see Akhtar, M.: Vol. 57, p. 159
Bocker, H., Knorre, W. A.: Antibiotica Research in Jena from Penicillin and Nourseothricin to Interferon. Vol. 70, p. 35
de Bont, J. A. M. see van der Werf, M. J.: Vol. 55, p. 147
van den Boom, D. see Jurinke, C.: Vol. 77, p. 57
Brainard, A. P. see Ho, N. W. Y.: Vol. 65, p. 163
Brazma, A., Sarkans, U., Robinson, A., Vilo, J., Vingron, M., Hoheisel, J., Fellenberg, K.: Microarray Data Representation, Annotation and Storage. Vol. 77, p. 113
Breuer, U. see Babel, W.: Vol. 71, p. 125
Broadhurst, D. see Shaw, A. D.: Vol. 66, p. 83
Bruckheimer, E. M., Cho, S. H., Sarkiss, M., Herrmann, J., McDonell, T. J.: The Bcl-2 Gene Family and Apoptosis. Vol 62, p. 75
Brüggemann, O.: Molecularly Imprinted Materials – Receptors More Durable than Nature Can Provide. Vol. 76, p. 127
Bruggink, A., Straathof, A. J. J., van der Wielen, L. A. M.: A 'Fine' Chemical Industry for Life Science Products: Green Solutions to Chemical Challenges. Vol. 80, p. 69
Buchert, J. see Suurnäkki, A.: Vol. 57, p. 261
Bungay, H. R. see Mühlemann, H. M.: Vol. 65, p. 193
Bungay, H. R., Isermann, H. P.: Computer Applications in Bioprocessin. Vol. 70, p. 109
Büssow, K. see Eickhoff, H.: Vol. 77, p. 103
Byun, S. Y. see Choi, J. W.: Vol. 72, p. 63

Cabral, J. M. S. see Fernandes, P.: Vol. 80, p. 115
Cahill, D. J., Nordhoff, E.: Protein Arrays and Their Role in Proteomics. Vol. 83, p. 177
Cantor, C. R. see Jurinke, C.: Vol. 77, p. 57
Cao, N. J. see Gong, C. S.: Vol. 65, p. 207
Cao, N. J. see Tsao, G. T.: Vol. 65, p. 243
Carnell, A. J.: Stereoinversions Using Microbial Redox-Reactions. Vol. 63, p. 57
Cash, P.: Proteomics of Bacterial Pathogens. Vol. 83, p. 93
Cen, P., Xia, L.: Production of Cellulase by Solid-State Fermentation. Vol. 65, p. 69
Chang, H. N. see Lee, S. Y.: Vol. 52, p. 27
Chauhan, V. S., Bhardwaj, D.: Current Status of Malaria Vaccine Development. Vol. 84, p. 143
Cheetham, P. S. J.: Combining the Technical Push and the Business Pull for Natural Flavours. Vol. 55, p. 1
Chen, Z. see Ho, N. W. Y.: Vol. 65, p. 163
Cho, S. H. see Bruckheimer, E. M.: Vol. 62, p. 75
Cho, G. H. see Choi, J. W.: Vol 72, p. 63
Choi, J. see Lee, S. Y.: Vol. 71, p. 183
Choi, J. W., Cho, G. H., Byun, S. Y., Kim, D.-I.: Integrated Bioprocessing for Plant Cultures. Vol. 72, p. 63
Christensen, B., Nielsen, J.: Metabolic Network Analysis – A Powerful Tool in Metabolic Engineering. Vol. 66, p. 209

Christians, F. C. see McGall, G.H.: Vol. 77, p. 21
Chui, G. see Drmanac, R.: Vol. 77, p. 75
Ciaramella, M. see van der Oost, J.: Vol. 61, p. 87
Contreras, B. see Sablon, E.: Vol. 68, p. 21
Conway de Macario, E., Macario, A. J. L.: Molecular Biology of Stress Genes in Methanogens: Potential for Bioreactor Technology. Vol. 81, p. 95
Cordero Otero, R.R. see Hahn-Hägerdal, B.: Vol. 73, p. 53
Cordwell S. J. see Nouwens, A.S.: Vol. 83, p. 117
Cornet, J.-F., Dussap, C. G., Gros, J.-B.: Kinetics and Energetics of Photosynthetic Micro-Organisms in Photobioreactors. Vol. 59, p. 153
da Costa, M. S., Santos, H., Galinski, E. A.: An Overview of the Role and Diversity of Compatible Solutes in Bacteria and Archaea. Vol. 61, p. 117
Cotter, T. G. see McKenna, S. L.: Vol. 62, p. 1
Croteau, R. see McCaskill, D.: Vol. 55, p. 107

Danielsson, B. see Xie, B.: Vol. 64, p. 1
DaPra, E. see Beyeler, W.: Vol. 70, p. 139
Darzynkiewicz, Z., Traganos, F.: Measurement of Apoptosis. Vol. 62, p. 33
Davey, H. M. see Shaw, A. D.: Vol. 66, p. 83
Dean, J. F. D., LaFayette, P. R., Eriksson, K.-E. L., Merkle, S. A.: Forest Tree Biotechnolgy. Vol. 57, p. 1
Debabov, V. G.: The Threonine Story. Vol. 79, p. 113
Demain, A.L., Fang, A.: The Natural Functions of Secondary Metabolites. Vol. 69, p. 1
Dhar, N. see Tyagi, A. K.: Vol. 84, p. 211
Diaz, R. see Drmanac, R.: Vol. 77, p. 75
Dochain, D., Perrier, M.: Dynamical Modelling, Analysis, Monitoring and Control Design for Nonlinear Bioprocesses. Vol. 56, p. 147
Dolfing, J. see Mogensen, A. S.: Vol. 82, p. 69
Drmanac, R., Drmanac, S., Chui, G., Diaz, R., Hou, A., Jin, H., Jin, P., Kwon, S., Lacy, S., Moeur, B., Shafto, J., Swanson, D., Ukrainczyk, T., Xu, C., Little, D.: Sequencing by Hybridization (SBH): Advantages, Achievements, and Opportunities. Vol. 77, p. 75
Drmanac, S. see Drmanac, R.: Vol. 77, p. 75
Du, J. see Gong, C. S: Vol. 65, p. 207
Du, J. see Tsao, G. T.: Vol. 65, p. 243
Dueser, M. see Raghavarao, K.S.M.S.: Vol. 68, p. 139
Dussap, C. G. see Cornet J.-F.: Vol. 59, p. 153
Dutta, N. N. see Ghosh, A. C.: Vol. 56, p. 111
Dutta, N. N. see Sahoo, G. C.: Vol. 75, p. 209
Dynesen, J. see McIntyre, M.: Vol. 73, p. 103

Eggeling, L., Sahm, H., de Graaf, A. A.: Quantifying and Directing Metabolite Flux: Application to Amino Acid Overproduction. Vol. 54, p. 1
Eggeling, L. see de Graaf, A. A.: Vol. 73, p. 9
Eggink, G., see Kessler, B.: Vol. 71, p. 159
Eggink, G., see van der Walle, G. J. M.: Vol. 71, p. 263
Ehrlich, H. L. see Rusin, P.: Vol. 52, p. 1
Eickhoff, H., Konthur, Z., Lueking, A., Lehrach, H., Walter, G., Nordhoff, E., Nyarsik, L., Büssow, K.: Protein Array Technology: The Tool to Bridge Genomics and Proteomics. Vol. 77, p. 103
Elias, C. B., Joshi, J. B.: Role of Hydrodynamic Shear on Activity and Structure of Proteins. Vol. 59, p. 47
Ellegaard, L. see Angelidaki, I.: Vol. 82, p. 1
Elling, L.: Glycobiotechnology: Enzymes for the Synthesis of Nucleotide Sugars. Vol. 58, p. 89
Eriksson, K.-E. L. see Kuhad, R. C.: Vol. 57, p. 45
Eriksson, K.-E. L. see Dean, J. F. D.: Vol. 57, p. 1

Faber, K. see Orru, R. V. A.: Vol. 63, p. 145
Fang, A. see Demain, A. L.: Vol. 69, p. 1
Farrace, M. G. see Autuori, F.: Vol. 62, p. 129
Farrell, R. L., Hata, K., Wall, M. B.: Solving Pitch Problems in Pulp and Paper Processes. Vol. 57, p. 197
Fellenberg, K. see Brazma, A.: Vol. 77, p. 113
Fernandes, P., Prazeres, D. M. F., Cabral, J. M. S.: Membrane-Assisted Extractive Bioconversions. Vol. 80, p. 115
Ferro, A., Gefell, M., Kjelgren, R., Lipson, D. S., Zollinger, N., Jackson, S.: Maintaining Hydraulic Control Using Deep Rooted Tree Systems. Vol. 78, p. 125
Fiechter, A.: Biotechnology in Switzerland and a Glance at Germany. Vol. 69, p. 175
Fiechter, A. see Ochsner, U. A.: Vol. 53, p. 89
Flechas, F. W., Latady, M.: Regulatory Evaluation and Acceptance Issues for Phytotechnology Projects. Vol. 78, p. 171
Foody, B. see Tolan, J. S.: Vol. 65, p. 41
Fréchet, J. M. J. see Xie, S.: Vol. 76, p. 87
Freitag, R., Hórvath, C.: Chromatography in the Downstream Processing of Biotechnological Products. Vol. 53, p. 17
Furstoss, R. see Orru, R. V. A.: Vol. 63, p. 145

Galinski, E. A. see da Costa, M. S.: Vol. 61, p. 117
Gàrdonyi, M. see Hahn-Hägerdal, B.: Vol. 73, p. 53
Gatfield, I. L.: Biotechnological Production of Flavour-Active Lactones. Vol. 55, p. 221
Gavala, H. N., Angelidaki, I., Ahring, B. K.: Kinetics and Modeling of Anaerobic Digestion Process. Vol. 81, p. 57
Gavala, H. N. see Skiadas, I. V.: Vol. 82, p. 35
Gefell, M. see Ferro, A.: Vol. 78, p. 125
Gemeiner, P. see Stefuca, V.: Vol. 64, p. 69
Gerlach, S. R. see Schügerl, K.: Vol. 60, p. 195
Ghose, T. K., Bisaria, V. S.: Development of Biotechnology in India. Vol. 69, p. 71
Ghosh, A. C., Mathur, R. K., Dutta, N. N.: Extraction and Purification of Cephalosporin Antibiotics. Vol. 56, p. 111
Ghosh, P. see Singh, A.: Vol. 51, p. 47
Gilbert, R. J. see Shaw, A. D.: Vol. 66, p. 83
Gill, R. T. see Stephanopoulos, G.: Vol. 73, p. 1
Gomes, J., Menawat, A. S.: Fed-Batch Bioproduction of Spectinomycin. Vol. 59, p. 1
Gong, C. S., Cao, N. J., Du, J., Tsao, G. T.: Ethanol Production from Renewable Resources. Vol. 65, p. 207
Gong, C. S. see Tsao, G. T.: Vol. 65, p. 243
Goodacre, R. see Shaw, A. D.: Vol. 66, p. 83
de Graaf, A. A., Eggeling, L., Sahm, H.: Metabolic Engineering for L-Lysine Production by *Corynebacterium glutamicum*. Vol. 73, p. 9
de Graaf, A. A. see Eggeling, L.: Vol. 54, p. 1
de Graaf, A. A. see Weuster-Botz, D.: Vol. 54, p. 75
de Graaf, A. A. see Wiechert, W.: Vol. 54, p. 109
Grabley, S., Thiericke, R.: Bioactive Agents from Natural Sources: Trends in Discovery and Application. Vol. 64, p. 101
Griengl, H. see Johnson, D. V.: Vol. 63, p. 31
Gros, J.-B. see Larroche, C.: Vol. 55, p. 179
Gros, J.-B. see Cornet, J. F.: Vol. 59, p. 153
Guenette M. see Tolan, J. S.: Vol. 57, p. 289
Gutman, A. L., Shapira, M.: Synthetic Applications of Enzymatic Reactions in Organic Solvents. Vol. 52, p. 87

Haagensen, F. see Mogensen, A. S.: Vol. 82, p. 69
Hahn-Hägerdal, B., Wahlbom, C.F., Gárdonyi, M., van Zyl, W.H., Cordero Otero, R.R., Jönsson, L.J.: Metabolic Engineering of *Saccharomyces cerevisiae* for Xylose Utilization. Vol. 73, p. 53
Haigh, J.R. see Linden, J.C.: Vol. 72, p. 27
Hall, D. O. see Markov, S. A.: Vol. 52, p. 59
Hall, P. see Mosier, N. S.: Vol. 65, p. 23
Hammar, F.: History of Modern Genetics in Germany. Vol. 75, p. 1
Hannenhalli, S., Hubbell, E., Lipshutz, R., Pevzner, P. A.: Combinatorial Algorithms for Design of DNA Arrays. Vol. 77, p. 1
Haralampidis, D., Trojanowska, M., Osbourn, A. E.: Biosynthesis of Triterpenoid Saponins in Plants. Vol. 75, p. 31
Häring, D. see Adam, E.: Vol. 63, p. 73
Harvey, N. L., Kumar, S.: The Role of Caspases in Apoptosis. Vol. 62, p. 107
Hasegawa, S., Shimizu, K.: Noninferior Periodic Operation of Bioreactor Systems. Vol. 51, p. 91
Hata, K. see Farrell, R. L.: Vol. 57, p. 197
Hecker, M.: A Proteomic View of Cell Physiology of Bacillus subtilis – Bringing the Genome Sequence to Life. Vol. 83, p. 57
van der Heijden, R. see Memelink, J.: Vol. 72, p. 103
Hein, S. see Steinbüchel, A.: Vol. 71, p. 81
Hembach, T. see Ochsner, U. A.: Vol. 53, p. 89
Henzler, H.-J.: Particle Stress in Bioreactor. Vol. 67, p. 35
Herrmann, J. see Bruckheimer, E. M.: Vol. 62, p. 75
Hill, D. C., Wrigley, S. K., Nisbet, L. J.: Novel Screen Methodologies for Identification of New Microbial Metabolites with Pharmacological Activity. Vol. 59, p. 73
Hiroto, M. see Inada, Y.: Vol. 52, p. 129
Ho, N. W. Y., Chen, Z., Brainard, A. P. Sedlak, M.: Successful Design and Development of Genetically Engineering Saccharomyces Yeasts for Effective Cofermentation of Glucose and Xylose from Cellulosic Biomass to Fuel Ethanol. Vol. 65, p. 163
Hoch, U. see Adam, W.: Vol. 63, p. 73
Hofman-Bang, J., Zheng, D., Westermann, P., Ahring, B. K., Raskin, L.: Molecular Ecology of Anaerobic Reactor Systems. Vol. 81, p. 151
Hoheisel, J. see Brazma, A.: Vol. 77, p. 113
Holló, J., Kralovánsky, U.P.: Biotechnology in Hungary. Vol. 69, p. 151
Honda, H., Liu, C., Kobayashi, T.: Large-Scale Plant Micropropagation. Vol. 72, p. 157
Hórvath, C. see Freitag, R.: Vol. 53, p. 17
Hou, A. see Drmanac, R.: Vol. 77, p. 75
Hubbell, E. see Hannenhalli, S.: Vol. 77, p. 1
Huebner, S. see Mueller, U.: Vol. 79, p. 137
Hummel, W.: New Alcohol Dehydrogenases for the Synthesis of Chiral Compounds. Vol. 58, p. 145

Ikeda, M.: Amino Acid Production Processes. Vol. 79, p. 1
Imamoglu, S.: Simulated Moving Bed Chromatography (SMB) for Application in Bioseparation. Vol. 76, p. 211
Inada, Y., Matsushima, A., Hiroto, M., Nishimura, H., Kodera, Y.: Chemical Modifications of Proteins with Polyethylen Glycols. Vol. 52, p. 129
Irwin, D. C. see Wilson, D. B.: Vol. 65, p. 1
Isermann, H. P. see Bungay, H. R.: Vol. 70, p. 109
Iyer, P. see Lee, Y. Y.: Vol. 65, p. 93

Jackson, S. see Ferro, A.: Vol. 78, p. 125
James, E., Lee, J. M.: The Production of Foreign Proteins from Genetically Modified Plant Cells. Vol. 72, p. 127

Jeffries, T. W., Shi, N.-Q.: Genetic Engineering for Improved Xylose Fementation by Yeasts. Vol. 65, p. 117
Jendrossek, D.: Microbial Degradation of Polyesters. Vol. 71, p. 293
Jenne, M. see Schmalzriedt, S.: Vol. 80, p. 19
Jin, H. see Drmanac, R.: Vol. 77, p. 75
Jin, P. see Drmanac, R.: Vol. 77, p. 75
Johnson, D. V., Griengl, H.: Biocatalytic Applications of Hydroxynitrile. Vol. 63, p. 31
Johnson, E. A., Schroeder, W. A.: Microbial Carotenoids. Vol. 53, p. 119
Johnsurd, S. C.: Biotechnolgy for Solving Slime Problems in the Pulp and Paper Industry. Vol. 57, p. 311
Johri, B. N., Sharma, A., Virdi, J. S.: Rhizobacterial Diversity in India and its Influence on Soil and Plant Health. Vol. 84, p. 49
Jönsson, L. J. see Hahn-Hägerdal, B.: Vol. 73, p. 53
Joshi, J. B. see Elias, C. B.: Vol. 59, p. 47
Jurinke, C., van den Boom, D., Cantor, C. R., Köster, H.: The Use of MassARRAY Technology for High Throughput Genotyping. Vol. 77, p. 57

Kaderbhai, N. see Shaw, A. D.: Vol. 66, p. 83
Karanth, N. G. see Krishna, S. H.: Vol. 75, p. 119
Karthikeyan, R., Kulakow, P. A.: Soil Plant Microbe Interactions in Phytoremediation. Vol. 78, p. 51
Kataoka, M. see Shimizu, S.: Vol. 58, p. 45
Kataoka, M. see Shimizu, S.: Vol. 63, p. 109
Katzen, R., Tsao, G. T.: A View of the History of Biochemical Engineering. Vol. 70, p. 77
Kawai, F.: Breakdown of Plastics and Polymers by Microorganisms. Vol. 52, p. 151
Kell, D. B. see Shaw, A. D.: Vol. 66, p. 83
Kessler, B., Weusthuis, R., Witholt, B., Eggink, G.: Production of Microbial Polyesters: Fermentation and Downstream Processes. Vol. 71, p. 159
Khosla, C. see McDaniel, R.: Vol. 73, p. 31
Khurana, J. P. see Tyagi, A. K.: Vol. 84, p. 91
Kieran, P. M., Malone, D. M., MacLoughlin, P. F.: Effects of Hydrodynamic and Interfacial Forces on Plant Cell Suspension Systems. Vol. 67, p. 139
Kijne, J. W. see Memelink, J.: Vol. 72, p. 103
Kim, D.-I. see Choi, J. W.: Vol. 72, p. 63
Kim, R. see Banks, M. K.: Vol. 78, p. 75
Kim, Y. B., Lenz, R. W.: Polyesters from Microorganisms. Vol. 71, p. 51
Kimura, E.: Metabolic Engineering of Glutamate Production. Vol. 79, p. 37
King, R.: Mathematical Modelling of the Morphology of Streptomyces Species. Vol. 60, p. 95
Kino-oka, M., Nagatome, H., Taya, M.: Characterization and Application of Plant Hairy Roots Endowed with Photosynthetic Functions. Vol. 72, p. 183
Kirk, T. K. see Akhtar, M.: Vol. 57, p. 159
Kjelgren, R. see Ferro, A.: Vol. 78, p. 125
Knorre, W. A. see Bocker, H.: Vol. 70, p. 35
Kobayashi, M. see Shimizu, S.: Vol. 58, p. 45
Kobayashi, S., Uyama, H.: *In vitro* Biosynthesis of Polyesters. Vol. 71, p. 241
Kobayashi, T. see Honda, H.: Vol. 72, p. 157
Kodera, F. see Inada, Y.: Vol. 52, p. 129
Kolattukudy, P. E.: Polyesters in Higher Plants. Vol. 71, p. 1
König, A. see Riedel, K: Vol. 75, p. 81
de Koning, G. J. M. see van der Walle, G. A. M.: Vol. 71, p. 263
Konthur, Z. see Eickhoff, H.: Vol. 77, p. 103
Kossen, N. W. F.: The Morphology of Filamentous Fungi. Vol. 70, p. 1
Köster, H. see Jurinke, C.: Vol. 77, p. 57
Krabben, P., Nielsen, J.: Modeling the Mycelium Morphology of Penicilium Species in Submerged Cultures. Vol. 60, p. 125

Kralovánszky, U. P. see Holló, J.: Vol. 69, p. 151
Krämer, R.: Analysis and Modeling of Substrate Uptake and Product Release by Procaryotic and Eucaryotik Cells. Vol. 54, p. 31
Kretzmer, G.: Influence of Stress on Adherent Cells. Vol. 67, p. 123
Krieger, N. see Mitchell, D. A.: Vol. 68, p. 61
Krishna, S. H., Srinivas, N. D., Raghavarao, K. S. M. S., Karanth, N. G.: Reverse Micellar Extraction for Downstream Processeing of Proteins/Enzymes. Vol. 75, p. 119
Kuhad, R. C., Singh, A., Eriksson, K.-E. L.: Microorganisms and Enzymes Involved in the Degradation of Plant Cell Walls. Vol. 57, p. 45
Kuhad, R. Ch. see Singh, A.: Vol. 51, p. 47
Kulakow, P. A. see Karthikeyan, R.: Vol. 78, p. 51
Kulakow, P. A. see Banks, M. K.: Vol. 78, p. 75
Kumagai, H.: Microbial Production of Amino Acids in Japan. Vol. 69, p. 71
Kumar, S. see Harvey, N. L.: Vol. 62, p. 107
Kunze, G. see Riedel, K.: Vol. 75, p. 81
Kwon, S. see Drmanac, R.: Vol. 77, p. 75

Lacy, S. see Drmanac, R.: Vol. 77, p. 75
Ladenstein, R., Antranikian, G.: Proteins from Hyperthermophiles: Stability and Enzamatic Catalysis Close to the Boiling Point of Water. Vol. 61, p. 37
Ladisch, C. M. see Mosier, N. S.: Vol. 65, p. 23
Ladisch, M. R. see Mosier, N. S.: Vol. 65, p. 23
LaFayette, P. R. see Dean, J. F. D.: Vol. 57, p. 1
Lammers, F., Scheper, T.: Thermal Biosensors in Biotechnology. Vol. 64, p. 35
Larroche, C., Gros, J.-B.: Special Transformation Processes Using Fungal Spares and Immobilized Cells. Vol. 55, p. 179
Latady, M. see Flechas, F. W.: Vol. 78, p. 171
Lazarus, M. see Adam, W.: Vol. 63, p. 73
Leak, D. J. see van der Werf, M. J.: Vol. 55, p. 147
Lee, J. M. see James, E.: Vol. 72, p. 127
Lee, S. Y., Chang, H. N.: Production of Poly(hydroxyalkanoic Acid). Vol. 52, p. 27
Lee, S. Y., Choi, J.: Production of Microbial Polyester by Fermentation of Recombinant Microorganisms. Vol. 71, p. 183
Lee, Y. Y., Iyer, P., Torget, R. W.: Dilute-Acid Hydrolysis of Lignocellulosic Biomass. Vol. 65, p. 93
Lehrach, H. see Eickhoff, H.: Vol. 77, p. 103
Lenz, R. W. see Kim, Y. B.: Vol. 71, p. 51
Licari, P. see McDaniel, R.: Vol. 73, p. 31
Lievense, L. C., van't Riet, K.: Convective Drying of Bacteria II. Factors Influencing Survival. Vol. 51, p. 71
Linden, J. C., Haigh, J. R., Mirjalili, N., Phisaphalong, M.: Gas Concentration Effects on Secondary Metabolite Production by Plant Cell Cultures. Vol. 72, p. 27
Lipshutz, R. see Hannenhalli, S.: Vol. 77, p. 1
Lipson, D. S. see Ferro, A.: Vol. 78, p. 125
Little, D. see Drmanac, R.: Vol. 77, p. 75
Liu, B. see Banks, M. K.: Vol. 78, p. 75
Liu, C. see Honda, H.: Vol. 72, p. 157
Lueking, A. see Eickhoff, H.: Vol. 77, p. 103
Lyberatos, G. see Pind, P. F.: Vol. 82, p. 135

MacLoughlin, P. F. see Kieran, P. M.: Vol. 67, p. 139
Macario, A. J. L. see Conway de Macario, E.: Vol. 81, p. 95
Malone, D. M. see Kieran, P. M.: Vol. 67, p. 139
Maloney, S. see Müller, R.: Vol. 61, p. 155
Mandenius, C.-F.: Electronic Noses for Bioreactor Monitoring. Vol. 66, p. 65

Markov, S. A., Bazin, M. J., Hall, D. O.: The Potential of Using Cyanobacteria in Photobioreactors for Hydrogen Production. Vol. 52, p. 59
Marteinsson, V. T. see Prieur, D.: Vol. 61, p. 23
Marx, A. see Pfefferle, W.: Vol. 79, p. 59
Mathur, R. K. see Ghosh, A. C.: Vol. 56, p. 111
Matsushima, A. see Inada, Y.: Vol. 52, p. 129
Mauch, K. see Schmalzriedt, S.: Vol. 80, p. 19
McCaskill, D., Croteau, R.: Prospects for the Bioengineering of Isoprenoid Biosynthesis. Vol. 55, p. 107
McDaniel, R., Licari, P., Khosla, C.: Process Development and Metabolic Engineering for the Overproduction of Natural and Unnatural Polyketides. Vol. 73, p. 31
McDonell, T. J. see Bruckheimer, E. M.: Vol. 62, p. 75
McGall, G.H., Christians, F.C.: High-Density GeneChip Oligonucleotide Probe Arrays. Vol. 77, p. 21
McGovern, A. see Shaw, A. D.: Vol. 66, p. 83
McGowan, A. J. see McKenna, S. L.: Vol. 62, p. 1
McIntyre, M., Müller, C., Dynesen, J., Nielsen, J.: Metabolic Engineering of the *Aspergillus*. Vol. 73, p. 103
McIntyre, T.: Phytoremediation of Heavy Metals from Soils. Vol. 78, p. 97
McKenna, S. L., McGowan, A. J., Cotter, T. G.: Molecular Mechanisms of Programmed Cell Death. Vol. 62, p. 1
McLoughlin, A. J.: Controlled Release of Immobilized Cells as a Strategy to Regulate Ecological Competence of Inocula. Vol. 51, p. 1
Memelink, J., Kijne, J. W., van der Heijden, R., Verpoorte, R.: Genetic Modification of Plant Secondary Metabolite Pathways Using Transcriptional Regulators. Vol. 72, p. 103
Menachem, S. B. see Argyropoulos, D. S. : Vol. 57, p. 127
Menawat, A. S. see Gomes J.: Vol. 59, p. 1
Menge, M. see Mukerjee, J.: Vol. 68, p. 1
Merkle, S. A. see Dean, J. F. D.: Vol. 57, p. 1
Meyer, H. E. see Sickmann, A.: Vol. 83, p. 141
Mirjalili, N. see Linden, J. C.: Vol. 72, p. 27
Mitchell, D.A., Berovic, M., Krieger, N.: Biochemical Engineering Aspects of Solid State Bioprocessing. Vol. 68, p. 61
Möckel, B. see Pfefferle, W.: Vol. 79, p. 59
Moeur, B. see Drmanac, R.: Vol. 77, p. 75
Mogensen, A. S., Dolfing, J., Haagensen, F., Ahring, B. K.: Potential for Anaerobic Conversion of Xenobiotics. Vol. 82, p. 69
Moore, J. C. see Arnold, F. H.: Vol. 58, p. 1
Moracci, M. see van der Oost, J.: Vol. 61, p. 87
Mosier, N. S., Hall, P., Ladisch, C. M., Ladisch, M. R.: Reaction Kinetics, Molecular Action, and Mechanisms of Cellulolytic Proteins. Vol. 65, p. 23
Mreyen, M. see Sickmann, A.: Vol. 83, p. 141
Mühlemann, H.M., Bungay, H.R.: Research Perspectives for Bioconversion of Scrap Paper. Vol. 65, p. 193
Mukherjee, J., Menge, M.: Progress and Prospects of Ergot Alkaloid Research. Vol. 68, p. 1
Mukhopadhyay, A.: Inclusion Bodies and Purification of Proteins in Biologically Active Forms. Vol. 56, p. 61
Mukhopadhyay, A., Basu, S. K.: Intracellular Delivery of Drugs to Macrophages. Vol. 84, p. 183
Mukhopadhyay, A. see Bhatia, P. K.: Vol. 64, p. 155
Mueller, U., Huebner, S.: Economic Aspects of Amino Acids Production. Vol. 79, p. 137
Müller, C. see McIntyre, M.: Vol. 73, p. 103
Müller, R., Antranikian, G., Maloney, S., Sharp, R.: Thermophilic Degradation of Environmental Pollutants. Vol. 61, p. 155
Müllner, S.: The Impact of Proteomics on Products and Processes. Vol. 83, p. 1

Nagatome, H. see Kino-oka, M.: Vol. 72, p. 183
Nagy, E.: Three-Phase Oxygen Absorption and its Effect on Fermentation. Vol. 75, p. 51
Necina, R. see Strancar, A.: Vol. 76, p. 49
Nielsen, J. see Christensen, B.: Vol. 66, p. 209
Nielsen, J. see Krabben, P.: Vol. 60, p. 125
Nielsen, J. see McIntyre, M.: Vol. 73, p. 103
Nisbet, L.J. see Hill, D.C.: Vol. 59, p. 73
Nishimura, H. see Inada, Y.: Vol. 52, p. 123
Nordhoff, E. see Cahill, D.J.: Vol. 83, p. 177
Nordhoff, E. see Eickhoff, H.: Vol. 77, p. 103
Nouwens, A. S., Walsh, B. J., Cordwell S. J.: Application of Proteomics to Pseudomonas aeruginosa. Vol. 83, p. 117
Nyarsik, L. see Eickhoff, H.: Vol. 77, p. 103

Ochsner, U. A., Hembach, T., Fiechter, A.: Produktion of Rhamnolipid Biosurfactants. Vol. 53, p. 89
O'Connor, R.: Survival Factors and Apoptosis: Vol. 62, p. 137
Ogawa, J. see Shimizu, S.: Vol. 58, p. 45
Ohta, H.: Biocatalytic Asymmetric Decarboxylation. Vol. 63, p. 1
Oliverio, S. see Autuori, F.: Vol. 62, p. 129
van der Oost, J., Ciaramella, M., Moracci, M., Pisani, F.M., Rossi, M., de Vos, W.M.: Molecular Biology of Hyperthermophilic Archaea. Vol. 61, p. 87
Orlich, B., Schomäcker, R.: Enzyme Catalysis in Reverse Micelles. Vol. 75, p. 185
Orru, R. V.A., Archelas, A., Furstoss, R., Faber, K.: Epoxide Hydrolases and Their Synthetic Applications. Vol. 63, p. 145
Osbourn, A. E. see Haralampidis, D.: Vol. 75, p. 31
Oude Elferink, S. J. W. H. see Stams, A. J. M.: Vol. 81, p. 31

Padmanaban, G.: Drug Targets in Malaria Parasites. Vol. 84, p. 123
Paul, G.C., Thomas, C.R.: Characterisation of Mycelial Morphology Using Image Analysis. Vol. 60, p. 1
Perrier, M. see Dochain, D.: Vol. 56, p. 147
Pevzner, P. A. see Hannenhalli, S.: Vol. 77, p. 1
Pfefferle, W., Möckel, B., Bathe, B., Marx, A.: Biotechnological Manufacture of Lysine. Vol. 79, p. 59
Phisaphalong, M. see Linden, J. C.: Vol. 72, p. 27
Piacentini, G. see Autuori, F.: Vol. 62, p. 129
Pind, P. F., Angelidaki, I., Ahring, B. K., Stamatelatou, K., Lyberatos, G.: Monitoring and Control of Anaerobic Reactors. Vol. 82, p. 135
Piredda, L. see Autuori, F.: Vol. 62, p. 129
Pisani, F.M. see van der Oost, J.: Vol. 61, p. 87
Podgornik, A. see Strancar, A.: Vol. 76, p. 49
Podgornik, A., Tennikova, T.B.: Chromatographic Reactors Based on Biological Activity. Vol. 76, p. 165
Pohl, M.: Protein Design on Pyruvate Decarboxylase (PDC) by Site-Directed Mutagenesis. Vol. 58, p. 15
Poirier, Y.: Production of Polyesters in Transgenic Plants. Vol. 71, p. 209
Pons, M.-N., Vivier, H.: Beyond Filamentous Species. Vol. 60, p. 61
Pons, M.-N., Vivier, H.: Biomass Quantification by Image Analysis. Vol. 66, p. 133
Prazeres, D. M. F. see Fernandes, P.: Vol. 80, p. 115
Prieur, D., Marteinsson, V. T.: Prokaryotes Living Under Elevated Hydrostatic Pressure. Vol. 61, p. 23
Prior, A. see Wolfgang, J.: Vol. 76, p. 233
Pulz, O., Scheibenbogen, K.: Photobioreactors: Design and Performance with Respect to Light Energy Input. Vol. 59, p. 123

Raghavarao, K. S. M. S., Dueser, M., Todd, P.: Multistage Magnetic and Electrophoretic Extraction of Cells, Particles and Macromolecules. Vol. 68, p. 139
Raghavarao, K. S. M. S. see Krishna, S. H.: Vol. 75, p. 119
Ramanathan, K. see Xie, B.: Vol. 64, p. 1
Raskin, L. see Hofman-Bang, J.: Vol. 81, p. 151
Reuss, M. see Schmalzriedt, S.: Vol. 80, p. 19
Riedel, K., Kunze, G., König, A.: Microbial Sensor on a Respiratory Basis for Wastewater Monitoring. Vol. 75, p. 81
van't Riet, K. see Lievense, L. C.: Vol. 51, p. 71
Roberts, S. M. see Allan, J. V.: Vol. 63, p. 125
Robinson, A. see Brazma, A.: Vol. 77, p. 113
Rock, S. A.: Vegetative Covers for Waste Containment. Vol. 78, p. 157
Roehr, M.: History of Biotechnology in Austria. Vol. 69, p. 125
Rogers, P. L., Shin, H. S., Wang, B.: Biotransformation for L-Ephedrine Production. Vol. 56, p. 33
Rossi, M. see van der Oost, J.: Vol. 61, p. 87
Rowland, J. J. see Shaw, A. D.: Vol. 66, p. 83
Roychoudhury, P. K., Srivastava, A., Sahai, V.: Extractive Bioconversion of Lactic Acid. Vol. 53, p. 61
Rusin, P., Ehrlich, H. L.: Developments in Microbial Leaching – Mechanisms of Manganese Solubilization. Vol. 52, p. 1
Russell, N. J.: Molecular Adaptations in Psychrophilic Bacteria: Potential for Biotechnological Applications. Vol. 61, p. 1

Sablon, E., Contreras, B., Vandamme, E.: Antimicrobial Peptides of Lactic Acid Bacteria: Mode of Action, Genetics and Biosynthesis. Vol. 68, p. 21
Sahai, V. see Singh, A.: Vol. 51, p. 47
Sahai, V. see Roychoudhury, P. K.: Vol. 53, p. 61
Saha-Möller, C. R. see Adam, W.: Vol. 63, p. 73
Sahm, H. see Eggeling, L.: Vol. 54, p. 1
Sahm, H. see de Graaf, A. A.: Vol. 73, p. 9
Sahoo, G. C., Dutta, N. N.: Perspectives in Liquid Membrane Extraction of Cephalosporin Antibiotics: Vol. 75, p. 209
Saleemuddin, M.: Bioaffinity Based Immobilization of Enzymes. Vol. 64, p. 203
Santos, H. see da Costa, M. S.: Vol. 61, p. 117
Sarkans, U. see Brazma, A.: Vol. 77, p. 113
Sarkiss, M. see Bruckheimer, E. M.: Vol. 62, p. 75
Sauer, U.: Evolutionary Engineering of Industrially Important Microbial Phenotypes. Vol. 73, p. 129
Scheibenbogen, K. see Pulz, O.: Vol. 59, p. 123
Scheper, T. see Lammers, F.: Vol. 64, p. 35
Schmalzriedt, S., Jenne, M., Mauch, K., Reuss, M.: Integration of Physiology and Fluid Dynamics. Vol. 80, p. 19
Schmidt, J. E. see Skiadas, I. V.: Vol. 82, p. 35
Schneider, K. see Beyeler, W.: Vol. 70, p. 139
Schomäcker, R. see Orlich, B.: Vol. 75, p. 185
Schreier, P.: Enzymes and Flavour Biotechnology. Vol. 55, p. 51
Schreier, P. see Adam, W.: Vol. 63, p. 73
Schroeder, W. A. see Johnson, E. A.: Vol. 53, p. 119
Schubert, W.: Topological Proteomics, Toponomics, MELK-Technology. Vol. 83, p. 189
Schügerl, K., Gerlach, S. R., Siedenberg, D.: Influence of the Process Parameters on the Morphology and Enzyme Production of Aspergilli. Vol. 60, p. 195
Schügerl, K. see Seidel, G.: Vol. 66, p. 115
Schügerl, K.: Recovery of Proteins and Microorganisms from Cultivation Media by Foam Flotation. Vol. 68, p. 191

Schügerl, K.: Development of Bioreaction Engineering. Vol. 70, p. 41
Schumann, W.: Function and Regulation of Temperature-Inducible Bacterial Proteins on the Cellular Metabolism. Vol. 67, p. 1
Schuster, K. C.: Monitoring the Physiological Status in Bioprocesses on the Cellular Level. Vol. 66, p. 185
Schwab, P. see Banks, M. K.: Vol. 78, p. 75
Scouroumounis, G. K. see Winterhalter, P.: Vol. 55, p. 73
Scragg, A. H.: The Production of Aromas by Plant Cell Cultures. Vol. 55, p. 239
Sedlak, M. see Ho, N. W. Y.: Vol. 65, p. 163
Seidel, G., Tollnick, C., Beyer, M., Schügerl, K.: On-line and Off-line Monitoring of the Production of Cephalosporin C by Acremonium Chrysogenum. Vol. 66, p. 115
Shafto, J. see Drmanac, R.: Vol. 77, p. 75
Sharma, A. see Johri, B. N: Vol. 84, p. 49
Sharma, M., Swarup, R.: The Way Ahead – The New Technology in an Old Society. Vol. 84, p. 1
Shamlou, P. A. see Yim, S. S.: Vol. 67, p. 83
Shapira, M. see Gutman, A. L.: Vol. 52, p. 87
Sharp, R. see Müller, R.: Vol. 61, p. 155
Shaw, A. D., Winson, M. K., Woodward, A. M., McGovern, A., Davey, H. M., Kaderbhai, N., Broadhurst, D., Gilbert, R. J., Taylor, J., Timmins, E. M., Alsberg, B. K., Rowland, J. J., Goodacre, R., Kell, D. B.: Rapid Analysis of High-Dimensional Bioprocesses Using Multivariate Spectroscopies and Advanced Chemometrics. Vol. 66, p. 83
Shi, N.-Q. see Jeffries, T. W.: Vol. 65, p. 117
Shimizu, K. see Hasegawa, S.: Vol. 51, p. 91
Shimizu, S., Ogawa, J., Kataoka, M., Kobayashi, M.: Screening of Novel Microbial for the Enzymes Production of Biologically and Chemically Useful Compounds. Vol. 58, p. 45
Shimizu, S., Kataoka, M.: Production of Chiral C3- and C4-Units by Microbial Enzymes. Vol. 63, p. 109
Shin, H. S. see Rogers, P. L.: Vol. 56, p. 33
Sickmann, A., Mreyen, M., Meyer, H. E.: Mass Spectrometry – a Key Technology in Proteome Research. Vol. 83, p. 141
Siedenberg, D. see Schügerl, K.: Vol. 60, p. 195
Singh, A., Kuhad, R. Ch., Sahai, V., Ghosh, P.: Evaluation of Biomass. Vol. 51, p. 47
Singh, A. see Kuhad, R. C.: Vol. 57, p. 45
Singh, R. P., Al-Rubeai, M.: Apoptosis and Bioprocess Technology. Vol. 62, p. 167
Skiadas, I. V., Gavala, H. N., Schmidt, J. E., Ahring, B. K.: Anaerobic Granular Sludge and Biofilm Reactors. Vol. 82, p. 35
Smith, J. S. see Banks, M. K.: Vol. 78, p. 75
Sohail, M., Southern, E. M.: Oligonucleotide Scanning Arrays: Application to High-Throughput Screening for Effective Antisense Reagents and the Study of Nucleic Acid Interactions. Vol. 77, p. 43
Sonnleitner, B.: New Concepts for Quantitative Bioprocess Research and Development. Vol. 54, p. 155
Sonnleitner, B.: Instrumentation of Biotechnological Processes. Vol. 66, p. 1
Southern, E. M. see Sohail, M.: Vol. 77, p. 43
Srinivas, N. D. see Krishna, S. H.: Vol. 75, p. 119
Srivastava, A. see Roychoudhury, P. K.: Vol. 53, p. 61
Stafford, D. E., Yanagimachi, K. S., Stephanopoulos, G.: Metabolic Engineering of Indene Bioconversion in *Rhodococcus sp.* Vol. 73, p. 85
Stamatelatou, K. see Pind, P. F.: Vol. 82, p. 135
Stams, A. J. M., Oude Elferink, S. J. W. H., Westermann, P.: Metabolic Interactions Between Methanogenic Consortia and Anaerobic Respiring Bacteria. Vol. 81, p. 31
Stark, D., von Stockar, U.: In Situ Product Removal (ISPR) in Whole Cell Biotechnology During the Last Twenty Years. Vol. 80, p. 149

Stefuca, V., Gemeiner, P.: Investigation of Catalytic Properties of Immobilized Enzymes and Cells by Flow Microcalorimetry. Vol. 64, p. 69
Steinbüchel, A., Hein, S.: Biochemical and Molecular Basis of Microbial Synthesis of Polyhydroxyalkanoates in Microorganisms. Vol. 71, p. 81
Stephanopoulos, G., Gill, R.T.: After a Decade of Progress, an Expanded Role for Metabolic Engineering. Vol. 73, p. 1
Stephanopoulos, G. see Stafford, D. E.: Vol. 73, p. 85
von Stockar, U., van der Wielen, L. A. M.: Back to Basics: Thermodynamics in Biochemical Engineering. Vol. 80, p. 1
von Stockar, U. see Stark, D.: Vol. 80, p. 149
Straathof, A. J. J. see Bruggink, A.: Vol. 80, p. 69
Strancar, A., Podgornik, A., Barut, M., Necina, R.: Short Monolithic Columns as Stationary Phases for Biochromatography. Vol. 76, p. 49
Suurnäkki, A., Tenkanen, M., Buchert, J., Viikari, L.: Hemicellulases in the Bleaching of Chemical Pulp. Vol. 57, p. 261
Svec, F.: Capillary Electrochromatography: a Rapidly Emerging Separation Method. Vol. 76, p. 1
Svec, F. see Xie, S.: Vol. 76, p. 87
Swanson, D. see Drmanac, R.: Vol. 77, p. 75
Swarup, R. see Sharma, M.: Vol. 84, p. 1

Taya, M. see Kino-oka, M.: Vol. 72, p. 183
Taylor, J. see Shaw, A. D.: Vol. 66, p. 83
Tenkanen, M. see Suurnäkki, A.: Vol. 57, p. 261
Tennikova, T. B. see Podgornik, A.: Vol. 76, p. 165
Thiericke, R. see Grabely, S.: Vol. 64, p. 101
Thomas, C. R. see Paul, G. C.: Vol. 60, p. 1
Thömmes, J.: Fluidized Bed Adsorption as a Primary Recovery Step in Protein Purification. Vol. 58, p. 185
Timmens, E. M. see Shaw, A. D.: Vol. 66, p. 83
Todd, P. see Raghavarao, K. S. M. S.: Vol. 68, p. 139
Tolan, J. S., Guenette, M.: Using Enzymes in Pulp Bleaching: Mill Applications. Vol. 57, p. 289
Tolan, J. S., Foody, B.: Cellulase from Submerged Fermentation. Vol. 65, p. 41
Tollnick, C. see Seidel, G.: Vol. 66, p. 115
Torget, R. W. see Lee, Y. Y.: Vol. 65, p. 93
Traganos, F. see Darzynkiewicz, Z.: Vol. 62, p. 33
Trojanowska, M. see Haralampidis, D.: Vol. 75, p. 31
Tsao, D. T.: Overview of Phytotechnologies. Vol. 78, p. 1
Tsao, G. T., Cao, N. J., Du, J., Gong, C. S.: Production of Multifunctional Organic Acids from Renewable Resources. Vol. 65, p. 243
Tsao, G. T. see Gong, C. S.: Vol. 65, p. 207
Tsao, G. T. see Katzen, R.: Vol. 70, p. 77
Tyagi, A. K., Dhar, N.: Recent Advances in Tuberculosis Research in India. Vol. 84, p. 211
Tyagi, A. K., Khurana, J. P.: Plant Molecular Biology and Biotechnology Research in the Post-Recombinant DNA Era. Vol. 84, p. 91

Ukrainczyk, T. see Drmanac, R.: Vol. 77, p. 75
Uyama, H. see Kobayashi, S.: Vol. 71, p. 241

VanBogelen, R. A.: Probing the Molecular Physiology of the Microbial Organism, Escherichia coli using Proteomics. Vol. 83, p. 27
Vandamme, E. see Sablon, E.: Vol. 68, p. 21
Verpoorte, R. see Memelink, J.: Vol. 72, p. 103
Viikari, L. see Suurnäkki, A.: Vol. 57, p. 261
Vilo, J. see Brazma, A.: Vol. 77, p. 113

Vingron, M. see Brazma, A.: Vol. 77, p. 113
Virdi, J. S. see Johri, B. N: Vol. 84, p. 49
Vivier, H. see Pons, M.-N.: Vol. 60, p. 61
Vivier, H. see Pons, M.-N.: Vol. 66, p. 133
de Vos, W.M. see van der Oost, J.: Vol. 61, p. 87

Wahlbom, C.F. see Hahn-Hägerdal, B.: Vol. 73, p. 53
Wall, M. B. see Farrell, R. L.: Vol. 57, p. 197
van der Walle, G. A. M., de Koning, G. J. M., Weusthuis, R. A., Eggink, G.: Properties, Modifications and Applications of Biopolyester. Vol. 71, p. 263
Walsh, B. J. see Nouwens, A.S.: Vol. 83, p. 117
Walter, G. see Eickhoff, H.: Vol. 77, p. 103
Wang, B. see Rogers, P. L.: Vol. 56, p. 33
Weichold, O. see Adam, W.: Vol. 63, p. 73
van der Werf, M. J., de Bont, J. A. M. Leak, D. J.: Opportunities in Microbial Biotransformation of Monoterpenes. Vol. 55, p. 147
Westermann, P. see Hofman-Bang, J.: Vol. 81, p. 151
Westermann, P. see Stams, A. J. M.: Vol. 81, p. 31
Weuster-Botz, D., de Graaf, A.A.: Reaction Engineering Methods to Study Intracellular Metabolite Concentrations. Vol. 54, p. 75
Weusthuis, R. see Kessler, B.: Vol. 71, p. 159
Weusthuis, R. A. see van der Walle, G. J. M.: Vol. 71, p. 263
Wiechert, W., de Graaf, A. A.: In Vivo Stationary Flux Analysis by ^{13}C-Labeling Experiments. Vol. 54, p. 109
van der Wielen, L. A. M. see Bruggink, A.: Vol. 80, p. 69
van der Wielen, L. A. M. see von Stockar, U.: Vol. 80, p. 1
Wiesmann, U.: Biological Nitrogen Removal from Wastewater. Vol. 51, p. 113
Williamson, N. M. see Allan, J. V.: Vol. 63, p. 125
Wilson, D. B., Irwin, D. C.: Genetics and Properties of Cellulases. Vol. 65, p. 1
Winson, M. K. see Shaw, A. D.: Vol. 66, p. 83
Winterhalter, P., Skouroumounis, G. K.: Glycoconjugated Aroma Compounds: Occurence, Role and Biotechnological Transformation. Vol. 55, p. 73
Witholt, B. see Kessler, B.: Vol. 71, p. 159
Wolfgang, J., Prior, A.: Continuous Annular Chromatography. Vol. 76, p. 233
Woodley, J. M.: Advances in Enzyme Technology – UK Contributions. Vol. 70, p. 93
Woodward, A. M. see Shaw, A. D.: Vol. 66, p. 83
Wrigley, S. K. see Hill, D. C.: Vol. 59, p. 73

Xia, L. see Cen, P.: Vol. 65, p. 69
Xie, B., Ramanathan, K., Danielsson, B.: Principles of Enzyme Thermistor Systems: Applications to Biomedical and Other Measurements. Vol. 64, p. 1
Xie, S., Allington, R. W., Fréchet, J. M. J., Svec, F.: Porous Polymer Monoliths: An Alternative to Classical Beads. Vol. 76, p. 87
Xu, C. see Drmanac, R.: Vol. 77, p. 75

Yanagimachi, K. S. see Stafford, D. E.: Vol. 73, p. 85
Yim, S. S., Shamlou, P. A.: The Engineering Effects of Fluids Flow and Freely Suspended Biological Macro-Materials and Macromolecules. Vol. 67, p. 83

Zheng, D. see Hofman-Bang, J.: Vol. 81, p. 151
Zhong, J.-J.: Biochemical Engineering of the Production of Plant-Specific Secondary Metabolites by Cell Suspension Cultures. Vol. 72, p. 1
Zollinger, N. see Ferro, A.: Vol. 78, p. 125
van Zyl, W. H. see Hahn-Hägerdal, B.: Vol. 73, p. 53

Subject Index

ABC transporter 239, 240
ACC deaminase 66
Acetamidase 220
Acinetobacter caloaceticus 68
Acquired immunity 158
Adjuvants 168
Agglutination reaction 57
Alfisols 59
AMP 252
Animal models 172
Anterobacter 54
Antibiotic resistance 220
Antibody-drug conjugates 183
Anticancer agents 201
Antifolate 195
Antifungals 58
Antigen, acidic/basic repeat (ABRA) 160
Antigenic polymorphism 170
Antisense oligonucleotides 197
Apical membrane antigen (AMA) 159
Apicoplast 123, 135
Apoptosis 251
Aromatic plants 19
Artemisinin 123
Atavaquone 128
Azolla 53
Azotobactin 63

Bacillus thuringiensis 43, 75
Barnase 114
- /barstar genes 12
Basmati rice 106, 113
BCG 242, 246
Biocontrol 49
Biodiesel 25
Bioenergy crops 26
Bioethanol 25
Biofertilizer 72, 73
-, mycorrhizal 21
Biofuels 24
Bioinformatics network 33
Bioluminescence 261
Biome maps 24
Biomonitoring system 24
Biopesticides 21
Bioprospecting 24
Bioremediation 26
Biosafety 38, 39
Biosensors 27
Breeding, marker-assisted 101
Bt toxin gene 12

Carbohydrate metabolic pathways 129
Carbohydrate-binding proteins 193
$CD4^+$ 245
Cell division 234, 249
Chaperonin-10/-60 258
Chickpea 69
Chloramphenicol acetyl transferase 218
Chloroplast biogenesis 99
Chloroquine 131
Chromophores 123
Ciprofloxacin 225
Circumsporozoite protein (CS) 150
Collagen 243
Colonization potential 62
Community dynamics 49
Coumarins 225
Crop improvement 91
Cropping, intensive 53
Cry1A(c) 43
Cyanobacteria 63
Cyclothialidines 225
Cytoadherence 138
Cytotoxic T lymphocyte (CTL) 151

Daunomycin 201
Deamination 225
Dehydrogenases 59
Deoxycoformycin 126
Designers' promoter 110
Diagnostic kits 19
Dihydrofolate reductase (DHFR) 125
Dihydropteroate synthase (DHPS) 125
DNA fingerprinting 19
DNA gyrase 224, 239

DNA helicases 95
DNA markers 91
DNA methyltransferases 228
DNA polymerase 225
DNA repair 223, 225
DNA vaccines 167, 256, 257
DOTS 211
Drug carrier 186
Drug delivery 183
Drug detecting/targeting 27
Drug efficacy 261
Drug resistance 134, 258
Dynamin 188

Edema 244
Efflux pumps 240, 241
Elastin 243
ELISA 254, 255
ELISPOT assay 264
Elongation factor G 229
Embryo transfer 22
Emergence promoting rhizobacteria 54
Endocytosis, receptor-mediated 187, 192
Endophytes 55
Endorhizosphere 53
Energy crops 25
Epidermal growth factor (EGF) 161
Erythrocyte binding antigen (EBA) 159
Erythrocyte membrane protein (EMP) 160
Ethylene 66
N-Ethylmaleimide sensitive fusion proteins (NSF) 190
Excision 226
Expression cassette/vector 218

Factor Xa 219
Fingerprinting 105
Flooded soil ecosystem 54
FLP 49
Folate pathway 125
Food/nutritional security 11
Food production, nutritionally balanced 97
Food vacuole 130
Framework maps 101
FtsE 249
FtsH 234, 235
Fusaric acid 59
FYM 71

Gall midge 103
Gametocytes 147
GDI displacement factor 190
Gene banks 19
Gene pyramiding 104
Gene therapy 29
Genetic clinics 29
Genetic disorders 29
Genetic diversity 105
Genetic Engineering Approval Committee (GEAC) 39
Genetic screens 260
Genome sequencing 108
Genome-wide analysis 101
Genomics, functional 29
Global burden 215
Glycosylation 235
Golgi network, trans (TGN) 189
Grain yield 70
Granuloma 243, 244
Green fluorescent protein 77
Growth promotion 49
Guanine nucleotide dissociation inhibitor 190

Hardening units 18
Healthcare 27
Heat-shock protein 236
Heme 131
Hemoglobin-like proteins 238
Hemozoin 123, 130
Heparin sulfate proteoglycans (HSPG) 151
Hepatitis virus 27, 233
Herbal product development 19
Hexosamines 243
HGPRT 126
Histidine kinase 249
HIV 215
HLA genes 263
Host cell signaling 251
Host immune response 246
Hot spots 24
Human Genome Project 29
Human resource development 31
Hybridization, subtractive 248
Hydrocarbon-yielding plants 25
Hydroloases, lysosomal 193
Hygromycin 217

ICGEB 131
IFN-γ 237, 244–247, 251, 255, 258, 264, 265
IGS 57
IL-2 258
IL-4 245
IL-10 255
Immunodiagnostics 28
Immunogenicity 218, 235
IMP 236
Indo US vaccine action programme 28
Industrial R & D 33

INH resistance 240
Insect resistance 113
Integrated nutrient management 21, 56
Integration signals 219
International Rice Genome Sequencing Programme (IRGSP) 12
ISR 58
IVET 58

Jatropha curcas 25
Jhangora 59
JNCASR 127

Kala-azar 195
Kanamycin 217

LDL 187
- receptor 188
Lectin 250
Leishmania donovani 195, 201
Leishmaniasis 196
Leprovac 28
Ligands, endocytosed 189
Light stimulation 95
Lipoarabinomannan 251
Lipopolysaccharides 237
Liposomes 258
Liver stage antigen (LSA) 153
Low-density lipoprotein (LDL) 187
Lymphadenitis 253
Lymphocyte responses 244

Macrophages 183
-, alveolar 251
- cancer cells 202
Malaria parasite 123
- -, life cycle 146
Male-sterile lines 114
Maltose binding protein 219
Mandua 59
Mannose receptors 193, 199
MDR-TB 265
Medicinal plants 19
Membrane permeability 239
Membrane scissors 188
Merozoite surface protein (MSP) 158
Merozoites 147
Methotrexate 195
Methylation 228
Methyltransferases 260
Microbicidal agents 243
Micropropagation Parks 16, 18
Microsites 58
Mitochondrial genome 96
Molecular fingerprinting 77
Molecular markers 101
Monocot transformation 111
Montanide ISA 720 161
MTX 195
Multidrug resistance 201
Multiple antigen peptide 154
Muramyl dipeptide 203
Mycobacteriophage 221, 222
Mycobacterium tuberculosis 195, 197, 211
Mycorrhiza 21
Mycorrhization helper bacteria 79

Naringenin 55
National Agricultural Policy (NAP) 11
National Bioethics Committee 29
National Biofertilizer Development Centre 20
National Bioscience Career Development 33
National Biotechnology Board (NBTB) 5
National Facility for Virus Diagnosis and Quality Control 19
Ndk 252
Neem 107
Neoplastic cells 201
Neopterin 248
Nitric oxide 150
Nitrogen-fixing bacteria 54
Nodulation, rhizobacteria 54

Organelle genome 91, 94
Organic agriculture 59
Overexpression 218

P solubilizers 49
Palindromic stretch 232
PBMCs 246
PBPs 241
PCR 56
PCR-based diagnostics 29
PGPR, transgenic 73
Pharmacogenomics 29
Pharmacophores 123
Phosphatases 66
Photomorphogenic mutants 95
Photosynthate 53
Phytochrome 96
Phytohormones 66
Phytosiderophores 54
PknG 252
Plant growth promoting rhizobacteria 54
Plant transgenics 91
Plasmodium falciparum 125
Plasmodium vivax 124
Plastid genome 94

Pleural fluid 253
PMP 236
Polyadenylating mRNA 247
Polyguanylic acid 197
Polymorphisms 222
Post genome initiative 29
Potato, transgenic high-protein 12
PQQ dependen GDH 68
Prawn culture, intensive 23
Pribnow-box 231
Prime-boost 167
Promoter probe vector 218
PR-proteins 74
Pseudomonas 54
PSM 66
Purine biosynthetic pathway 125
Pyoverdine 62
Pyrimidine biosynthetic pathway 127

Quality assessment 84
Quinolones 225
Quorum sensing 78

Rab proteins 190
RecA protein 227, 228, 259
Receptor-mediated endocytosis 187, 192
Receptors, macrophages 192
Regulatory elements 91
RepB 220, 221
Reporter gene 218, 220
Response regulator 249
RFLP 57
Rhamnolipids 60
Rhizoplane 53, 56
Rhizosphere 53
Rhizosphere-induced genes 59
Rhoptry associated antigen (RAP) 160
Ribosome recycling factor 229
Rice 12
- genome 108
Rice-wheat 52
Ricin 201
Rifampicin 258
RNA polymerase 232, 261
Root exudates 57

Salt-tolerant bacteria 55
Scavenger receptor 193
Schizogony 147
Schizont 147
Secondary metabolites 19
Serine repeat antigen (SERA) 160
shibire mutant 188
Shuttle vector 218
Siderophores 62
Siderotyping 62
Signal transduction 96
SmpB 249
SOD 236
Software 20
SOS response 227
Sporozoites 147
–, irradiated 149
SSB protein 226
r-Strategist 55
Stresses, abiotic 98, 112
–, biotic 98
Supercoiling 225
Surface antigen 233
Surface plasmon resonance 226

TB vaccine 255–257
T-cell proliferation 237
TGN motif 232
Th1 responses 237, 255
Thrombospondin related adhesive protein (TRAP) 150
Tissue culture pilot plant 17
TNF-α 204, 251
Topoisomerases 94, 223, 224
Transcription 230
- terminators 232
Transgenic plants 38
Transgenics 110
Transposition 222
Tuberculosis 195
Tuberculosis, meningitis 253
–, pulmonary 254

Ujhani 59
UPGMA 57
Uracil DNA glycosylase 225

Vaccination, BCG 245
Vaccine development, cholera 112
Vaccines 27, 216, 218, 255
–, anti-viral 205
–, TB 255–257
Vaccinia virus 168
Virus diagnosis 19
VPN 34

White spot shrimp virus 23
Women's Biotechnology Park 45

X-ray crystallography 258

Yield increasing bacteria 54

Be the first to know

with the new online notification service

Springer Alert

You decide how we keep you up to date on new publications:

- Select a specialist field within a subject area
- Take your pick from various information formats
- Choose how often you'd like to be informed

And receive customised information to suit your needs

Register now

http://www.springer.de/alert

and then you are one click away from a world of chemistry information!

Come and visit Springer's Chemistry Online Library

http://www.springer.de/chem

Springer

008818_001x_1c

GPSR Compliance
The European Union's (EU) General Product Safety Regulation (GPSR) is a set of rules that requires consumer products to be safe and our obligations to ensure this.

If you have any concerns about our products, you can contact us on

ProductSafety@springernature.com

In case Publisher is established outside the EU, the EU authorized representative is:

Springer Nature Customer Service Center GmbH
Europaplatz 3
69115 Heidelberg, Germany

www.ingramcontent.com/pod-product-compliance
Lightning Source LLC
Chambersburg PA
CBHW070249260926
51849CB00001BA/88
* 9 7 8 3 5 4 0 0 0 6 0 9 1 *